Advanced Structured Materials

Volume 215

Common engineering materials are reaching their limits in many applications, and new developments are required to meet the increasing demands on engineering materials. The performance of materials can be improved by combining different materials to achieve better properties than with a single constituent, or by shaping the material or constituents into a specific structure. The interaction between material and structure can occur at different length scales, such as the micro, meso, or macro scale, and offers potential applications in very different fields.

This book series addresses the fundamental relationships between materials and their structure on overall properties (e.g., mechanical, thermal, chemical, electrical, or magnetic properties, etc.). Experimental data and procedures are presented, as well as methods for modeling structures and materials using numerical and analytical approaches. In addition, the series shows how these materials engineering and design processes are implemented and how new technologies can be used to optimize materials and processes.

Advanced Structured Materials is indexed in Google Scholar and Scopus.

Azman Ismail · Fatin Nur Zulkipli ·
Ahmad Fitriadhy ·
Sayyid Zainal Abidin Syed Ahmad ·
Andreas Öchsner
Editors

Exploring Diversity in Engineering and Technology for Knowledge and Innovation

Springer

Editors
Azman Ismail
Maritime Engineering Technology
and Centre for Women Advancement
and Leadership, Malaysian Institute
of Marine Engineering Technology
Universiti Kuala Lumpur
Lumut, Perak, Malaysia

Ahmad Fitriadhy
Program of Naval Architecture, Faculty
of Ocean Engineering Technology
Universiti Malaysia Terengganu
Kuala Nerus, Malaysia

Andreas Öchsner
Faculty of Mechanical Engineering
Esslingen University of Applied Science
Esslingen am Neckar, Baden-Württemberg,
Germany

Fatin Nur Zulkipli
Information Science Studies, College
of Computing, Informatics
and Mathematics
Universiti Teknologi MARA
Machang, Kelantan, Malaysia

Sayyid Zainal Abidin Syed Ahmad
Program of Maritime Technology
and Naval Architecture, Faculty of Ocean
Engineering Technology
Universiti Malaysia Terengganu
Kuala Nerus, Malaysia

ISSN 1869-8433 ISSN 1869-8441 (electronic)
Advanced Structured Materials
ISBN 978-3-031-64329-3 ISBN 978-3-031-64330-9 (eBook)
https://doi.org/10.1007/978-3-031-64330-9

This Springer imprint is published by the registered company Springer Nature Switzerland AG
The registered company address is: Gewerbestrasse 11, 6330 Cham, Switzerland

Preface

These research papers cover a wide range of topics, including sediment transportation in rivers, efficient adsorption processes, renewable energy, structural resilience, and environmental studies. Researchers explore various areas including the stability of fishing vessels, the impact of mining on riverbeds, traditional navigation methods, medical procedures, advanced materials, and the utilization of artificial intelligence in data analysis. These comprehensive papers also address structural analysis, chloride-induced damage assessment, and the potential of soy biodiesel in marine engines. It concludes with discussions on probabilistic models, sustainable aquaculture, planning for renewable energy, and the integration of technology in environmental monitoring. This diverse compilation underscores the unwavering pursuit of knowledge and innovation across scientific and engineering disciplines, offering promise for a brighter and more technologically advanced future.

Lumut, Malaysia — Azman Ismail
Machang, Malaysia — Fatin Nur Zulkipli
Kuala Nerus, Malaysia — Ahmad Fitriadhy
Kuala Nerus, Malaysia — Sayyid Zainal Abidin Syed Ahmad
Esslingen am Neckar, Germany — Andreas Öchsner

Contents

Chapter 1
Determination of Sediment Transport Rate at Dungun River, Terengganu

Nur Ayuni Zaiddin, Mohd Sofiyan Sulaiman, Nur Sara Hazeqqa Johari, Muhammad Hafizzudin Muhammad Ali, Kishan Gunesegeran, and Edi Mawardi

Abstract Excessive erosion and sedimentation have become the greatest factors that influence the effect on river morphology in Malaysia. Recently, the sand mining activities from rivers in the area of Dungun, Terengganu State, have increased alarmingly due to commercial demand in the construction market. The excessive river sand mining may lead to anthropogenic stresses that render water quality degradation, alteration in flow, and sediment regimes. The main objective of this paper is to observe the longitudinal cumulative mass change along the Dungun River system. The hydrologic engineering center's-river analysis system (HEC-RAS) version 6.0 was used to simulate quasi-unsteady sediment transport analysis. The HEC-RAS model is able to simulate riverbed change as well as the area that susceptible toward

N. A. Zaiddin · M. S. Sulaiman (✉) · N. S. H. Johari · M. H. M. Ali · K. Gunesegeran · E. Mawardi
Faculty of Ocean Engineering Technology, Universiti Malaysia Terengganu, 21030 Kuala Nerus, Terengganu Darul Iman, Malaysia
e-mail: sofiyan@umt.edu.my

N. A. Zaiddin
e-mail: ayunizaiddin@gmail.com

N. S. H. Johari
e-mail: nursarahazeqqajohari@gmail.com

M. H. M. Ali
e-mail: muhdhafizzudinali@gmail.com

K. Gunesegeran
e-mail: kishan.gunesegeran@newinti.edu.my

E. Mawardi
e-mail: edimawardi@utu.ac.id

K. Gunesegeran
Faculty of Engineering and Quantity Surveying, Inti International University, 71800 Nilai, Negeri Sembilan, Malaysia

E. Mawardi
Fakultas Teknik, Universitas Teuku Umar, Kabupaten Aceh Barat, Aceh 23681, Indonesia

A. Ismail et al. (eds.), *Exploring Diversity in Engineering and Technology for Knowledge and Innovation*, Advanced Structured Materials 215,
https://doi.org/10.1007/978-3-031-64330-9_1

either erosion or deposition. After a series of simulation, it was found that the net cumulative of erosion occurred at the upstream with an estimated sediment loss of 130,000 tons/year while the net deposition occurred at the downstream area with an estimated sediment load of 300,000 tons/year. This finding is crucial for mitigation purpose which might help navigation activities at the downstream of Dungun River.

Keywords Sediment transport · HEC-RAS 6.0 · Dungun River · Sedimentation · Erosion

1.1 Introduction

Rivers behave as a dynamic equilibrium system because they are made up of a dynamic combination of water, sediment, aquatic organisms, and riparian vegetation. The shape of rivers changes due to the occurrence of erosion, deposition, and sediment transport dynamics. All rivers carry eroded sediments in three different forms which are known as bed load, suspended load, or wash load. The sediment transport in rivers depends on several factors such as sediment type and particle size, size of drainage area, land use cover and vegetation adjacent to the catchment areas, climate changes pattern and temperature, flood events, and basin slope [1, 2]. The first laboratory-scale research of sediment transport began in 1998 and studied the starting of particle motion in a rigid rectangular channel [3]. For researchers in Malaysia, these small steps will result in a major leap in their potential to broaden their understanding of sediment transport study. Most researchers in Malaysia focused their research on the total bed material load, which included both the bed load and the suspended load [4].

Based on the current situation, human activities contribute more impact on the change of river channel compared to natural occurrences such as floods, droughts, and landslides [5, 6]. Sand mining is believed to alter flow regimes and river morphology due to the hungry water phenomenon [7]. Sand is a resource that is primarily used in the construction industry, specifically for road construction, building sites, brick making, and restoring disturbed environments [8]. A site visit at the Dungun River basin revealed abundant quantities of graded sand, which can be mined easily and economically for a wide range of purposes (see Fig. 1.1). Rapid development in recent years has increased the demand for river sand as a building material. This has resulted in a ubiquity of river sand mining activities, which has given rise to a slew of issues that require immediate attention from authorities. Sand mining permit procedures are frequently expressed in regulatory terms without taking technical account of how they can affect the ecosystem [9].

By clearing away the sediment from the active channel bed, rivers experience severe deposition, bank erosion, disruption of the sediment mass balance at the downstream, and channel adjustments that extend over the extract site itself. These factors all disrupt the river system by the natural process of sediment transport [1]. Degradation of the riverbed, encroachment into the river's buffer zone, and a deterioration in river water quality are all effects of sand mining. These physical impacts arise as

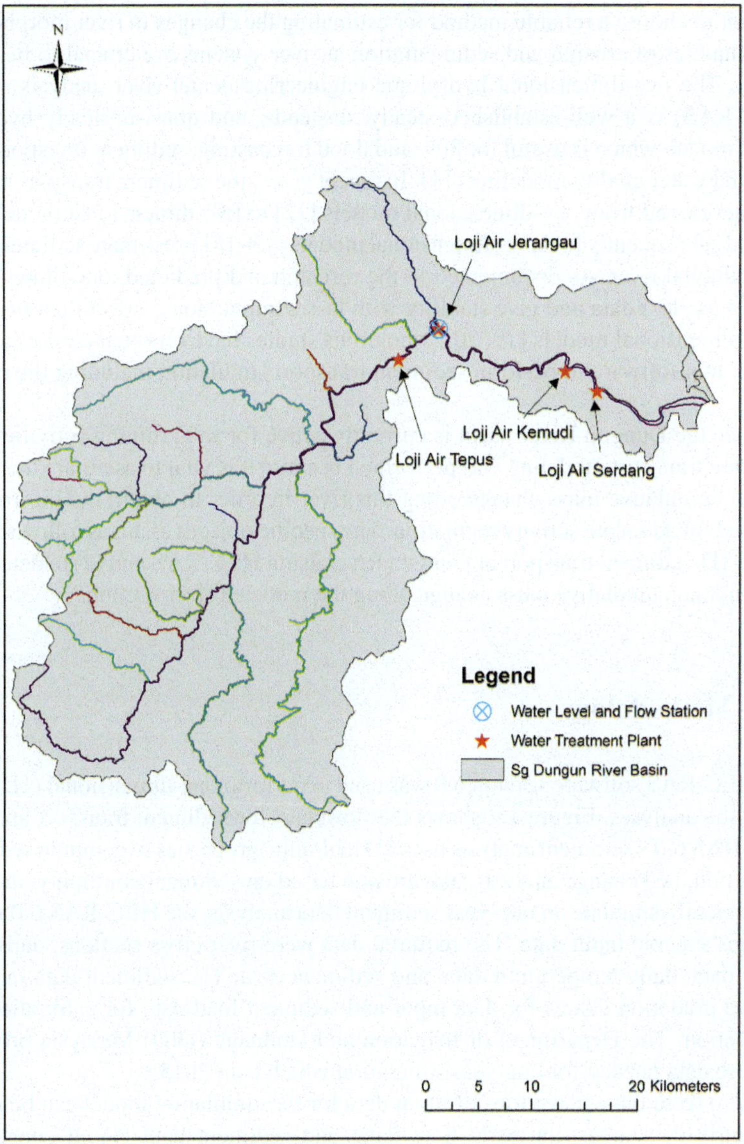

Fig. 1.1 Dungun River basin in Terengganu State

sand, gravel, and other material production rates rise and eventually reach levels that are greater than what can be produced by their natural processes. The sustainability of the environment and the river itself might be threatened when over-mining occurs frequently [10].

Prior to choose a reliable method for estimating the changes in river morphology, the dynamics of erosion and sedimentation at river systems are crucial to be determined. The one-dimensional hydrologic engineering center-river analysis system (HEC-RAS) is a well-established steady, unsteady, and quasi-unsteady hydrodynamic model, which is useful for flow and flood forecasting, sediment transport analysis, and water quality modeling [11]. In recent years, the sediment transport models have developed from one-dimensional models [12] to two-dimensional models [13, 14] and subsequently to three-dimensional models [15–18] to estimate sedimentation in an alluvial river. As documented in the research and predicted, one-dimensional models use less data and give stability with less computational effort than two- and three-dimensional models [19, 20]. Numerous studies have reported on the capabilities of this software to perform sediment transport simulation including the one by [21].

Since the Dungun River basin is currently active for sand mining activities, 1D-sediment transport modeling was performed because it is vital to ascertain the longitudinal cumulative mass change along this river in order to obtain the information required for adequate activity regulation. The specific objectives are as follows: (1) to model 1D-sediment transport at Dungun River using HEC-RAS and (2) to determine longitudinal cumulative mass change along the modeled river section.

1.2 Methodology

The HEC-RAS software version 6.0 was used to perform one-dimensional (1D) river hydraulic analyses. Figure 1.2 shows the flowchart for sediment transport analysis. HEC-RAS 6.0's sediment analysis uses 1D hydraulic properties to compute sediment deposition, bed change analysis, and erosion based on sediment continuity analysis. To efficiently simulate an one-year sediment data analysis, the HEC-RAS 6.0 model requires specific input data. The required data were river cross sections, daily flow series data, daily temperature data, and sediment data. The sediment data includes the bed gradation values for data input and sediment load data for calibration and verification. The Department of Irrigation and Drainage (DID) Malaysia provided all these data needed for one-year simulation which is in 2015.

There were three categories of input data for the simulation model can be made; geometric data, quasi-unsteady flow data, and sediment data [1]. It consists of boundary conditions, with flow hydrograph as the upstream boundary condition. The first step in sediment transport modeling was generation of the river geometry using the geospatial layout system in HEC-RAS Mapper (Fig. 1.3). Stream centerlines, cross sections, storage areas, 2D flow areas, and hydraulic structures may all be included in a model's geometry. There are 73 cross sections involved for the defined river reach as shown in Fig. 1.4.

In the quasi-unsteady flow editor, flow series and normal depth data are required to complete the boundary condition type section. HEC-RAS separates the flow series into three sections which are the flow duration, computational increments, and flow

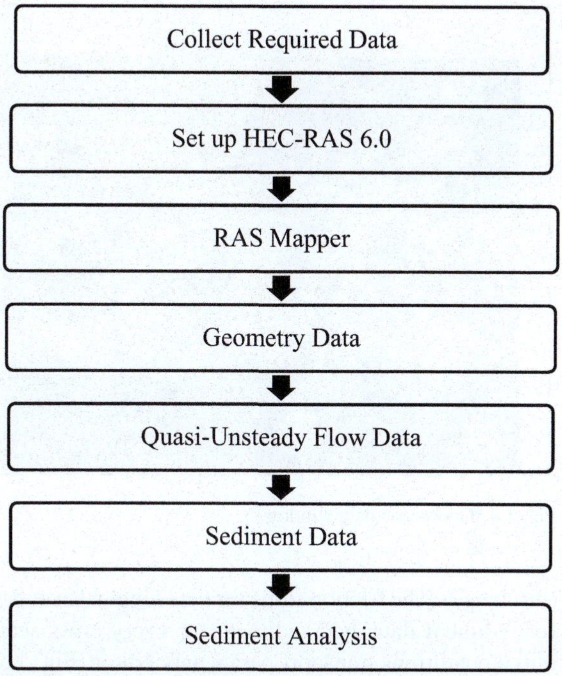

Fig. 1.2 Flowchart for sediment transport analysis

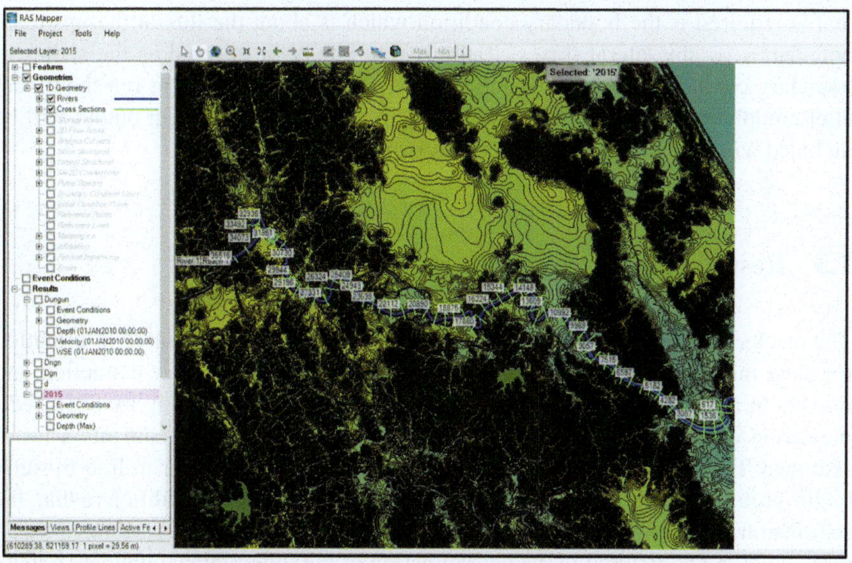

Fig. 1.3 HEC-RAS mapper editor

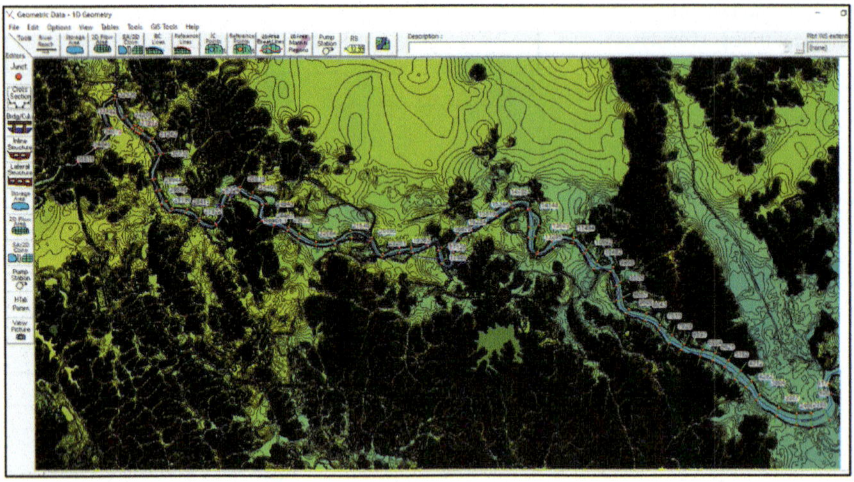

Fig. 1.4 Geometric data window

(discharge). The friction slope for the Dungun River is set at 0.0003 m/m. Meanwhile for sediment data, bed sediment for every cross section is entered directly in the initial conditions transport parameters editor (Fig. 1.5) as each cross section needs a sediment reservoir or a sediment thickness below the bed. Another sediment data that is required is the boundary condition which is about the flux at the upstream boundaries or any lateral fluxes. Rating curve has been selected as the sediment boundary condition types. The final stage of the simulation was to run the model after completing all the required data. Therefore, the final sediment output will be included with 73 cross sections from upstream to the downstream.

1.3 Results and Discussion

The results from HEC-RAS simulation can be used to provide required information for sand mining purposes such as level of deposition, location for extraction and maximum allowable mining depth [10]. Based on a previous study by [22], sedimentation of canals become a major issue since the main river contains a lot of sediment. HEC-RAS software was used to simulate the main river in Iraq to study on the problem of sediment transport caused by river erosion. In addition to that, the sediment transport simulation output using the Laursen (Copeland) sediment transport equation has resulted in the determination of the longitudinal cumulative mass change of the Dungun River in 2015. The following output from the simulation is shown in Figs. 1.6 and 1.7.

After a series of simulation, it was found that net cumulative of erosion in 2015 occurred at the upstream with an estimated sediment loss of 130,000 tons/year while

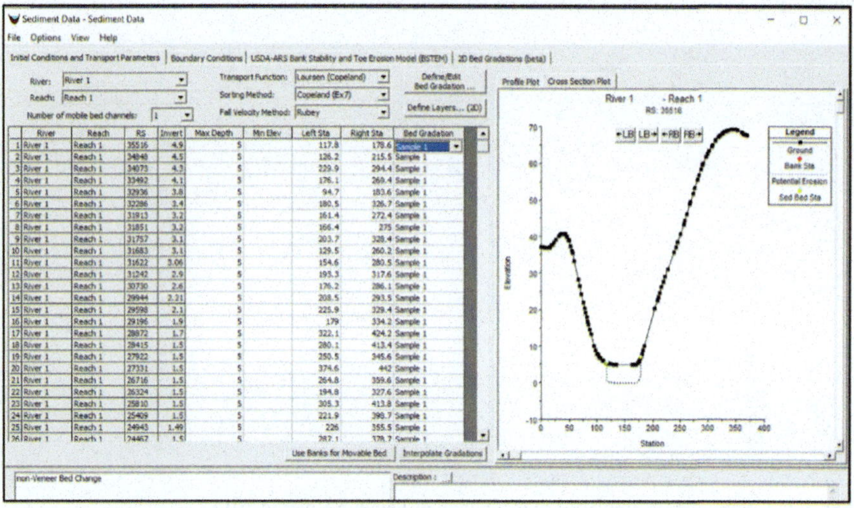

Fig. 1.5 Initial conditions transport parameters editor

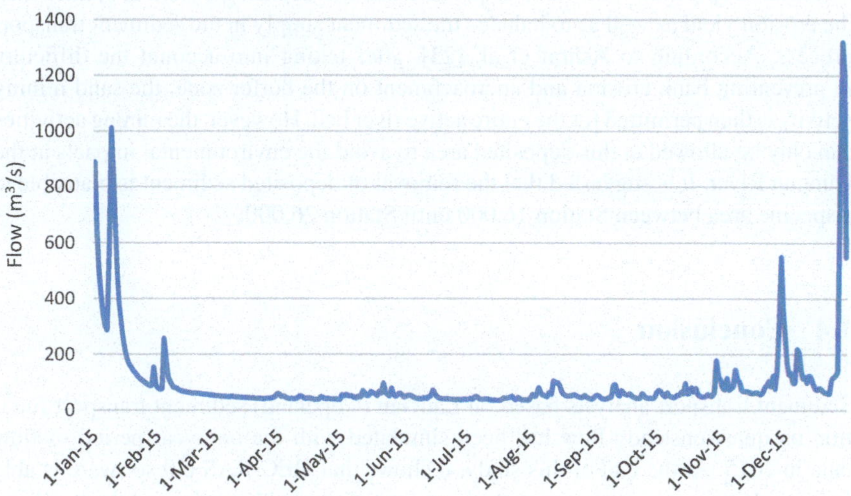

Fig. 1.6 Flow of Dungun River in 2015

net deposition occurred at the downstream area with an estimated sediment load of 300,000 tons/year. During floods, the majority of sediment is transported and it normally grows as a power function of flow, meaning that doubling flow typically results in more than doubling in sediment transport [7]. Sand extraction's feasibility will be determined by the depth and volume of deposition, taking into account the rivers' capacity to replenish the sediment.

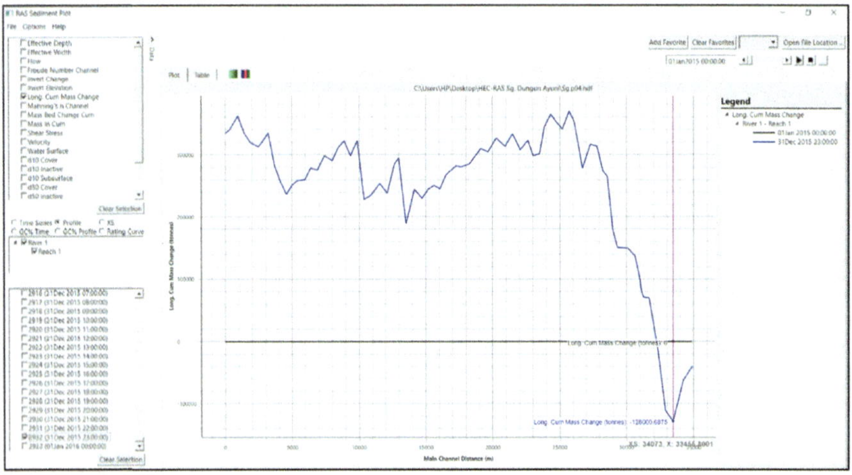

Fig. 1.7 Longitudinal cumulative mass change of Dungun River in 2015

Hence, deposition removal is required when the deposition yield is greater than the erosion yield as well as to balance the sediment supply in the sediment transport process. According to Ashraf et al. [23], after taking into account the difficulty in preventing bank erosion and encroachment on the buffer zone, the sand mining activity is then permitted for the entire active river bed. However, the mining activities can only be allowed at this deposited area to avoid the environmental impacts at the Dungun River. It is suggested that the removal of deposited sediment is plausible at a specific area between Station 16,000 until Station 26,000.

1.4 Conclusion

Sediment transport analysis based on Laursen (Copeland) sediment transport function on quasi-unsteady flow has been simulated with the one-year period of flow data in 2015. At the end of this study, it shows that HEC-RAS 6.0 software is able to produce the net erosion and deposition for longitudinal profile; and the bed level changes within each cross section. This model is useful to study bed and riverbank changes in large or complex river systems such as the Dungun River. The outcome from this study must be verified using the measured sediment load or surveyed cross section which are not covered in this paper. The future works should incorporate the calibration and validation element in HEC-RAS modeling.

Acknowledgements This research was fully funded by the Fundamental Research Grant Scheme [FRGS/1/2020/TK0/UMT/02/9] under the Ministry of Higher Education (MOHE) Malaysia.

References

1. Joshi, N., Lamichhane, G.R., Rahaman, M.M., Kalra, A., Ahmad, S.: Application of HEC-RAS to study the sediment transport characteristics of Maumee River in Ohio. In: World Environmental and Water Resources Congress 2019: Hydraulics, Waterways, and Water Distribution Systems Analysis, pp. 257–267. American Society of Civil Engineers Reston, VA (2019)
2. Sulaiman, M.S., Sinnakaudan, S.K., Ng, S.F., Strom, K.: Occurrence of bed load transport in the presence of stable clast. Int. J. Sedim. Res. **32**, 195–209 (2017)
3. Ab Ghani, A., Salem, A., Abdullah, R., Yahaya, A., Zakaria, N.: Incipient motion of sediment particles over loose deposited beds in a rigid rectangular channel. In: Proceedings of the Eighth International Conference on Urban Storm Drainage (1999)
4. Sulaiman, M.S., Sinnakaudan, S.K., Azhari, N.N., Abidin, R.Z.: Behavioral of sediment transport at lowland and mountainous rivers: a special reference to selected Malaysian rivers. Environ. Earth Sci. **76**, 300 (2017)
5. Surian, N., Cisotto, A.: Channel adjustments, bedload transport and sediment sources in a gravel-bed river, Brenta River, Italy. Earth Surf. Process. Landf. **32**, 1641–1656 (2007)
6. Sulaiman, M.S., Nazaruddin, A., Salleh, N.M., Abidin, R.Z., Miniandi, N.D., Yusoff, A.H.: Landslide occurrences in Malaysia based on soil series and lithology factors. Int. J. Adv. Sci. Technol. **28**, 1–26 (2019)
7. Kondolf, G.M.: PROFILE: hungry water: effects of dams and gravel mining on river channels. Environ. Manage. **21**, 533–551 (1997)
8. Arbogast, B.F., Knepper, D.H., Langer, W.H.: The Human Factor in Mining Reclamation. US Department of the Interior, US Geological Survey (2000)
9. Ismail, W., Ibrahim, M., Maz, M.R.R., Najib, S., Syazwan, M.: Study on the river sand removal capacity: case study at Bota Kiri, Sungai Perak. Technology **10**, 1149–1157 (2019)
10. Teo, F.Y., Noh, M.M., Ghani, A.A., Zakaria, N., Chang, C.: River sand mining capacity in Malaysia. In: Proceedings of the 37th IAHR World Congress, pp. 538–546 (2017)
11. Timbadiya, P., Patel, P., Porey, P.: HEC-RAS based hydrodynamic model in prediction of stages of lower Tapi River. Hydraul. Eng. **17**, 110–117 (2011)
12. Wu, W., Wang, S.S.Y.: One-dimensional explicit finite-volume model for sediment transport. J. Hydraul. Res. **46**, 87–98 (2008)
13. Zainalfikry, M.K., Ab Ghani, A., Zakaria, N.A., Chan, N.W.: HEC-RAS one-dimensional hydrodynamic modelling for recent major flood events in Pahang River. In: AWAM International Conference on Civil Engineering, pp. 1099–1115. Springer (2019)
14. Guan, M., Liang, Q.: Software, a two-dimensional hydro-morphological model for river hydraulics and morphology with vegetation. Environ. Modell. Softw. **88**, 10–21 (2017)
15. Wu, W.M., Rodi, W., Wenka, T.: 3D numerical modeling of flow and sediment transport in open channels. J. Hydraul. Eng. ASCE **126**, 4–15 (2000)
16. Burkow, M., Griebel, M.: A full three dimensional numerical simulation of the sediment transport and the scouring at a rectangular obstacle. Comput. Fluids **125**, 1–10 (2016)
17. Török, G.T., Baranya, S., Rüther, N.: 3D CFD modeling of local scouring, bed armoring and sediment deposition. Water **9**, 56 (2017)
18. Sawadogo, O., Basson, G.R., Schneiderbauer, S.: Physical and coupled fully three-dimensional numerical modeling of pressurized bottom outlet flushing processes in reservoirs. Int. J. Sedim. Res. **34**, 461–474 (2019)
19. Werner, M.G.F.: Impact of grid size in gis based flood extent mapping using a 1D flow model. Phys. Chem. Earth **26**, 517–522 (2001)
20. Horritt, M., Bates, P.: Evaluation of 1D and 2D numerical models for predicting river flood inundation. Hydrology **268**, 87–99 (2002)
21. Gibson, S., Brunner, G., Piper, S., Jensen, M.: Sediment transport computations in HEC-RAS, eighth federal interagency sedimentation conference (8thFISC), pp. 57–64. Reno, NV (2006)
22. Ali Omran, Z., Shamkhi Jaber, W.: Simulation of sediment transport in Al-Hilla River in Iraq using the HEC-RAS software. J. Kerbala Univ. **13**, 8–18 (2017)

23. Ashraf, M.A., Maah, M.J., Yusoff, I., Wajid, A., Mahmood, K.: Sand mining effects, causes and concerns: a case study from Bestari Jaya, Selangor, Peninsular Malaysia. J. Sci. Res. Essays **6**, 1216–1231 (2011)

Chapter 2
Kinetics, Isotherms, and Thermodynamic Studies on the Efficient Adsorption of Ionizable Organic *β*-Agonist (Ractopamine) in Aqueous Solution onto Mechanically Synthesized Copper (II) Isonicotinate Metal–Organic Frameworks [Cu (INA)$_2$]-MOF

Usman Armaya'u, Marinah Mohd Ariffin, Saw Hong Loh, Wan Mohd Afiq Wan Mohd Khalik, and Hanis Mohd Yusoff

Abstract The adsorptive performance of mechanochemically synthesized copper (II) isonicotinate metal–organic frameworks ([Cu (INA)$_2$]-MOFs) on ractopamine in water is reported. The MOF characteristic was revealed by Fourier transform infrared (FTIR), showing all the projected functional groups. The X-ray diffraction (XRD) diffractogram shows an excellent crystalline size of 39.78 nm. Morphology, shape, and size revealed from field emission scanning electron microscopy (FESEM) with energy dispersive X-ray (EDX) show a porous nature, revealing a rectangular bar-like shape with a rough surface that aids the retention of its porous nature with the elemental distribution graph. Ionizable organic compounds (like ractopamine with secondary amide and phenolic groups) contain acidic or basic functional groups that

U. Armaya'u · M. M. Ariffin (✉) · S. H. Loh · W. M. A. W. M. Khalik · H. M. Yusoff
Faculty of Science and Marine Environment, Universiti Malaysia Terengganu, 21030 Kuala Nerus, Terengganu Darul Iman, Malaysia
e-mail: erin@umt.edu.my

U. Armaya'u
e-mail: p4222@pps.umt.edu.my

S. H. Loh
e-mail: lohsh@umt.edu.my

W. M. A. W. M. Khalik
e-mail: wan.afiq@umt.edu.my

H. M. Yusoff
e-mail: hanismy@umt.edu.my

U. Armaya'u
Faculty Faculty of Applied Sciences, Al-Qalam University, Katsina, Nigeria

A. Ismail et al. (eds.), *Exploring Diversity in Engineering and Technology for Knowledge and Innovation*, Advanced Structured Materials 215,
https://doi.org/10.1007/978-3-031-64330-9_2

exist in the ionic state under the appropriate pH value for better removal. Evaluating the pH influence revealed the MOF's performed better at an alkaline pH while the isotherm and kinetics fit pseudo-second-order and Langmuir models based on the statistical fits of R^2, RMSE, and AIC, respectively. An excellent removal efficiency of 98% under 40 min was obtained with an adsorbate concentration of 40 mg/L and a dosage of 4 mg/g. The thermodynamics study confirmed the spontaneous, endothermic, and random nature of the process. The MOF's high reusability even after six cycles proves the [Cu $(INA)_2$]-MOF as an efficient adsorbent for the removal of ractopamine and in fact, for all β-agonists.

Keywords Ractopamine · β-agonists · Adsorption · Removal · Metal–organic frameworks

2.1 Introduction

Ractopamine, (RAC) is known as a synthetic β-agonist compound that is a member of the phenolic group that is structurally similar to the standard term for β-receptor stimulants which is a phenyl β-ethanolamine compounds with substituents at the terminal amino group on the aromatic ring [1]. Each of the RAC molecule's constituent parts which include the benzylic hydroxyl group, amino functional group, and two phenolic groups has the potential to affect the molecule when subjected to a wide range of different environmental conditions (Fig. 2.1). Four different stereoisomers can be found in ractopamine, which has two chiral centers in its structure. Free ractopamine is named under standard IUPAC naming as 4-[3-[[2-hydroxy-2-(4-hydroxyphenyl)ethyl]amino]butyl]phenol with a molecular formula $C_{18}H_{23}NO_3$ and a molecular weight of 301.386 g mol^{-1} with properties such as water-soluble (31 g/L; log K_{ow} 2.4; pK_a 9, mobile in the environment and can be water-borne from animal wastes [2, 3]. Pharmacological activity of these compounds is exceptionally potent. Tocolytic and bronchodilator drugs are the primary applications in both human and veterinary medicine, respectively. On the other hand, due to the anabolic effects that are induced in animals when higher doses are given to them, [4], ractopamine is used as a feed additive that is utilized in different trade names in over 20 different nations, which are the USA, Mexico, Canada, and Japan, with the goals of enhancing feed efficiency, accelerating the rate of weight gain, and increasing the leanness of the carcass [2]. The addition of β-agonists to the diets of animals raised for human consumption is not without its critics and controversies. RAC use has been shown to have negative effects on the heart, thyroid, urethra, and prostate [5], and it was found to be responsible for cardiac toxicity in greyhounds at levels that were two to four times higher than the recommended dietary quantities for swine [6] as the approved dosages of these drugs as additives in complete feed is around 2–25 mg/kg [7]. A recent comprehensive review disclosed a higher rise of reported cases of β-agonist detections in different environmental water [8] as these compounds can make their way into water sources and easy to leach from land dumping areas due to their polar nature, into groundwater by illegal or unintentional spills of unmetabolized β-agonists, flooded rivers from farm fields carrying the dung of animals that have been detected with β-agonists, and sewage systems [9–11], these make it difficult to remove using conventional methods. The analysis for these compounds in water is still difficult and complicated, even though progress has been made in identifying effective approaches to find β-agonists in various sample types detected at the lowest level. Their pK_a has a significant bearing on their ability to form bonds with proteins, as well as their lipophilicity, solubility, and permeability. In addition to this, it has a significant impact on the parameters of their absorption, adsorption, transportation, metabolism, and excretion. According to certain reports, their concentrations typically reach 20 mg/L in some of the drinking water that has been treated in accordance with the conventional procedures [12, 13]. According to the Hakk [2] reports, free ractopamine, which has a $\log K_{ow}$ of 2.4, a pK_a of 9.4, and an acceptable solubility of 31 g/L, might spread in water from animal feces and is easy to travel around and found in plenty in the environment. RAC has been selected as a model β-agonists for this research due to its harmful effects on the environment and its toxicity to people and aquatic creatures, as shown in Fig. 2.1.

However, reported adsorbents for RAC removal and for the removal of β-agonist in general were not many, but adsorbents based to detect clenbuterol utilizing molecularly imprinted polymers coated with particles as fluorescence were reported by Tang in 2015 [14], and a study using molecularly imprinted polymers (MIP) was also reported by Tang in 2016 [15]. The adsorbent of granular activated carbon (GAC) was also reported by de Ridder [16]. However, the efficiency of these adsorbents was limited as they required a longer equilibrium period in addition to being difficult to regenerate. On the other hand, elimination of β-agonists was accomplished by methods such as oxidation [17] and photo-degradation [18]. However, oxidation and photocatalytic degradation were found to produce intermediates with a higher toxicity than the pollutants; consequently, researchers consider adsorption

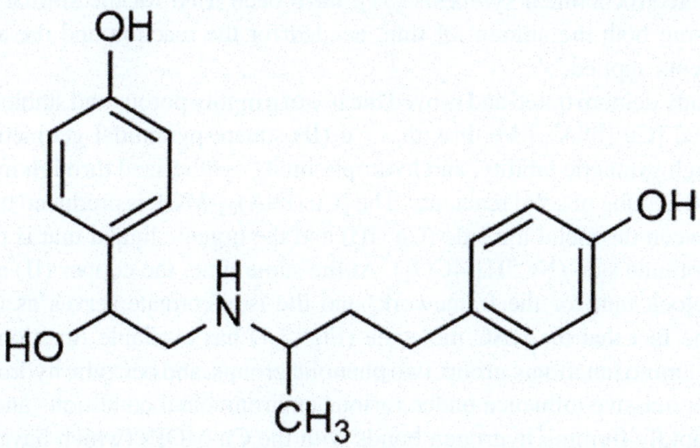

Fig. 2.1 Chemical structure of the model ractopamine

as the physical process with the most advantageous, easy, promising, economical, and environmentally compatible method for removing pollutants during wastewater recovery [19, 20].

A literature review revealed by Yang [21] and Lv [22], both of which used aluminum-based MOFs (BUT-19) and both of which took hours to days during the synthesis process, are the only two obtainable studies so far that are available to us that have carried out removal studies of β-agonist in environmental waters using newly discovered MOFs materials. These studies were carried out by using hydrothermal synthesis to create fluorescent zirconium-based MOFs (Ui0-66) and solvothermal synthesis.

MOFs are of great interest because of the unusual qualities that they possess, which include high crystallinity, pore sizes with homogenous distribution, ultra-high surface area, and volume tunability. Their microporous structures, which include the surface areas, were reported to be of about 9000 $m^2\ g^{-1}$ with specific volumes of about 2 $cm^3\ g^{-1}$ with wide range of pore sizes and topologies, reported by Neshastehgar [23], Zhao [24], and Zango [25]. So far, the remarkable adaptability and practical applications of MOFs, in addition to their simple, direct, and uncomplicated synthesis, which differs from that of other typical porous solids, have been some of the many interesting reasons that researchers in academia and industry have been drawn to them [26].

MOFs are usually produced by the extensive procedures of either hydrothermal or solvothermal methods, but these processes need a longer reaction time and the use of a significant quantity of organic solvents, both of which generate serious environmental problems. With the discovery of the first mechanochemical synthesis by Pichon in 2006 [27], chemists have been paying a lot of attention to mechanochemistry over the course of the past decade. The search for novel, environmentally friendly synthetic techniques has refueled a resurgence of mechanochemical synthesis (MS) in the manufacture of metal–organic frameworks (MOFs) [28]. Alternative methods of synthesis, such as microwave-assisted synthesis [29], sonochemical synthesis [30], and electrochemical synthesis [31], have been tried for the similar objective of decreasing both the amount of time needed for the reaction and the amount of large solvents applied.

This study demonstrated and proved the use of a highly porous and stable Cu-based MOF called [Cu $(INA)_2$]-MOF with a Cu (II) square-pyramidal geometry, exceptionally high synthetic lability, and hydrophobicity synthesized through mechanical mixing and heating of solid reactants. The [Cu $(INA)_2$]-MOF is produced by creating a link between the metal ion side (Cu (II)) and the organic ligand that is present on the isonicotinate side ($NC5H_4$-$4CO_2$). At the same time, the copper (II) acts as the building block unit for the framework, and the isonicotinate serves as the linker. Despite the fact that the RAC molecule (Fig. 2.1) has available functional groups such as its amino functional group, two phenolic groups, and benzylic hydroxyl group that have a massive influence under various environmental conditions and are suitable for rapidly forming hydrogen bonds with the Cu-MOFs (which have available hydroxyl, carboxyl, and amino groups), the adsorption process is still anticipated to be sped up as a result of all of these factors [2, 3].

For the purposes of this study, the solvent-free synthesis method, also known as mechano-synthesis, has been selected over other solvent-based synthesis methods. This is due to the solvent-free synthesis method's significant advantages over the solvent-based methods of synthesis, among which are listed below: (i) there is no need to use hazardous and toxic solvents; (ii) there is no need to worry about discarding of solvent; (iii) short processing time and money-saving; (iv) the environment is protected from unwanted chemicals (environmentally friendly); (v) less time needed, and simple synthesis, as well as (vi) higher/quantitative yield is achieved, and (vii) further purification after synthesis is not needed.

2.2 Methodology

2.2.1 Chemicals and Materials

All the chemical substances used in this study were used as received in analytical quality. Isonicotinic acid (98%), copper acetate monohydrate (98.5%), and ractopamine sulfate were obtained from Sigma Aldrich (St. Louis, MO, USA). Chemicals supplied by IT Tech Research Selangor in Malaysia are methanol (99.5%), sodium hydroxide, and hydrochloric acid. It should be noted that RAC is a hazardous material that should be handled with extreme caution. As a result, the adsorption studies were carried out in a well-equipped laboratory environment. The solvent involved in this study was discharged adequately.

2.2.1.1 *Synthesis of [Cu $(INA)_2$]-MOF*

A solvent-free method was used to synthesize [Cu$(INA)_2$]-MOF following Tella's [32] procedure with slight modification. An amount of 0.25 g (1 mmol) copper acetate monohydrate (Cu $(CH_3COO)2H_2O$) and 0.246 g (2 mmol) isonicotinic acid ($C_5H_4N(CO_2H)$) were mixed using a mortar and pestle, then pulverized for 5 min. The product obtained was further confirmed by the characteristic odor of acetate and the color change has been observed from green to blue. The product was heated to 150 °C at the duration of 12 h to remove the glacial acetic acid entrapped in the pores of the [Cu $(INA)_2$]-MOF. After cooling in a desiccator, the products were stored and sealed in a suitable sample vial. This sample was used without additional treatment throughout the next series of the experimental method.

2.2.1.2 Characterization

A PerkinElmer FTIR spectrometer scanned from 500 to 4000 cm^{-1} for functional group identification. Thermal stability was assessed using a Shimadzu TGA-50 analyzer in N_2 atmosphere and was utilized as a thermogravimetric analyzer. A scanning electron microscope (SEM/EDX) Zeiss Supra was used to analyze the surface features, morphology, and topology, as well as the elemental analysis of the MOFs material. For an in-depth surface morphology, topography, microstructure, and elemental evidence of the synthesized material a field emission scanning electron microscopy (FESEM/EDX) model Zeiss Supra 55 VP was used. The crystallinity of the material was determined using a X-ray Diffractometer (XRD) of a known model, Bruker D8 Advance.

2.2.2 *Removal Experiment*

2.2.2.1 Preparation of Ractopamine Solution

Ractopamine (2 mg) was prepared in methanol into a 100 mL volumetric flask to get a final concentration of 20 mg L^{-1}. The stock solution was stored at 4 °C in a refrigerator prior to use. Freshly prepared working solutions with milli-*Q* water were carried out every day by diluting the necessary stock solution to acquire the requisite amount. It is possible to separate polar components from non-polar components by combining methanol and water together as compared to utilizing either methanol or water alone. A decrease in surface tension, a reduction in the polarity of the water, an increase in density, and the facilitation of the collection of chemicals at the interface are some of the advantages that can result from employing a mixture of water and methanol.

2.2.2.2 Batch Adsorption Study

Stock solution was diluted with milli-*Q* water and the working solution was adjusted to the appropriate concentration; batch adsorption studies were carried out with a 40 mL solution. [Cu $(INA)_2$]-MOF was suspended in a 150 ml conical flask containing a RAC solution with initial concentrations of 10–60 mg L^{-1}, 1–6 mg g^{-1}. The flask was shaken using an incubator shaker (brand: stacked shaker SKE 8000) at a temperature controlled at 25 °C for the duration of 10–70 min at a speed of 150 revolutions per minute (rpm). Using a quartz cuvette and a Shimadzu UV-1800 UV–Vis spectrophotometer, a sample solution aliquot of 2 mL was filtered using a nylon syringe membrane with a thickness of 0.45 μl at 10 min of time intervals. The absorbance of the analyte was determined by determining how much of the filtrate solution contained the analyte.

The effects of pH were initially investigated by adjusting the pH of the solution with 0.1 N NaOH or M HCl to a various range of five different pHs which are pH3, pH5, pH7, pH11 to pH13, and the effects of temperature were investigated by adjusting the temperature to a various range starting from 25 °C followed by 30 °C, 35 °C, 40 °C, 45 °C and up to 50 °C.

The percentage of removal (%R), the amount of adsorbate taken up at a specific time interval (q_t), and the quantity of adsorbate taken up at equilibrium (q_e) were all determined by applying Eqs. (2.1)–(2.3) according to [33], as shown below:

$$\%R = \frac{(C_0 - C_t)}{C_0} \times 100 \tag{2.1}$$

$$q_t = \frac{(C_0 - C_t)V}{W} \tag{2.2}$$

$$q_e = \frac{(C_0 - C_e)V}{W} \tag{2.3}$$

where C_0 (mg L^{-1}) is the initial concentration of RAC, C_t (mg L^{-1}) is the concentration at a time (minute), and C_e (mg L^{-1}) is the RAC equilibrium concentration. The adsorbent's dry weight, measured in *g*, is denoted by the symbol *W*, while the volume of the solution under study is denoted by *V* (*L*).

The results of the adsorption measurements, each of which was carried out in triplicate, were used to calculate the average values. In the batch studies that were carried out, the most successful combination parameters are at pH11, at the temperature of 25 °C with RAC starting concentration at 50 mg L^{-1}, adsorbent dosage of 0.004 g L^{-1}, and contact length of 50 min.

2.2.2.3 Regeneration Study

In this work, both the reusability of the adsorbent as well as its ability to regenerate were investigated. After the typical runs of the adsorption study, the study was carried out by separating the supernatant and washing the MOFs residue with 50 mL of methanol of 100% purity followed by distilled water. After that, the MOF was dried for 3 h using vacuum at a 100 °C, and the product was kept in a vial inside a desiccator to be used in the next experimental procedure. For reusability of the experiment, a total of six cycles in a row was conducted, and the regenerated [Cu $(INA)_2$]-MOF was subjected to the same ideal conditions that had been determined during the batch adsorption tests to investigate its potential for RAC adsorption.

2.2.3 *Adsorption Kinetics*

Kinetics research provides a comprehensive explanation toward the mechanism of adsorption and the process of rate-controlling phase. RAC adsorption data onto the [Cu $(INA)_2$]-MOF are applied using three important formulas or model approaches known as pseudo-first-order, pseudo-second-order, and intraparticle diffusion, expressed by the following Eqs. (2.4)–(2.6), respectively [33].

$$\ln(q_e - q_t) = \ln q_e - K_1 t \quad (2.4)$$

$$\frac{t}{q_t} = \frac{1}{K_2 q_{e^2}} + \frac{t}{q_e} \quad (2.5)$$

$$q_t = K_P t^{1/2} + C \quad (2.6)$$

where q_e and q_t are the amounts of RAC adsorbed at equilibrium and at a specific time, respectively, whereas K_1 and K_2 are the pseudo-first-order and pseudo-second-order rate constants, respectively. K_p is the intraparticle diffusion rate constants and C is the concentration.

2.2.4 Adsorption Isotherms

The equilibrium relationship between the adsorbate and the adsorbent, which determines how well the MOFs bind to the RAC, was determined by isotherm modeling of Langmuir, Freundlich, and Temkin [33]. This equilibrium relationship is described by the following respective linear equations in Eqs. (2.7)–(2.9).

$$\frac{C_e}{q_e} = \frac{1}{K_L q_m} + \frac{1}{q_m} C_e \quad (2.7)$$

$$\log q_e = \log K_F + \frac{1}{n} \log C_e \quad (2.8)$$

$$q_e = \frac{RT}{b_T} \ln A_T + \left(\frac{RT}{b_T}\right) \ln C_e \quad (2.9)$$

where q_m is the adsorption capacity, K_L and K_F are the Langmuir and Freundlich constants, respectively, $1/n$ is the adsorption intensity, b_T (kJ/mol) is the Temkin constant relating to the heat of adsorption, and A_T is the equilibrium binding constant corresponding to the highest binding energy (maximum) while R is the universal gas constant and T is the absolute temperature.

Additionally, these two models evaluate the best for the adsorption process [33]. The coefficient of determination (R^2), adjusted coefficient of determination (R^2 adj.), root means square error (RMSE), and Akaike information criterion (AIC) as can be determined fitted with statistical regression analysis using Eqs. (2.10)–(2.12) as follows:

$$R^2 = 1 \frac{\Sigma\left(q_{e\,exp} - q_{e\,cal}\right)^2}{\Sigma\left(q_{e\,exp}\right)^2} \quad (2.10)$$

$$\mathrm{RMSE} = \sqrt{\sum_{n=1}^{i}\left(q_{\mathrm{e\,exp}} - q_{\mathrm{e\,model}}\right)^2} \tag{2.11}$$

$$\mathrm{AIC} = n\ln\left(\frac{\mathrm{SSE}}{n}\right) + 2n_{\mathrm{p}} + \frac{2n_{\mathrm{p}}\left(n_{\mathrm{p}}+1\right)}{n\left(n_{\mathrm{p}}+1\right)} \tag{2.12}$$

where the experimental adsorption capacity is denoted by $q_{\mathrm{e\,exp}}$ and the model adsorption capacity, denoted by $q_{\mathrm{e\,model}}$. n is the number of observations, and p is the number of parameters. SSE is the sum of the square errors that were collected. In general, a higher R^2 value implies that the model is more linear, and a lower RMSE and AIC value indicates that the model fits the data very well.

2.2.5 *Thermodynamics Study*

The thermodynamics analysis was carried out based on temperature differences, with information on the adsorption process nature and with the involved parameters that govern it. Based on the adsorption process, parameters for the thermodynamic study which are the Gibbs free energy change ($\Delta G°$), entropy change ($\Delta S°$), and standard enthalpy change ($\Delta H°$) were determined using the following Eqs. (2.13)–(2.15) as stated by van't Hoff [34].

$$\Delta G^{\circ} = -\mathrm{RT}\ln K_{\mathrm{d}} \tag{2.13}$$

$$K_{\mathrm{d}} = \frac{q_{\mathrm{e}}}{C_{\mathrm{e}}} \tag{2.14}$$

$$\Delta G^{\circ} = \Delta H^{\circ} - T\Delta S \tag{2.15}$$

2.3 Results and Discussion

2.3.1 *Characterization*

The FTIR spectrum in Fig. 2.2 shows the infrared spectra for the MOF in the visible range. Table 2.1 shows a summary of the major functional groups and their peaks.

Figure 2.3 depicts the thermogravimetric analysis (TGA) curve with the essential information of the weight loss percentage against the temperature. TGA demonstrated a decrease in weight throughout the range of temperatures. At the temperature of 240 °C the first weight loss can be seen. Additionally, the thermogram

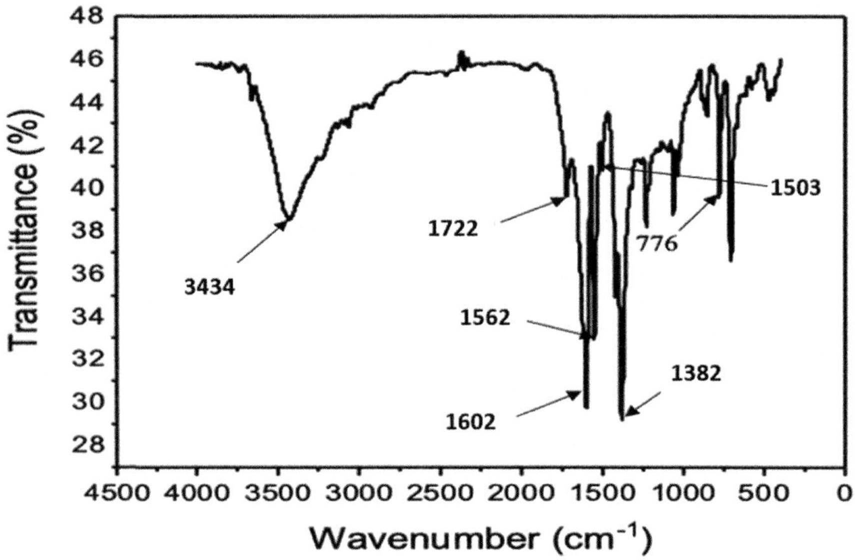

Fig. 2.2 FTIR spectrum of [Cu $(INA)_2$]-MOF

Table 2.1 FTIR summary

S. No.	Functional group	Peak (cm^{-1})
1	^{-}OH	3434
2	COO—asymmetric stretching	1562
3	COO—symmetric stretching	1503
4	Complete coordination of the isonicotinic acid	1722
5	Cu–O	776
6	C=N	1384

curves revealed that high level of thermal stability properties for [Cu $(INA)_2$]-MOFs compound, as the material showed an 82% in weight when heated to 246.38 °C. In 2001, Liu [35] also had the same successful result.

X-ray diffraction (XRD) was used to investigate the crystallinity of the sample. The XRD pattern previously described by Pichon [27], Lu and Babb [35], and Tella [32] correlated to high-intensity peaks at angles 2θ 10.57° and 21.25° in the diffractogram of the MOFs and the value of the Miller indices (hkl) were (100) and (200) respectively as shown in Fig. 2.4. The present MOF was found to have a crystalline size of 39.78 nm when the Scherrer equation [see Eq. (2.16)] [36, 37] was used to compute the sizes of sub-μm crystallites in a diffraction pattern.

$$D = 1 + \frac{K\lambda}{\beta \cos\theta}, \tag{2.16}$$

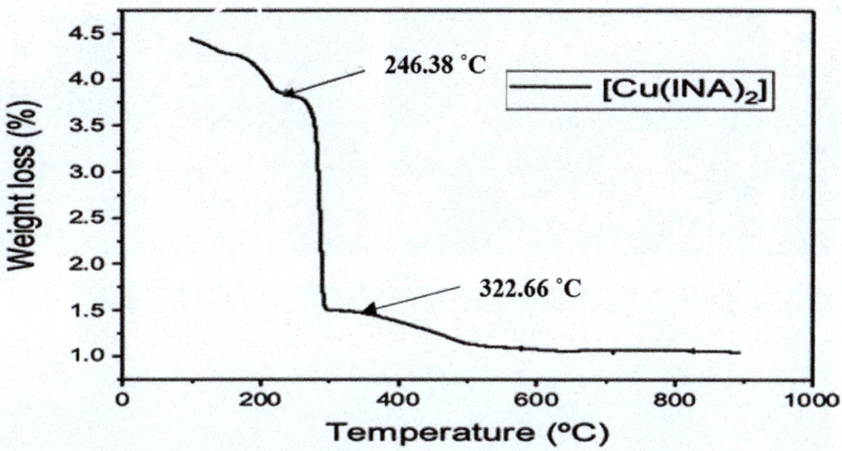

Fig. 2.3 TGA graph of [Cu (INA)$_2$]-MOF

where D is the average size of ordered (crystalline) domains, K is a dimensionless shape factor, the wavelength is represented by λ, while peak broadening at half maximum intensity (FWHM) by β, the Bragg constant is after deducting the θ and instrumental line broadening.

SEM images at different magnifications can be seen in Fig. 2.5 which shows morphological images of the microstructure and element properties of the MOFs from the secondary or backscattered electron signal to make detailed, high-resolution

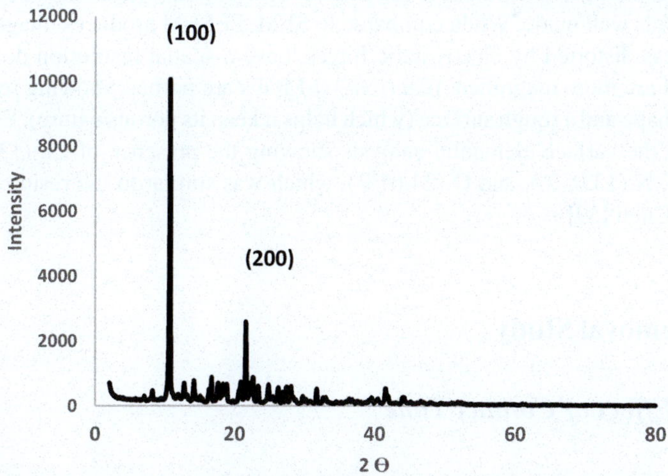

Fig. 2.4 XRD of [Cu (INA)$_2$]-MOF with the Miller indices (hkl), which define the reciprocal axial intercepts of a plane of atoms with the unit cell; more precisely, the (hkl) plane of atoms intercepts the unit cell

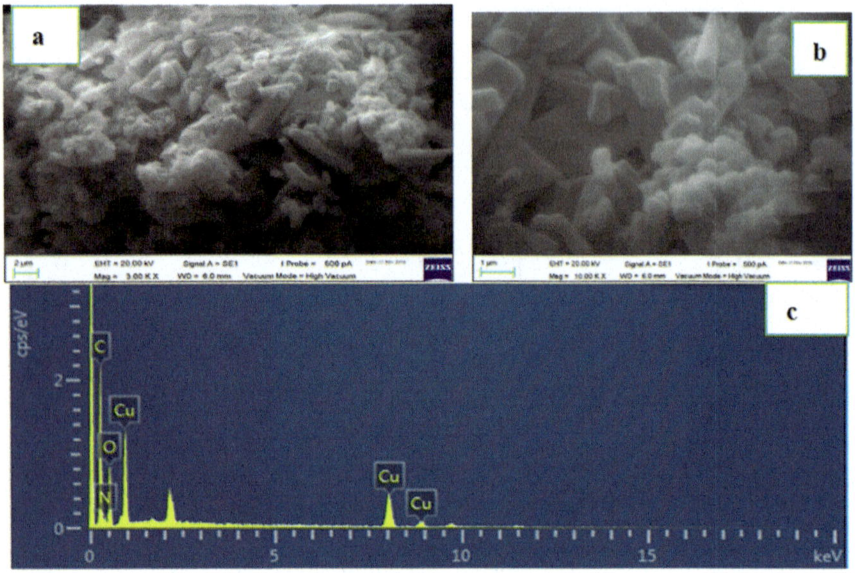

Fig. 2.5 SEM images of [Cu (INA)$_2$]-MOF material at a different magnification of **a** 3.00 K X at WD of 6.0 mm, **b** 5.00 K X at WD of 6.0 mm, **c** EDX spectrums of the [Cu(INA)$_2$]-MOF

images of the sample and additional EDX apparatus to identify the elements present in the [Cu (INA)$_2$]-MOFs material.

Figure 2.6 represents the FESEM images of MOFs from the mechanically generated method showing crystalline properties with edges, which indicated that the material was well made. When compared to SEM, FESEM produces images that are clearer, less distorted by electrostatic forces, have a spatial resolution down to 11/2 nm, and are more magnified. [Cu (INA)$_2$]-MOFs are porous, showing rectangular bar-like shape and a rough surface (which helps it keep its porous nature). EDX result presented the surface elemental analysis showing the presence of Cu (13.93%), C (48.41%), N (13.05%), and O (24.61%), which was similar to the results from Shi [38] and Chen [39].

2.4 Removal Study

2.4.1 Effect of Contact Time

Adsorption time before, during, and after equilibrium is always important, and it is always one of the most important things to look at when performing an adsorption experiment. This study finds that the best conditions for the adsorption of RAC on the synthesized adsorbent are an adsorbate with a concentration of 40 mg L^{-1},

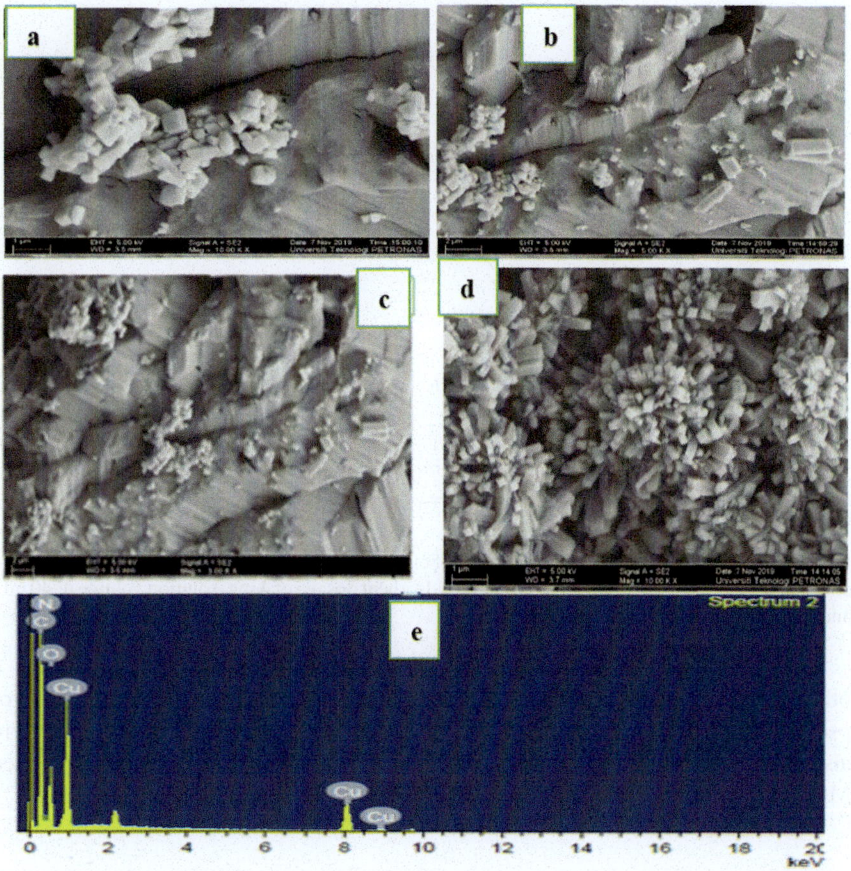

Fig. 2.6 FESEM images of [Cu $(INA)_2$]-MOF at the same point in a different magnification, **a** 3.00 K X at 3.5 mm, **b** 5.00 K X at 3.5 mm, **c** 10.00 K X at 3.5 mm, **d** 5.00 K X at 3.7 mm magnifications, **e** EDX spectrums of the [Cu $(INA)_2$]-MOF

and 4 mg of adsorbent dose, a pH of 11 at 25 °C. The adsorption of RAC onto [Cu $(INA)_2$]-MOF shown in Fig. 2.7 illustrates how the extent and influence the contact time have on the RAC study, as it was removed effectively at a high rate for the first 10 min of the process. After that, it continued to rise at a slower rate for the next 40 min reaching a steady level. This meant that 392 mg g^{-1} of RAC was removed. The quick removal of the pollutant was attributed to the good interactions between RAC molecules and the [Cu $(INA)_2$]-MOFs. These interactions were made possible by the presence of certain functional groups (hydroxyl groups ($^{-}$OH) and nitrogen ($^{-}$N)) and the aromatic nature of the MOF's surface, both of which are good adsorption sites [34], as well as the availability of unoccupied spaces in the surface of the MOF [40]. The adsorption was kept going for some more minutes to make sure that [Cu $(INA)_2$]-MOF would work well in the equilibrium state. Several

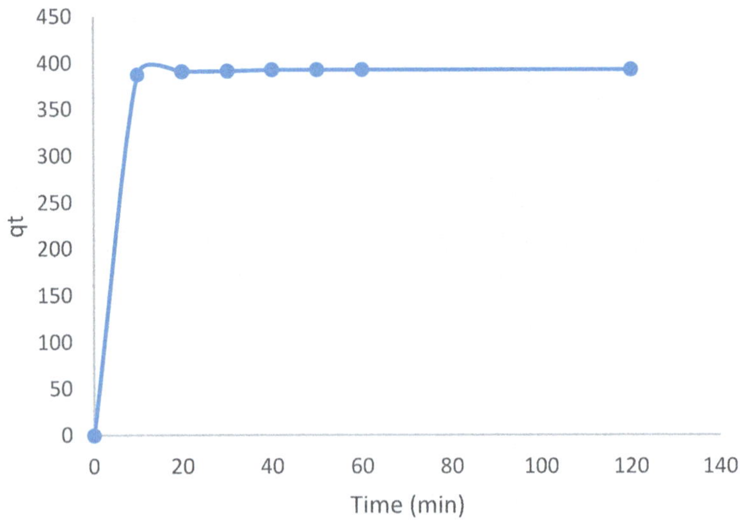

Fig. 2.7 Effect of contact time for adsorption of clenbuterol onto [Cu $(INA)_2$]-MOF with adsorbate concentration of 40 mg L^{-1} under pH of 11, an adsorbent dosage of 4 mg, solution volume of 40 mL, and temperature of 25 °C

other adsorption experiments [41, 42] showed the same observation. Linear plots of pseudo-first-order, pseudo-second-order, and intraparticle adsorption models were also used in the adsorption kinetics experiments to learn more about how the studied MOFs behave when they are used in adsorption analysis.

2.4.2 *Effect of pH*

The effect of different pH levels on RAC adsorption onto the [Cu $(INA)_2$]-MOF was investigated at 3, 5, 7, 9, 11, and 13 levels. As can be seen in Fig. 2.8, the findings indicate that the pH of the solution has a substantial impact on the efficiency of the MOF on RAC adsorption. Because RAC is phenolic-type β-agonist in which pH ranges of 10–11 are reportedly ideal for their extraction and adsorption [22, 43], its species are generally positively charged in solution at pH's ranging from 2 to 10. This is because RAC has an acidity constant (pK_a) of 9.4. The RAC adsorption process on Cu-MOFs is favored by this phenomenon, as evidenced by the fact that the adsorption efficiency increased as the pH of the solution increased up to pH 11 proven by a good percentage removal of approximately 92% and a very high adsorption capacity of more than 392.8 mg g^{-1} [20]. The attraction established between the negatively charged surface of the adsorbent and the RAC molecules is what causes a considerable increase in adsorption to take place when the pH of the solution is increased. According to zeta potential experiments carried out earlier in

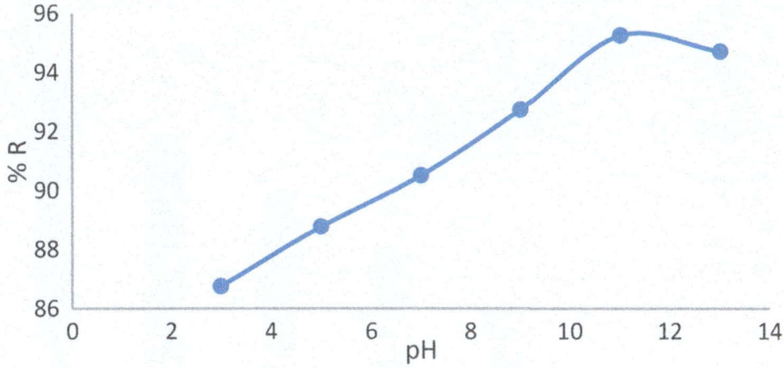

Fig. 2.8 Effect of pH on the adsorption of ractopamine onto [Cu $(INA)_2$]-MOF

our lab (result not reported), [Cu $(INA)_2$]-MOFs have a negatively charged surface at all pH levels. Therefore, better adsorption efficacy was found at a pH of 11. On the other hand, as the pH rises above 11, the amine group on the adsorbent may begin to progressively deprotonate, and there will be no H-donor site on them. Because of this, the adsorbent becomes deprotonated, which leads to a significant reduction in the efficacy of the adsorption process at pH levels higher than 11 [44, 45].

2.4.3 Effect of Ractopamine Initial Concentration

Adsorption experiments were carried out with an adsorbent dose of 4 mg g^{-1} to evaluate the influence of RAC initial concentrations ranging from 10 to 60 mg L^{-1}. As can be observed in Fig. 2.9, the effect of the initial RAC concentration on adsorption became more significant as higher concentrations of RAC were used. This is because of an increase in the amount of RAC adsorption that takes place at various starting doses. When there was a higher concentration of RAC molecules in the solution, the active sites of the MOF were surrounded by a significantly higher number of RAC molecules, and the adsorption was both more efficient and easier to do. Tella [32] and Han [46] found the same result. The findings of our research demonstrated that the RAC removal adsorption capacity of the MOFs was greater than 392.8 mg/L when working with a 40 mL solution containing an adsorbate concentration of about 40 mg L^{-1}, with 4 mg of adsorbent dose, and at pH11 at a temperature of 25 °C. Additionally, a significant removal percentage of more than 98.0% was achieved. It was found that the quantity of RAC that was adsorbed onto [Cu $(INA)_2$]-MOF was directly proportional to the amount of adsorbate that was utilized. This result was also obtained by Hamza [42].

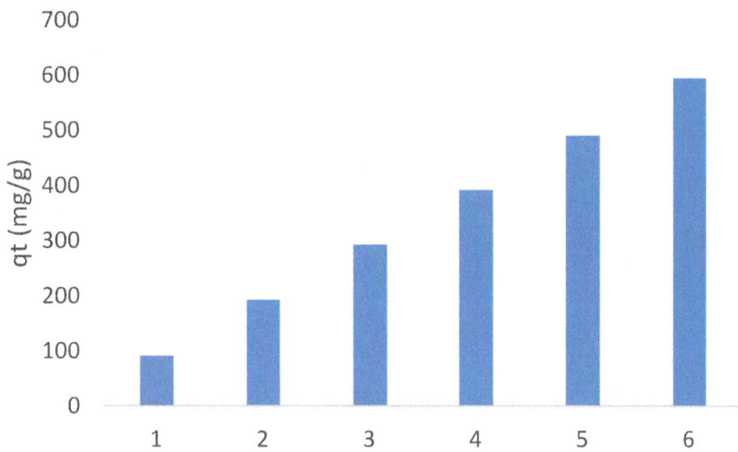

Fig. 2.9 Effect of ractopamine concentrations for adsorption onto [Cu $(INA)_2$]-MOF, adsorbate concentration of 40 mg L^{-1}, pH of 11, dosage of 4 mg, solution volume of 40 mL at 25 °C

2.4.4 *Effect of [Cu $(INA)_2$]-MOF Dosage*

For the adsorbent to be considered useful and worthy the materials need to be capable of achieving maximum adsorption efficacy even at the smallest dose practically possible and in the shortest length of time [7]. The effect of the adsorbent's dose on RAC adsorption was studied using 40 ml solution volume, a concentration of 40 mg L^{-1}, maintained at 25 °C, under an alkaline pH of 11. A dosage of 4 mg g^{-1} of [Cu $(INA)_2$]-MOF was determined to be effective at 40 min of equilibrium contact time. The MOFs % removal was extremely effective and the same above 96% around 4, 5, and 6 mg g^{-1} but 4 mg g^{-1} was used because it had a larger adsorption capacity (with lower dose) as compared to 5 6 and 6 6 mg g^{-1}.

2.4.5 *Kinetics of the Adsorption Studies*

The interaction between the adsorbate and the surface of the MOF is what decides whether there is a chance of adsorption of the analyte RAC onto the mechano-synthesized [Cu $(INA)_2$]-MOF. After the runs of adsorption process and running the three most prominent models, the pseudo-second-order model has the most robust support and best fit for the adsorption data. These statistical values are presented in Table 2.2. In addition to this, the values of q_e that were measured (in mg g^{-1}), which were included in the results, match the experimental values. Tang [15] also obtained similar results when he used [Cu $(INA)_2$]-MOF in his adsorption studies. The pseudo-second-order model was selected as the model for the adsorption of the

β-agonist on the [Cu $(INA)_2$]-MOF since it had an exceptional level of agreement with all the equilibrium concentrations and statistical values that were observed.

2.4.6 Isotherm Studies

The adsorption data for the surface contact of the [Cu $(INA)_2$]-MOF with the RAC molecules at the process equilibrium were reported in Table 2.3 and were examined utilizing three different major isotherm models, namely the Langmuir, Freundlich, and Temkin isotherm models. When compared to the other models, Langmuir has the greater R^2 of 1, adj. R^2 of 0.899, RMSE of 0.001, and AIC value of − 129.171 (the graphs are presented). Based on the statistical information presented above, and the best-fit model for the current inquiry is Langmuir model which best represents the experimental data. Tang [38] identified a similar scenario when he used a covalent imprinting approach to explore the capacity of a molecularly imprinted polymer to determine β-agonist compound in a portable water sample. The Langmuir finding also reveals homogeneous adsorption sites on the adsorption surface, implying the formation of a RAC monolayer.

2.4.7 Effect of Temperature and Thermodynamic Studies

The temperature studies and thermodynamics analysis for the adsorption of RAC toward the [Cu $(INA)_2$]-MOFs were carried out at varied temperature ranges of 25, 30, 35, 40, 45, to 50 °C, which is equivalent to 298, 303, 308, 313, 318, to 323 K. The batch adsorption experiments were performed in a 40 mL solution containing an adsorbate concentration of 40 mg L^{-1} under optimal alkaline (Fig. 2.10). The values from the percentage removal decreased from 81 to 75%, which corresponds to the adsorption capacity reduction from 373 down to 307 mg g^{-1} (Fig. 2.10), indicating that the mechanism was an endothermic reaction [47] and that the best adsorption temperature is 25 °C (lowest); this has been further confirmed by the number of adsorption temperatures carried out at different studies from researchers [20]. When the temperature increases, the contact between the molecules of the adsorbate and the molecules of the adsorbent decreases, which is a clear indication that physisorption is occurring [48]. In addition, the increase in temperature caused a reduction in the values of all the thermodynamic parameters (Table 2.4) for the Gibbs free energy change ($\Delta G°$), indicating that the process is endothermic. It can be deduced from the presence of this that there is an unfavorable spontaneous process occurring when the temperature is high [7]. Positive entropy changes ($\Delta S°$) were also seen, which provided more evidence that there was some degree of randomness in the system and the amount of disorder.

Table 2.2 Adsorption kinetics parameters for the removal of ractopamine onto the [Cu $(INA)_2$] MOFs

Pseudo-first-order	(mg g^{-1})	$q_{e, exp}$ (mg L^{-1})	$q_{e, cal}$ (mg L^{-1})	K_1 $(min)^{-1}$	R^2	R^2adj	RMSE	AIC
	40	392.86	1.6273	0.014	0.3344	0.542	14.038	27.864
Pseudo-second-order	(mg g^{-1})	$q_{e, exp}$ (mg L^{-1})	$q_{e\ cal}$ (g mg^{-1})	K_2 (g mg^{-1} min^{-1})	R^2	R^2adj	RMSE	AIC
	40	392.86	400	0.03125	1	1	0.048	-40.81
Intraparticle diffusion	(mg g^{-1})		Kp (mg^{-1} g^{-1} $min^{1/2}$) C		R^2	R^2adj	RMSE	AIC
	1		3.3234	274.13	0.2044	0.280	15.441	39.963

Table 2.3 Isotherm parameters for adsorption of ractopamine onto $[Cu(INA)_2]$-MOF

Isotherm model	Parameter	$Cu(INA)_2$ ractopamine
Langmuir	q_m (mg L^{-1})	400
	K_L (L mg^{-1})	250
	R_L	9.9
	R^2	0.9996
	R^2adj	0.114
	RMSE	0.018
	AIC	− 46.452
Freundlich	K_F (mg g^{-1})	102.62
	N	0.02889
	R^2	0.1158
	R^2adj	0.093
	RMSE	0.020
	AIC	− 45.193
Temkin	A (mg g^{-1})	237.36
	b_T (kJmol^{-1})	3.617
	R^2	0.1242
	R^2adj	− 0.095
	RMSE	34.743
	AIC	44.143

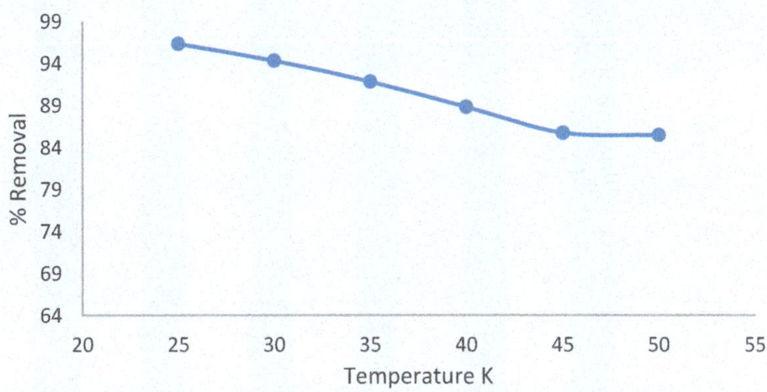

Fig. 2.10 Effect of temperature for ractopamine adsorption onto [Cu $(INA)_2$]-MOF

2.4.8 Adsorbent Regeneration

Indicative of their potential as an adsorbent for the purposes of removal is the simplicity with which it is possible to recover the adsorbent following active batch

Table 2.4 Thermodynamic parameters for the adsorption of ractopamine onto ([Cu (INA)$_2$])-MOFs

Temp (K)	$\Delta G°$ (kJ/mol)	$\Delta H°$ (kJ/mol)	$\Delta S°$ (J/mol/K)
298	– 0.00803	0.2156	0.0277
303	– 0.00818		
308	– 0.00832		
313	– 0.00846		
318	– 0.00859		
323	– 0.00873		

adsorption. As a consequence of this, the synthesized [Cu (INA)$_2$]-MOF was regenerated after active adsorption using the method described in Sect. 2.3, and the adsorption of RAC was carried out in a 40 mL solution with an adsorbate concentration of 40 mg L^{-1} at an ideal alkaline pH of 11 under 40 min, as was previously determined. Adsorption effectiveness during the first cycle and each of the succeeding six cycles was in the region of 98%, and the maximum adsorption capacity reached more than 392 mg g^{-1} throughout all the cycles (Fig. 2.11). The higher values obtained indicated that it is possible to attribute them to extra heating of the adsorbent after every adsorption cycle (before every recycle study), which increases the porosity of the solid adsorbent. Additionally, it was supported by the data obtained that the [Cu (INA)$_2$]-MOF had high reusability for the adsorption of RAC.

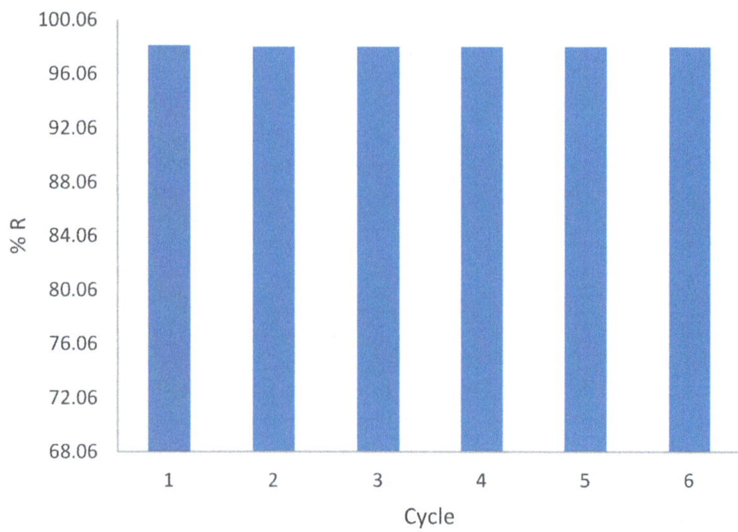

Fig. 2.11 Regeneration and reusability of the [Cu (INA)$_2$]-MOF for the adsorption of ractopamine

Table 2.5 Comparison of different adsorbents used for the adsorption of β-agonist with our studied adsorbent

Adsorbent materials	Equilibrium time (min)	Q_{max} (mg g^{-1})	%Removal	Analytes	References
MIP@UCPs prove	20	2.18	81.7	Clenbuterol	[14]
Activated carbon (GAC)	200	Unprovided	32–98	Clenbuterol	[16]
MIP@CIP	20	7.34	74.0	Clenbuterol	[15]
Ui-66 MOF	60	160	80.0	Clenbuterol	[21]
Al(III)-MOF (BUT-19)	35	362	51.74	Ractopamine	[22]
[Cu $(INA)_2$] MOF	40	400	98	Ractopamine	This study

2.4.9 *Comparison Studies*

A comparison of the relevant characteristics of adsorbents used for the removal of RAC in water with our reported [Cu $(INA)_2$]-MOF is presented in Table 2.5. Because of the paucity of relevant data on the removal of RAC in the published literature, as well as the removal of other β-agonists in general. Data in Table 2.5 indicate the fact that our study has a significantly higher adsorption capacity of over 392 mg g^{-1} and a significant percentage removal of 98% over the period of equilibrium time. Furthermore, in addition to the benefit of conveniently regenerating the adsorbent, which has not been addressed in many of the articles reviewed, the maximum adsorption capacity (Q_{max}) of 400 mg g^{-1} discovered by our research, other studies that were quoted did not provide this value, and the values that they did provide were lower than the one discovered by our study. However, precise comparisons are not possible because the scientists used such a wide variety of experimental and laboratory conditions in their work. A clear good adsorption capacity can be seen in [22] work but still not as much excellent adsorption capacity and percent removal as the recent work and the adsorption process followed by Lv [22] like the solvent synthesis is not fully as the recent work that used greener synthesis.

2.5 Conclusion

In this research, both the efficient synthesis of [Cu $(INA)_2$]-MOFs via a mechanochemical method and an investigation into its characterization are discussed as well as the adsorption study was successfully achieved. The FTIR examination found evidence of the existence of active functional groups in the MOF, while the TG analysis found evidence of increased thermal stability in the framework, with

534.32 °C temperature at which the framework decomposed. The good crystallinity of the MOFs was observed at 38.9 nm. The [Cu $(INA)_2$]-MOFs were able to effectively remove RAC after only 40 min, demonstrating the fast removal of the compound. The alkaline pH of 11 was selected because it produces the best adsorption results during the pH optimization process. Additionally, increasing the adsorbate (analyte) concentration from 10 to 60 mg L^{-1} improved adsorption efficiency, and 40 mg L^{-1} was regarded as the optimal condition for adsorption. On the other hand, an adsorbent dosage of 4 mg g^{-1} produced an excellent adsorption effectiveness as well. The statistical fits of R^2, RMSE, and AIC favored the pseudo-second-order kinetics of the process as the most appropriate explanation for the data. Isotherm data from the process pointed to the Langmuir model as the one that had the best statistical fit, as determined by fitting of statistical models. The investigation into thermodynamics revealed that an endothermic process was occurring and that temperature increase affects both the adsorption efficiency and capacity, while the reusability study shows the ability of the MOFs to maintain its high adsorption efficiency for up to six cycles and more. [Cu $(INA)_2$]-MOFs had an efficiency of 98% and an adsorption capacity of 392 mg L^{-1}. [Cu $(INA)_2$]-MOFs had a remarkable ability to remove RAC from environmental waters in a short period of time, according to the findings acquired from our study. This research recommended the [Cu $(INA)_2$]-MOFs for application in the removal of all β-agonists as they all have similar general physical and chemical characteristics which can include their chemical reactions under basic pH.

Acknowledgements The authors would like to express their profound gratitude to the financial institutions, namely Universiti Malaysia Terengganu and Al-Qalam University Katsina, Nigeria, for their overwhelming assistance in whatever capacity.

References

1. Gu, C., Ren, P., Zhang, F., Zhao, G., Shen, J., Zhao, B.: Detection of six β-agonists by three multiresidue immunosensors based on an anti-bovine serum albumin-ractopamine-clenbuterol-salbutamol antibody. ACS Appl. Mater. Interfaces (2020). https://doi.org/10.1021/acsomega.0c00249
2. Hakk, H., Shelver, W.L., Casey, F.X.M.: Fate and transport of the β-adrenergic agonist ractopamine hydrochloride in soil–water systems. J. Environ. Sci. (China), 40–48 (2015). https://doi.org/10.1016/j.jes.2015.11.026
3. Preechakasedkit, P., Ngamrojanavanich, N., Khongchareonporn, N., Chailapakul, O.: Novel ractopamine–protein carrier conjugation and its application to the lateral flow strip test for ractopamine detection in animal feed. J. Zhejiang Univ. B **20**(2), 193–204 (2019). https://doi.org/10.1631/jzus.B1800112
4. Cun, L., et al.: Simultaneous determination of clenbuterol, salbutamol and ractopamine in milk by reversed-phase liquid chromatography tandem mass spectrometry with isotope dilution. J. Chromatogr. A **1217**(50), 7873–7877 (2010). https://doi.org/10.1016/j.chroma.2010.10.055
5. Catalano, D., et al.: Physiopathological changes related to the use of ractopamine in swine: clinical and pathological investigations. Livest. Sci. **144**(1–2), 74–81 (2012). https://doi.org/10.1016/j.livsci.2011.10.016

6. Yaeger, M.J., Mullin, K., Ensley, S.M., Ware, W.A., Slavin, R.E.: Myocardial toxicity in a group of greyhounds administered ractopamine. Vet. Pathol. **49**(3), 569–573 (2012). https://doi.org/10.1177/0300985811424752
7. Zhang, L.Y., Chang, B.Y., Dong, T., He, P.L., Yang, W.J., Wang, Z.Y.: Simultaneous determination of salbutamol, ractopamine, and clenbuterol in animal feeds by SPE and LC–MS. J. Chromatogr. Sci. **47**(4), 324–328 (2009). https://doi.org/10.1093/chromsci/47.4.324
8. Armaya'u, U., Ariffin, M.M., Loh, S.H., Mohd Khalik, W.M.A.W., Yusoff, H.M.: β-agonist in the environmental waters: a review on threats and determination methods. Green Chem. Lett. Rev. **15**(1), 232–251 (2022). https://doi.org/10.1080/17518253.2022.2032843
9. Zhang, W., Wang, P., Su, X.: Current advancement in analysis of β-agonists. TrAC Trends Anal. Chem. **85**, 1–16 (2016). https://doi.org/10.1016/j.trac.2016.08.011
10. Jaimes-Correa, J.C., Snow, D.D., Bartelt-Hunt, S.L.: Seasonal occurrence of antibiotics and a beta agonist in an agriculturally-intensive watershed. Environ. Pollut. **205**, 87–96 (2015). https://doi.org/10.1016/j.envpol.2015.05.023
11. Rezazadeh, M., Yamini, Y., Seidi, S., Maryam Rezazadeh, S.S., Yamini, Y.: Electrically assisted liquid-phase microextraction for determination of β2-receptor agonist drugs in wastewater. J. Sep. Sci. **35**(4), 571–579 (2012). https://doi.org/10.1002/jssc.201100869
12. Rodil, R., Quintana, J.B., Concha-Graña, E., López-Mahía, P., Muniategui-Lorenzo, S., Prada-Rodríguez, D.: Emerging pollutants in sewage, surface and drinking water in Galicia (NW Spain). Chemosphere **86**(10), 1040–1049 (2012). https://doi.org/10.1016/j.chemosphere.2011.11.053
13. Quintana, J.B., Rodil, R., Cela, R.: Reaction of β-blockers and β-agonist pharmaceuticals with aqueous chlorine. Investigation of kinetics and by-products by liquid chromatography quadrupole time-of-flight mass spectrometry. Anal. Bioanal. Chem. **403**(8), 2385–2395 (2012). https://doi.org/10.1007/s00216-011-5707-7
14. Tang, Y., et al.: Upconversion particles coated with molecularly imprinted polymers as fluorescence probe for detection of clenbuterol. Biosens. Bioelectron. **71**, 44–50 (2015). https://doi.org/10.1016/j.bios.2015.04.005
15. Tang, Y., et al.: Determination of clenbuterol in pork and potable water samples by molecularly imprinted polymer through the use of covalent imprinting method. Food Chem. **190**, 952–959 (2016). https://doi.org/10.1016/j.foodchem.2015.06.067
16. de Ridder, D.J., et al.: Influence of natural organic matter on equilibrium adsorption of neutral and charged pharmaceuticals onto activated carbon. Water Sci. Technol. **63**(3), 416–423 (2011). https://doi.org/10.2166/wst.2011.237
17. Lan, Y., Coetsier, C., Causserand, C., Serrano, K.G.: An experimental and modelling study of the electrochemical oxidation of pharmaceuticals using a boron-doped diamond anode. Chem. Eng. J. **333**(2017), 486–494 (2018). https://doi.org/10.1016/j.cej.2017.09.164
18. Yan, S., et al.: Development of fluorescence surrogates to predict the photochemical transformation of pharmaceuticals in wastewater effluents. Environ. Sci. Technol. **51**(5), 2738–2747 (2017). https://doi.org/10.1021/acs.est.6b05251
19. Sakkas, V.A., et al.: Heterogeneous photocatalytic degradation of the pharmaceutical agent salbutamol in aqueous titanium dioxide suspensions. Appl. Catal. B Environ. **77**(1–2), 135–144 (2007). https://doi.org/10.1016/j.apcatb.2007.07.017
20. Selkälä, T., et al.: Rapid uptake of pharmaceutical salbutamol from aqueous solutions with anionic cellulose nanofibrils: the importance of pH and colloidal stability in the interaction with ionizable pollutants. Chem. Eng. J. **350**, 378–385 (2018). https://doi.org/10.1016/j.cej.2018.05.163
21. Yang, H., et al.: Determination and removal of clenbuterol with a stable fluorescent zirconium(IV)-based metal organic framework. Microchim. Acta **186**(7), 454 (2019). https://doi.org/10.1007/s00604-019-3586-3
22. Lv, J., et al.: Effective removal of clenbuterol and ractopamine from water with a stable Al(III)-based metal-organic framework. Inorg. Chem. **60**(3), 1814–1822 (2021). https://doi.org/10.1021/acs.inorgchem.0c03296

23. Neshastehgar, M., Rahmani, P., Shojaei, A., Molavi, H.: Enhanced adsorption removal performance of UiO-66 by rational hybridization with nanodiamond. Microporous Mesoporous Mater. **296**(2019), 110008 (2020). https://doi.org/10.1016/j.micromeso.2020.110008
24. Zhao, R., Ma, T., Zhao, S., Rong, H., Tian, Y., Zhu, G.: Uniform and stable immobilization of metal-organic frameworks into chitosan matrix for enhanced tetracycline removal from water. Chem. Eng. J. **382**, 122893 (2020). https://doi.org/10.1016/j.cej.2019.122893
25. Zango, Z.U., et al.: A critical review on metal-organic frameworks and their composites as advanced materials for adsorption and photocatalytic degradation of emerging organic pollutants from wastewater. Polymers (Basel) **12**(11), 2648 (2020). https://doi.org/10.3390/polym12112648
26. Li, J.R., Sculley, J., Zhou, H.C.: Metal-organic frameworks for separations. Chem. Rev. **112**(2), 869–932 (2012). https://doi.org/10.1021/cr200190s
27. Pichon, A., Lazuen-Garay, A., James, S.L.: Solvent-free synthesis of a microporous metal-organic framework. CrystEngComm **8**(3), 211–214 (2006). https://doi.org/10.1039/b513750k
28. Janiak, C.: Engineering coordination polymers towards applications. J. Chem. Soc. Dalt. Trans. **3**(14), 2781–2804 (2003). https://doi.org/10.1039/b305705b
29. Ni, Z., Masel, R.I.: Rapid production of metal-organic frameworks via microwave-assisted solvothermal synthesis. J. Am. Chem. Soc. **128**(38), 12394–12395 (2006). https://doi.org/10.1021/ja0635231
30. Abdollahi, N., Masoomi, M.Y., Morsali, A., Junk, P.C., Wang, J.: Sonochemical synthesis and structural characterization of a new Zn(II) nanoplate metal–organic framework with removal efficiency of Sudan red and Congo red. Ultrason. Sonochem. **45**(Feb), 50–56 (2018). https://doi.org/10.1016/j.ultsonch.2018.03.001
31. Li, Z., Cui, J., Liu, Y., Li, J., Liu, K., Shao, M.: Electrosynthesis of well-defined metal-organic framework films and the carbon nanotube network derived from them toward electrocatalytic applications. ACS Appl. Mater. Interfaces **10**(40), 34494–34501 (2018). https://doi.org/10.1021/acsami.8b12854
32. Tella, A.C., Owalude, S.O., Ojekanmi, C.A., Oluwafemi, O.S.: Synthesis of copper-isonicotinate metal-organic frameworks simply by mixing solid reactants and investigation of their adsorptive properties for the removal of the fluorescein dye. New J. Chem. **38**(9), 4494–4500 (2014). https://doi.org/10.1039/c4nj00411f
33. Isiyaka, H.A., Jumbri, K., Sambudi, N.S., Zango, Z.U., Saad, B., Mustapha, A.: Removal of 4-chloro-2-methylphenoxyacetic acid from water by MIL-101(Cr) metal-organic framework: kinetics, isotherms and statistical models. R. Soc. Open Sci. **8**(1), 201553 (2021). https://doi.org/10.1098/rsos.201553
34. Saruchi, Kumar, V.: Adsorption kinetics and isotherms for the removal of rhodamine B dye and Pb^{+2} ions from aqueous solutions by a hybrid ion-exchanger. Arab. J. Chem. **12**(3), 316–329 (2019). https://doi.org/10.1016/j.arabjc.2016.11.009
35. Lu, J.Y., Babb, A.M.: An extremely stable open-framework metal-organic polymer with expandable structure and selective adsorption capability. Chem. Commun. **2**(13), 1340–1341 (2002). https://doi.org/10.1039/b200213b
36. Gencalp Irizalp, S., Saklakoglu, N.: Laser peening of metallic materials, vol. 1–3 (2017). https://doi.org/10.1016/B978-0-12-803581-8.09160-8
37. Rabiei, M., Palevicius, A., Monshi, A., Nasiri, S., Vilkauskas, A., Janusas, G.: Comparing methods for calculating nano crystal size of natural hydroxyapatite using X-ray diffraction. Nanomaterials **10**(9), 1–21 (2020). https://doi.org/10.3390/nano10091627
38. Shi, G., Xu, W., Wang, J., Yuan, Y., Chaemchuen, S., Verpoort, F.: A Cu-based MOF for the effective carboxylation of terminal alkynes with CO_2 under mild conditions. J. CO_2 Util. **39**(May), 101177 (2020). https://doi.org/10.1016/j.jcou.2020.101177
39. Chen, Y., Li, L., Li, J., Ouyang, K., Yang, J.: Ammonia capture and flexible transformation of M-2(INA) (M=Cu Co, Ni, Cd) series materials. J. Hazard. Mater. **306**, 340–347 (2016). https://doi.org/10.1016/j.jhazmat.2015.12.046
40. Tella, A.C., et al.: Synthesis of zinc-carboxylate metal-organic frameworks for the removal of emerging drug contaminant (amodiaquine) from aqueous solution. J. Environ. Sci. (China) **64**(June), 264–275 (2018). https://doi.org/10.1016/j.jes.2017.06.015

41. Zango, Z.U., et al.: Adsorption of chrysene in aqueous solution onto MIL-88(Fe) and NH_2-MIL-88(Fe) metal-organic frameworks: Kinetics, isotherms, thermodynamics and docking simulation studies. J. Environ. Chem. Eng. **8**(2), 103544 (2020). https://doi.org/10.1016/j.jece.2019.103544
42. Isiyaka, H.A., et al.: Adsorption of dicamba and MCPA onto MIL-53(Al) metal–organic framework: response surface methodology and artificial neural network model studies. RSC Adv. **10**(70), 43213–43224 (2020). https://doi.org/10.1039/D0RA07969C
43. Spisso, B.F., Lopes, C.C., Marques, M.A., Neto, F.R.: Determination of beta-agonists in bovine urine : comparison of two extraction/clean-up procedures for high-resolution gas chromatography-mass spectrometry analysis. J. Anal. Toxicol. **24** (2000)
44. Zhou, L., Sleiman, M., Ferronato, C., Chovelon, J.M., de Sainte-Claire, P., Richard, C.: Sulfate radical induced degradation of $B2$-adrenoceptor agonists salbutamol and terbutaline: Phenoxyl radical dependent mechanisms. Water Res. **123**, 715–723 (2017). https://doi.org/10.1016/j.watres.2017.07.025
45. Zhou, L., Wang, Q., Zhang, Y., Ji, Y., Yang, X.: Aquatic photolysis of $\beta 2$-agonist salbutamol: kinetics and mechanism studies. Environ. Sci. Pollut. Res. **24**(6), 5544–5553 (2017). https://doi.org/10.1007/s11356-016-8207-7
46. Han, R., Zou, W., Yu, W., Cheng, S., Wang, Y., Shi, J.: Biosorption of methylene blue from aqueous solution by fallen phoenix tree's leaves. J. Hazard. Mater. **141**(1), 156–162 (2007). https://doi.org/10.1016/j.jhazmat.2006.06.107
47. Zango, Z.U., Sambudi, N.S., Jumbri, K., Bakar, N.H.H.A., Saad, B.: Removal of pyrene from aqueous solution using fe-based metal-organic frameworks. IOP Conf. Ser. Earth Environ. Sci. **549**(1), 012061 (2020). https://doi.org/10.1088/1755-1315/549/1/012061
48. Tella, A., Nwosu, F., Adimula, V.: Removal of Anthracene from solution using $[Cu(INA)_2]$ metal-organic frameworks synthesized by a solvent free method. Niger. J. Mater. Sci. Eng. **7**, 26 (2016)

Chapter 3
A Computational Study on the Performance and Emission Characteristics of Soy Biodiesel Fuel in Marine Diesel Engines

Che Wan Mohd Noor, Abdin Shakirin Zufayri, Rizalman Mamat, Ahmad Fitriadhy, Mohammad Fadhli Ahmad, and Sheikh Alif Ali

Abstract Biodiesel is a promising alternative to conventional diesel fuel due to its renewable and environmentally friendly nature. However, the use of biodiesel in engines can lead to decreased performance and increased emissions. This study aims to investigate the effect of soybean methyl ester (SME) biodiesel fuel on the performance and emissions of a marine diesel engine compared to diesel no. 2 fuel. The experiment was conducted using a computational software, Diesel RK, with a Cummins NT-855 M engine as the engine parameter. Biodiesel blends with varying percentages of SME (5, 20, 40, and 100%) were tested along with diesel no. 2 fuel. The engine parameters tested were brake torque, specific fuel consumption, and brake thermal efficiency, while the emission parameters were the fraction of wet NO_x and specific particulate matter. The results showed that the brake torque produced by biodiesel blends was lower by 1.48–26.9% compared to diesel no. 2 fuel. Moreover, specific fuel consumption and brake thermal efficiency for biodiesel

C. W. M. Noor (✉) · A. S. Zufayri · A. Fitriadhy · M. F. Ahmad · S. A. Ali
Faculty of Ocean Engineering Technology and Informatics, Universiti Malaysia Terengganu, 21030, Kuala Nerus, Terengganu, Malaysia
e-mail: che.wan@umt.edu.my

A. S. Zufayri
e-mail: abdinfaye@gmail.com

A. Fitriadhy
e-mail: a.fitriadhy@umt.edu.my

M. F. Ahmad
e-mail: fadhli@umt.edu.my

S. A. Ali
e-mail: sheikh_alif@umt.edu.my

C. W. M. Noor · R. Mamat
Centre for Automotive Engineering, Universiti Malaysia Pahang, 26600 Pekan, Pahang, Malaysia
e-mail: rizalman@ump.edu.my

A. Ismail et al. (eds.), *Exploring Diversity in Engineering and Technology for Knowledge and Innovation*, Advanced Structured Materials 215,
https://doi.org/10.1007/978-3-031-64330-9_3

blends were higher by 4.36% to 59% and 11.32% to 53.79%, respectively, compared to diesel no. 2 fuel. Additionally, the emission of NO_x and particulate matter from biodiesel was higher by 8.06% and 81.37%, respectively, compared to diesel no. 2 fuel. Overall, the study provides insight into the effects of SME biodiesel blends on marine diesel engine performance and emissions, which can help in the development of more efficient and sustainable fuel options for the marine industry.

Keywords Soy biodiesel · Marine engine · Engine performance · Engine emissions · Diesel RK

3.1 Introduction

The diesel engine plays a crucial role in various sectors, such as agriculture, maritime transport, automotive, and air transport. However, its continuous usage without proper regard for the environment has had significant impacts. The exponential increase of greenhouse gases in the atmosphere has led to a health crisis and climate change, resulting in acid rain and ozone depletion [1]. With the growing pattern of industrialization and transformation, the demand for world energy is also increasing at a faster pace. Many countries import diesel fuel to meet their energy needs, which are primary sources of carbon monoxide (CO), carbon dioxide (CO_2), and sulfur oxide (SOx) emissions [2].

Resource scarcity has also been a persistent issue for fossil fuels, particularly for farm-based fuels. Thus, alternative, safe, and sustainable energy sources are being researched to improve the performance of internal combustion engines and minimize greenhouse gas emissions due to decreased oil supplies, increased environmental consciousness, and stringent emission laws [3]. Biodiesel is one such alternative that can be made with vegetable oils or animal fats consisting of long-chain alkyl (methyl, propyl, or ethyl) esters [4]. In the marine sector, diesel engines are used for marine vessels, and they are much larger than standard diesel engines. These engines have super long-stroke engines compared to regular diesel engines, and they typically use heavy fuel oil diesel and standard diesel fuel. Moreover, the components in both engines are different due to their characteristics.

Several researchers have researched the performance and emission characteristics of marine diesel engines with biodiesel fuel, a renewable source of fuel. The latest study by [5], using fatty acid methyl ester (PFAME) biodiesel on a Cummins NT-855M engine, reported that the use of PFAME biodiesel has the advantage of lowering toxic emissions by 7.8% and 9.1% for nitrogen oxide and carbon dioxide, respectively, which are favorable to the atmosphere in the long run. They also found that the engine combustion improved with the presence of oxygen in the biodiesel, contributing to an increase in peak pressure and heat release by 0.18%–2.73% and 0.88%–4.49%, respectively.

In order to contribute to this field, this study aims to determine the performance and emission of a marine diesel engine using soybean methyl ester biodiesel. The study uses the Diesel RK software simulation, which is a low-cost and time-efficient method for multiple engine operation tests [6, 7]. The software's customizability also allows for the specification of various engine types. Other finding also agreed that the software is user-friendly, freely available, and has been validated for use in dual-fueled engine simulation [8]. This study's results are expected to be useful for numerical researchers as it uses Diesel RK software to run the experiment.

3.2 Methodology

In this computational study, the effect of different fuels on a marine diesel engine was modeled using a computational software. Diesel RK software was used in this study to simulate and analyze the engine test conditions. To set up the simulation, the engine specifications, such as the geometry, injection system, and combustion characteristics, were entered into the software. The software was also configured with the fuel properties and the test conditions, such as the engine speed and load.

In this simulation, two types of fuel were used, namely the standard diesel fuel, diesel no. 2, and soybean methyl ester (SME) biodiesel. The standard diesel fuel is composed of 100% diesel fuel and can also be referred to as B0. The SME biodiesel fuel was used in varying mixtures of 5, 20, 40, and 100% volume to create B5, B20, B40, and B100 SME biodiesel fuels. Table 3.1 provides the fuel properties data for the tested fuels.

The engine used in this study was six-cylinder, in-line, four-stroke Cummins NT-855M. The engine is commonly used as a prime mover in mid-size vessels or as an auxiliary engine in large-sized vessels. The engine specification details are listed in Table 3.2.

The simulation was run to predict the engine performance and emissions under each test condition. The engine was operated at varying speeds ranging from 1000 to 1600 rpm in order to predict its performance parameters, including torque, specific

Table 3.1 Fuel properties data

Fuels	B5	B20	B40	B100	Diesel no. 2 (B0)
Substance (diesel/palm oil), %	95/5	80/20	60/40	0/100	100/0
Low heating value of fuel, MJ/kg	39.89	41.18	39.89	36.22	42.5
Density of fuel at 323 K, kg/m^3	852	841	852	885	830
Dynamic viscosity coefficient at 323 K, Pa s	0.00367	0.00334	0.003687	0.00463	0.00300
Cetane no	49.37	48.68	49.37	51.3	48

Table 3.2 Engine specification

Engine model	Cummins NT-855M
Type	Four stroke, direct injection
Number of cylinders	In line-six-cylinder
Bore × stroke	139 mm × 152 mm
Displacement volume	14 L
Maximum torque	1068 Nm
Maximum power	201 kW
Compression ratio	14.5:1
Cooling system	Water-cooled

fuel consumption, brake thermal efficiency, as well as emissions such as NO_x, CO, and particulate matter.

3.3 Results and Discussion

The performance and emission characteristics of an engine using various blends of biodiesel and diesel fuels were determined using the Diesel RK simulation software. The results are presented and discussed below based on the engine output parameters.

3.3.1 Brake Torque

Figure 3.1 illustrates the variations of engine brake torque and rpm, which show a linear pattern in this experiment. However, the B100 blend exhibits decreasing values compared to other fuels due to its fuel properties, particularly its lower heating efficiency than diesel no. 2. The higher density and viscosity of SME biodiesel also hinder the mixing of air and fuel particles in the combustion chamber due to lower velocity and volatility, reducing combustion efficiency. Diesel RK data shows that biodiesel blends of B5, B20, and B40 do not produce significant changes in brake torque, except for B100, which starts with a low value of torque and gradually increases with speed, while diesel no. 2 shows slight increases and decreases in torque value. This pattern is due to B100's lowest heating value, highest density, and highest viscosity among the fuels. B100 exhibits a gradual increase of 2.79–11.49% throughout the experiment, while the other fuels show slight increases of 0.32–1.07% for B5, 1.31–2.17% for B20, 0.49–1.31% for B40, and 0.52–1.64% for diesel no. 2 fuel. Diesel no. 2 produced the highest torque, 1.56% more than the highest torque produced by biodiesel fuels. Among the different percentages of biodiesel blends, B20 produced the highest torque.

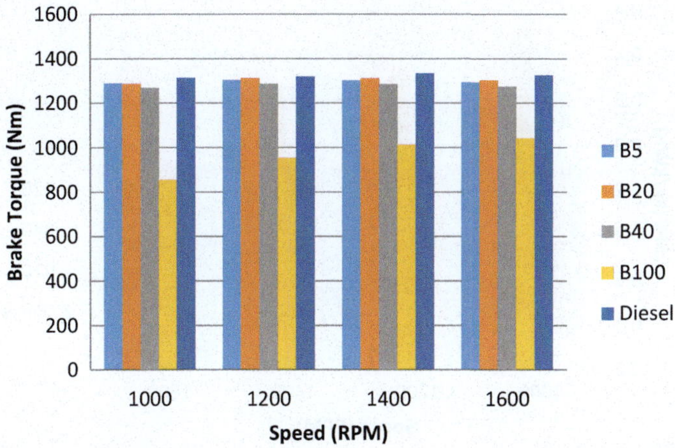

Fig. 3.1 Brake torque versus speed

3.3.2 Specific Fuel Consumption

The specific fuel consumption (SFC) is a measure of fuel flow required to produce a unit of power output. Figure 3.2 shows the trend of SFC at various speeds. As expected, SME biodiesel requires more fuel than diesel no. 2 to produce the same amount of power due to its properties, such as higher density and viscosity. Within the measured range of initial and maximum speeds, the SFC for SME biodiesel was significantly higher than diesel no. 2 fuel due to its lower calorific value caused by fuel-borne oxygen. The trend is similar for all fuel blends except B100, which exhibits a decreasing pattern. Figure 3.2 shows that B100 has the highest fuel consumption, consuming 47.4%, 52.4%, 44.8%, and 59% more fuel than B5, B20, B40, and diesel no. 2, respectively. As the speed increases, the fuel consumption gradually decreases due to the fuel's increased efficiency. For all other fuel blends except B100, there is no significant change, with an increase of only 0.44–0.88% for B5, 0.46–2.27% for B20, 0.43–1.29% for B40, 0.48–1.9% for diesel no. 2, while B100 exhibits an increase of 2.89–21.86%.

3.3.3 Brake Thermal Efficiency

The brake thermal efficiency (BTE) is a crucial engine performance parameter that represents the proportion of fuel energy converted into usable power output. Figure 3.3 depicts the difference in thermal brake efficiency between diesel no. 2 fuel and SME biodiesel fuel at increasing speeds. The figure illustrates that SME biodiesel fuel has a better performance expectation than diesel no. 2. This result can be affected by the properties of the fuel, such as its heating value and viscosity. The

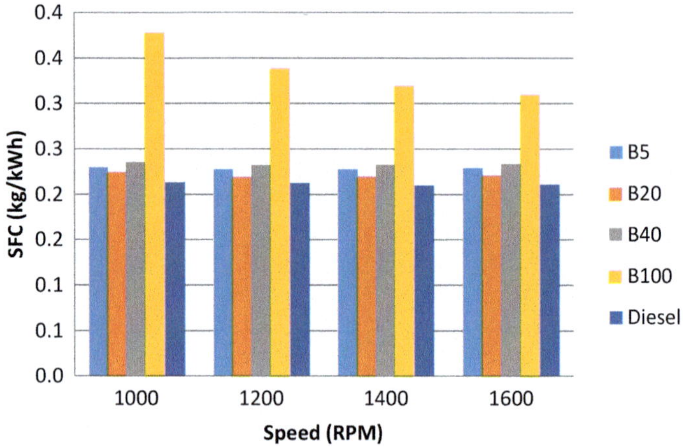

Fig. 3.2 Specific fuel consumption versus speed

figure shows an increasing trend of BTE with an increase in speed. The percentage increases were 23.2% to 38.9%, 22.15% to 35.44%, 24.35% to 38.96%, and 30.6% to 48.98% for B5, B20, B40, and B100, respectively, while diesel no. 2 yields 19.9% to 31.84%. Among the other fuels, B100 shows the highest percentage, while B20 has the least percentage among the other biodiesel fuels. Meanwhile, the highest percentage produced by diesel no. 2 is less than 19.27%, 10.16%, 18.28%, and 34.97% than B5, B20, B40, and B100, respectively.

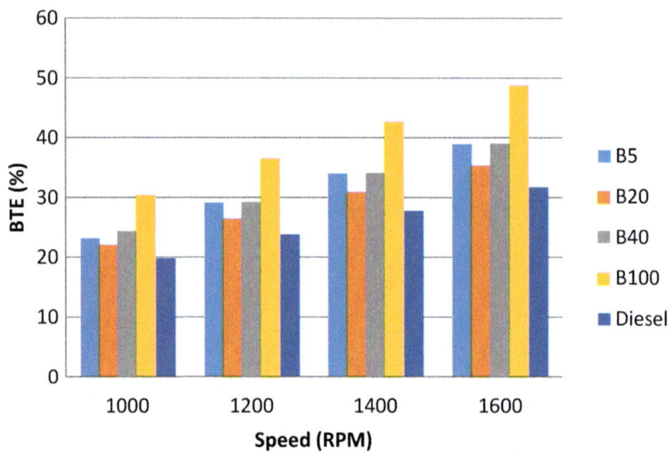

Fig. 3.3 Brake thermal efficiency versus speed

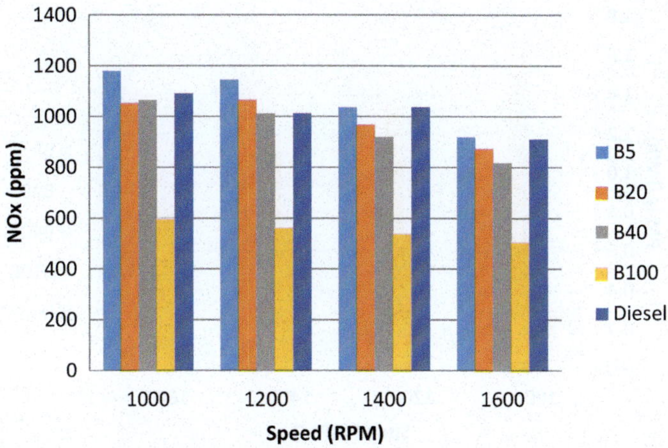

Fig. 3.4 Nitrogen oxide versus speed

3.3.4 Nitrogen Oxide Emissions

Nitrogen oxide emission (NO_x) represents a category of substances comprising nitrogen oxides that are created by the reaction of nitrogen and oxygen gases during combustion in the exhaust system of internal combustion engines. Figure 3.4 illustrates the NO_x emissions produced by SME biodiesel and diesel no. 2 at various speeds during the experiment. As the biodiesel blend increased, the amount of NO_x emissions decreased, which was expected since the higher oxygen and cetane number in the fuel would result in lower NO_x emissions. The presence of PFAME in biodiesel also contributes to the reduction of NO_x emissions due to higher levels of oxygen and cetane than petroleum diesel [5]. The NO_x emissions pattern gradually decreased as the temperature during combustion decreased, and the trend continued with higher speed, with a reduction of 12.76–28.26% for B5, 10.82–21.84% for B20, 12.37–30% for B40, 6.32–17.88% for B100, and 10.96–19.55% for diesel no. 2 fuel. B100 showed the lowest NO_x emissions among the biodiesel fuels and was still lower than diesel no. 2. The reason for this is that biodiesel has poor spray characteristics due to its higher density and viscosity, resulting in improper combustion in the engine. B5 had higher oxygen content than diesel no. 2, leading to a higher value of NO_x emissions.

3.3.5 Particulate Matter

Particulate matter (PM) denotes a solid and liquid particle suspended in the air, which can be harmful. Incomplete combustion of hydrocarbons (HCs) due to insufficient oxygen supply to the combustion chamber can cause the production of PM. Figure 3.5

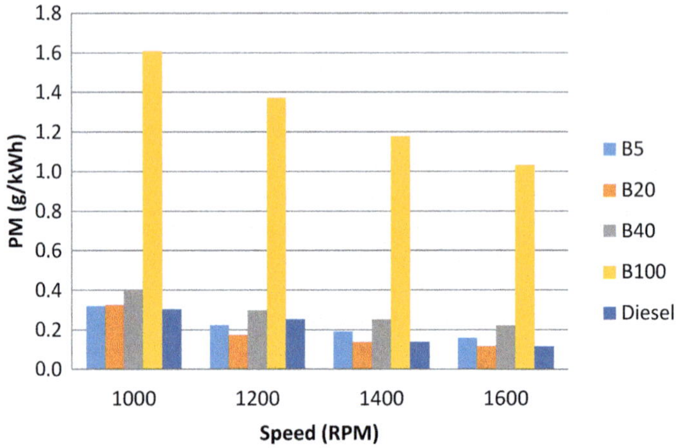

Fig. 3.5 Particulate matter versus speed

illustrates the emission pattern of PM at different engine speeds. The PM emission decreases gradually with increasing speed due to better combustion inside the engine. This results in lower PM emissions compared to the initial levels at engine start-up. Based on Fig. 3.5, B100 produces the highest PM emission at 1.61 g/kWh, whereas B5, B20, B40, and diesel no. 2 produce 0.32 g/kWh, 0.33 g/kWh, 0.40 g/kWh, and 0.30 g/kWh, respectively. The graph also shows a percentage decline in PM emission with increasing speed, ranging from 15.79 to 50%, 14.29–63.63%, 12–45%, 11.86–35.4%, and 14.29–60% throughout the experiment.

3.4 Conclusion

This study successfully investigated the effects of varying percentages of SME biodiesel using the Diesel RK computer simulation model. The following conclusions can be drawn from the results:

i. The brake torque for B5, B20, and B40 showed no significant change among the biodiesel percentages, with B20 SME producing slightly higher torque than other biodiesel percentages. Diesel no. 2 produced higher torque due to its high heating value compared to biodiesel fuel.
ii. All biodiesel blends had higher specific fuel consumption than diesel fuel due to their need for more fuel to produce power, with B100 having the highest consumption.
iii. The brake thermal efficiency produced by the biodiesel blend was higher than diesel fuel, with B100 having the highest percentage of 48.96%. As the speed of the engine increased, the percentage of BTE also increased.

iv. NO_x and PM emissions decreased with an increase in speed for biodiesel blends and diesel fuel. NO_x released by biodiesel did not vary much compared to diesel fuel despite having a higher oxygen content and cetane number. However, B100 blend produced the highest amount of PM compared to other biodiesel blends and diesel fuel.

v. B20 biodiesel blend is the best choice as it produces higher torque and better brake thermal efficiency than diesel fuel. It also had lower specific fuel consumption and lower emissions of NO_x and particulate matter than other biodiesel blends and diesel fuel.

Overall, the results suggest that using SME biodiesel blends as an alternative fuel in diesel engines can improve their performance and reduce harmful emissions. However, further research is needed to optimize the fuel properties and engine parameters for optimal performance and emissions reduction.

Acknowledgements The authors would like to acknowledge financial support from the Ministry of Higher Education of Malaysia under the Fundamental Research Grant Scheme (Ref. No.: FRGS/1/2020/WAB05/UMT/02/2 and Vote No. 59626).

References

1. Kapadia, H., Brahmbhatt, H., Dabhi, Y., Chourasia, S.: Investigation of emulsion and effect on emission in CI engine by using diesel and bio-diesel fuel: a review. Egypt. J. Pet. **28**(4), 323–337 (2019). https://doi.org/10.1016/j.ejpe.2019.06.004
2. Malvade, A.V., Satpute, S.T.: Production of palm fatty acid distillate biodiesel and effects of its blends on performance of single cylinder diesel engine. Procedia Eng. **64**, 1485–1494 (2013). https://doi.org/10.1016/j.proeng.2013.09.230
3. Temizer, İ, Cihan, Ö., Eskici, B.: Numerical and experimental investigation of the effect of biodiesel/diesel fuel on combustion characteristics in CI engine. Fuel **270**, 117523 (2020). https://doi.org/10.1016/j.fuel.2020.117523
4. Gautam, R., Kumar, S.: Performance and combustion analysis of diesel and tallow biodiesel in CI engine. Energy Rep. **6**, 2785–2793 (2020). https://doi.org/10.1016/j.egyr.2020.09.039
5. Noor, C.W.M., Mamat, R., Noor, M.M., Samo, K.B., Nik, W.B.W.: Effects of low proportion palm fatty acids methyl ester blends on the performance and combustion of marine diesel engine. J. Biobased Mater. Bioenergy **12**(2), 153–160 (2018). https://doi.org/10.1166/jbmb.2018.1766
6. Al-Dawody, M.F., Bhatti, S.K.: Experimental and computational inves-tigations for combustion, performance and emission parameters of a diesel engine fueled with soybean biodiesel-diesel blends. Energy Procedia **52**, 421–430 (2014). https://doi.org/10.1016/j.egypro.2014.07.094
7. Alghafis, A., Raouf, E.A.: Optimization of injection timing and injection duration of a diesel engine running on pure biodiesel SME (Soya Methyl Ester). Open J. Appl. Sci. **10** (07), 486–502 (2020). https://doi.org/10.4236/ojapps.2020.107034
8. Rajak, U., Prerana, N., Tikendra, N.V., Ibham, V., Ümit, A.: Numerical and experimental investigation of hydrogen enrichment in a dual-fueled CI engine: a detailed combustion, performance, and emission discussion. Int. J. Hydrogen Energy **47**(76), 32741–32752 (2022). https://doi.org/10.1016/j.ijhydene.2022.07.144

Chapter 4
An Excellent Relationship-Based Model Development in Deriving the Short-Term Probability Distribution of Offshore Structural Responses

Sayyid Zainal Abidin Syed Ahmad, Mohd Azlan Musa, Muhammad Faris Roslan, Mohd Khairi Abu Husain, Noor Irza Mohd Zaki, and Nurul 'Azizah Mukhlas

Abstract Accurate calculation of the structural responses has been extensively used based on a time-domain approach. Due to its ability to compute the nonlinear association between wave forces and the structural responses, the most method regularly used for predicting the probability distribution of the extreme responses in such statistical random condition is the Monte Carlo time simulation (MCTS) method. However, this method suffers from excessive sampling variability, and hence, a large number of simulated response records are required to reduce the sampling variability to acceptable levels. As well, a simple technique for derivation of the probability distribution of extreme responses is still a limited source. Findings show that a more efficient technique by an efficient time simulation (ETS) method possesses an advantage of the correlation between the extreme values of surface elevation (input) and

S. Z. A. S. Ahmad (✉) · M. A. Musa · M. F. Roslan
Faculty of Ocean Engineering Technology, Universiti Malaysia Terengganu, 21030 Kuala Nerus, Malaysia
e-mail: s.zainal@umt.edu.my

M. A. Musa
e-mail: mohdazlan@umt.edu.my

M. F. Roslan
e-mail: p5029@pps.umt.edu.my

M. K. A. Husain · N. I. M. Zaki
Faculty of Artificial Intelligence, Universiti Teknologi Malaysia, 54100 Kuala Lumpur, Malaysia
e-mail: mohdkhairi.kl@utm.my

N. I. M. Zaki
e-mail: noorirza.kl@utm.my

N. '. Mukhlas
School of Civil Engineering, Universiti Teknologi Malaysia, 80050 Johor Bahru, Malaysia
e-mail: nurulazizah@utm.my

A. Ismail et al. (eds.), *Exploring Diversity in Engineering and Technology for Knowledge and Innovation*, Advanced Structured Materials 215,
https://doi.org/10.1007/978-3-031-64330-9_4

their corresponding responses (output). In this article, the extended adaptation of ETS procedures provides an excellent correlation between input–output variables for developing the ETS-regression (ETS-reg) models. This proposed ETS-reg model is then verified by the extensive MCTS method and compared to another association of relationship-based models, namely the modified finite-memory nonlinear system (MFMNS) and its improved version of MFMNS (eMFMNS) models. In conclusion, the different models-based relationships are examined to the high design waves. Thus, the most accurate regression-based models in forecasting the return period of 100-year extreme responses are discussed.

Keywords Time-domain · Probability distribution · Monte Carlo-based simulation · Structural responses · Efficient time simulation · Finite-memory system

4.1 Introduction

In fact, the offshore installations were subjected to unpredictable wave loads [1]. Computer numerical analysis could be used to calculate the offshore hydrodynamic properties, i.e. motion, acceleration and loads [2]. Numerically (refer Sect. 3.2.2), the calculation sequences started using generic wave energy spectrum [3]. In wave hydrodynamics, the water kinematics are replicated using the Airy wave theory (AWT). The hydrodynamic calculation due to the motion of particle kinematic and its acceleration is observed in the forces of Morison wave loads. The Morison model is then utilized to compute the hydrodynamical forces activating on a slender cylindrical structure, which is applied to replicate the magnitude of base shear and overturning moment.

The Monte Carlo-based simulation technique is used as reference in the simulation due to an incredibly precise approach for estimating the offshore responses in such wave and current circumstance. Also, the Monte Carlo-based simulation is acknowledged as the ability approach to determine such structural nonlinearities in predicting wave responses [4]. However, the involved nonlinearity estimation was excessively time-consuming, and several researchers presented many methods in order to improve time consumption.

One such approach is an efficient-based time simulation (ETS) technique, which recommends an additional efficient practice without compromising the accuracy [5]. In that manner, an expanded version of the ETS method (refer Sect. 3.3.5) has the potential to progress because of its fine relationship among input of sea surface elevation and output of wave-induced response records, which gives the right solution for computationally demanding issues.

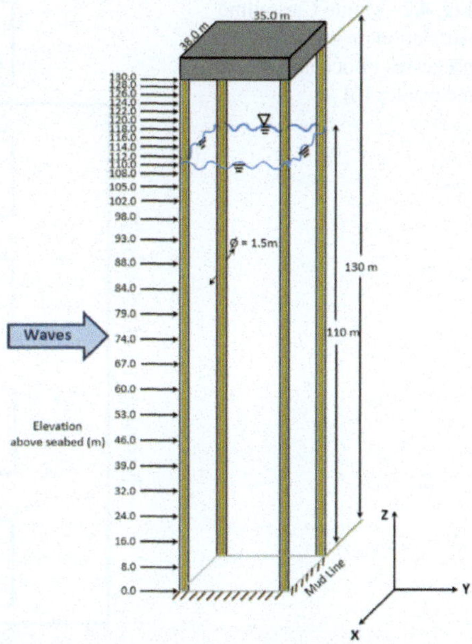

Fig. 4.1 Schematic diagram of structural models

4.2 Model Specifications and Wave Modelling Analysis

This section is divided into three subsections: the platform description, the illustration of the platform model and the computer numerical analysis by stochastic process.

4.2.1 Offshore Structural Model and its Specifications

The four-legged fixed offshore structures in the ocean are used as the basis for calculating structural forces-induced responses, as shown in Fig. 4.1. In this investigation, a jacket modelling structure that was in the North Atlantic Sea at a depth of 110 m [6]. Thirty-point loads, which add up to a total amount of nodal loads, indicate the distributed load on all of the legs [7].

4.2.2 Ocean Wave Modelling Analysis

A stochastic analysis is carried out utilizing a numerical computer model to compute the load-induced responses, which is performed based on the ocean wave energy via the Pierson–Moskowitz spectrum. A wave spectrum based on a fast Fourier transform (FFT) technique is utilized for generating the real-time ocean wave simulation. As

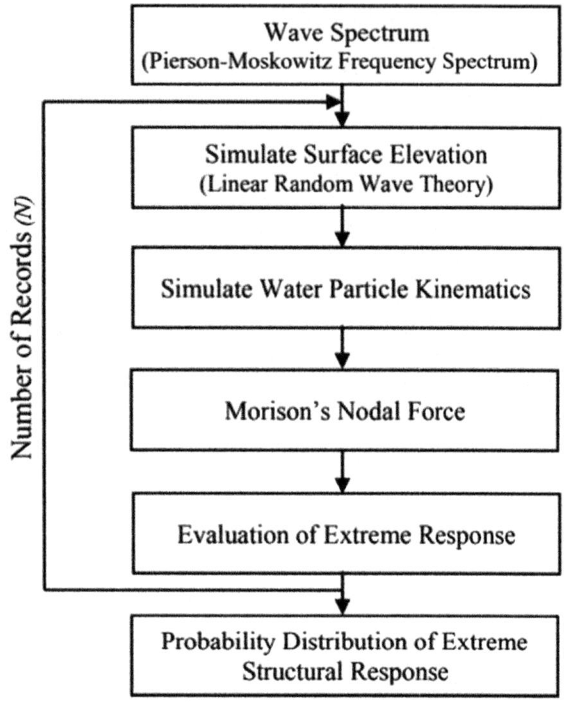

Fig. 4.2 Monte Carlo time simulation on numerical processes of ocean wave mechanics [6]

shown in Fig. 4.2, the computation procedures start with the wave spectrum using the FFT method to deal with the AWT, wave-induced fluid motion of particle kinematics and their corresponding wave loads in order to evaluate offshore structural responses [8].

4.3 Short-Term Probability Distribution of 100-Year Extreme Responses

The compilation of physical data was analysed numerically based on ocean modelling. In order to test the accuracy and efficiency, the short-term analysis regarding the 100-year return period of response values is predicted among established ETS-reg, MFMNS and eMFMNS models and its corresponding MCTS procedure.

4.3.1 MCTS Procedure

In this analysis, the conventional MCTS process was used to define the accuracy and efficiency level. According to [9], the MCTS method is verified as one of the most fundamental approaches to be used in predicting the extreme structural responses.

4.3.2 MFMNS Model

The earlier research has shown the established FMNS model efficiently replicates the structural responses [10]. The FMNS model is an efficient solution and enhances the computational effort in comparison with the MCTS method while maintaining its precision. The FMNS model is not discussed in this paper, but a modified nonlinear finite-memory (MFMNS) model is preferred due to the similar approach between them, and the MFMNS model is sufficient for the quasi-static analysis [11].

4.3.3 eMFMNS Model

The additional extreme value residue (EVR) term in eMFMNS model development influences the models. An adjustment to the MFMNS model known as eMFMNS is introduced to enhance the accurateness of the estimation of 100-year responses [12].

4.3.4 ETS Method

The previous implementation of ETS method was made by [13], which delivered a fine relationship among sea surface elevation and the comparable wave-induced structural response. The intent of the suggested ETS technique minimizes a huge amount of simulated records without losing precision [13].

4.3.5 Development of Efficient Time Simulation-Regression (ETS-reg) Procedures

In this study, the manifestation of a recommended model acknowledged as an ETS-based regression or ETS-reg model is a useful discovery. This type of model is established based on the excellent relationship between the variables from the ETS method [14].

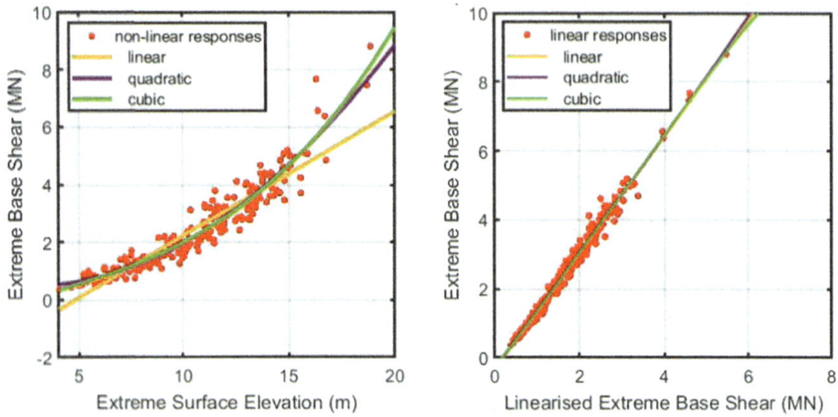

(a) Relationship-based surface elevation (b) Relationship-based linearised responses

Fig. 4.3 **a** ETS-reg$_{SE}$ model based on relationship-surface elevation **b** ETS-reg$_{LR}$ model based on relationship-linearized responses. For the cases of total base shear responses, $H_s = 15$ m, $T_z = 13.75$ s, $U = 0$ m/s

4.3.5.1 ETS-reg$_{SE}$ Model

According to [15], the presence of a proposed method known as an ETS-reg model is a useful discovery. As displayed in Fig. 4.3a, the developed ETS-reg$_{SE}$ model is based upon the surface elevation (SE) from the ETS method as the input for determining the extreme structural responses.

4.3.6 ETS-reg$_{LR}$ Model

In line with [16], the ETS-reg$_{LR}$ model is the advancement of the ETS-reg$_{SE}$ model in the prediction of the 100-year extreme response values. Due to the limitation on certain wave conditions, the ETS-reg$_{SE}$ is not practicable while the nonlinearity behaviour from surface elevation affected the model accuracy. The improved ETS-reg$_{LR}$ model relies on the linearized responses (LR) as the input variable in model development, as illustrated in Fig. 4.3.

4.4 Comparison and Validation of the Regression Models

4.4.1 Level of Accuracy of Extreme Response Prediction by ETS-Regression Models

By providing a good result, the level of accuracy of models needs to be initially examined. Those models are validated with the benchmark MCTS procedure to verify the level of accuracy. The simulated data is plotted by Gumbel distribution for the 100-year short-term extreme responses. Table 4.2 demonstrates that the ETS-reg models are used in predicting the 100-year responses (P_{100}) for high sea states without current impact. Besides, the comparison and validation of the proposed ETS-reg$_{LR}$ model are examined to the variation of relationship-based regression models.

As observed, the extreme responses of ETS-reg models are compared and validated with the MCTS method, as displayed in Fig. 4.4. From Table 4.2, the 100-year extreme response value predicted by the benchmark method of MCTS is 11.1459 ($\approx$ 11.15) MN and the ETS-reg$_{LR}$ models is 11.3220 MN, which is better than the ETS-reg$_{SE}$ model under a predicted value of 10.0723 MN. Thus, the prediction of the 100-year probability distribution of extreme response made by the ETS method was 11.55 MN, where the improved version of ETS method offers more accuracy than the ETS method. The ETS-reg$_{LR}$ model is considered the higher level of accuracy, which provides a minimum error value of around 1.58%. On the other hand, the most robust model goes to the MFMNS model producing only 0.99% relative error.

In summary, the accurate result could be achieved after modification has been done by using the revised input parameters. Thus, the prediction of 100-year responses by the ETS-reg$_{LR}$ model is better compared to the ETS-reg$_{SE}$ model and its predecessor of ETS method.

4.5 Conclusion

This paper has concluded a comprehensive analysis of predicted load-induced responses using the relationship-based model development via the ETS-reg$_{SE}$ and ETS-reg$_{LR}$ models. Also, another regression model, such as MFMNS and eMFMNS models were included in this analysis, as mentioned in Table 4.1. For testing the capability of the proposed version of ETS-reg, the case study was on the basis of wave conditions of high waves combined with the nonexistence of current impact. As a result, the proposed ETS-reg$_{LR}$ model provides the lower percentage error in estimating the 100-year return period of responses in comparison with the previous approaches, particularly the ETS method and another version of ETS method through ETS-reg$_{SE}$ model.

In fact, the calculation of structural response needs a complicated procedure in the numerical analysis. Besides, the stochastic process by the wide-ranging MCTS

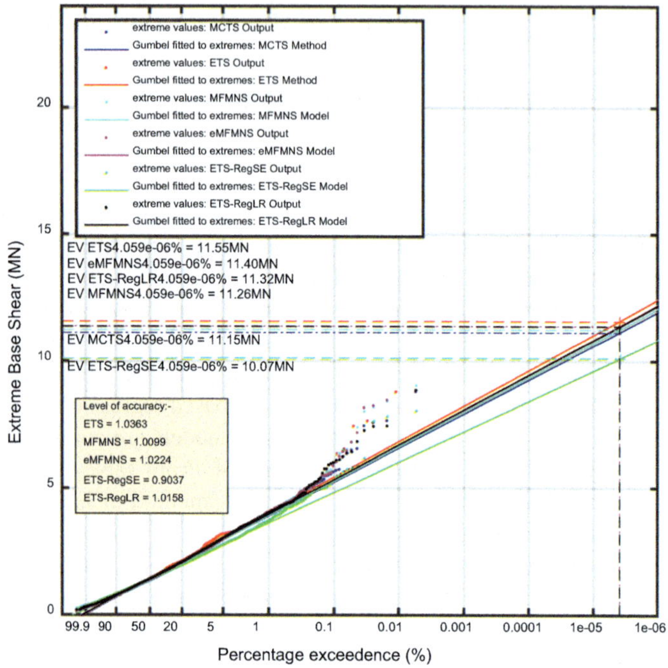

Fig. 4.4 Comparison the 100-year extreme responses between MCTS method and other relationship-based model development; quasi-static base shear responses, $H_s = 15$ m, $T_z = 13.75$ s and $U = 0.00$ m/s

Table 4.1 Types of ETS-reg models regarding the coefficient of determination for base shear responses $H_s = 15$ m, $T_z = 13.75$ s, $U = 0$ m/s

ETS-regression models	Relationship-SE	Relationship-LR	ETS-Reg$_{SE}$	ETS-Reg$_{LR}$
	R		r^2	
ETS-Reg$_{\mathrm{L}}$	0.9408	0.9931	0.8851	0.9862
ETS-Reg$_{\mathrm{P}}$	–	–	0.9284	0.9863
ETS-Reg$_{\mathrm{C}}$	–	–	0.9296	0.9864

Table 4.2 MCTS and ETS methods regarding association of other regression models in calculating the P_{100}-year short-term responses for base shear quasi-static responses and without current. $H_s =$ 15 m (P_{100}MCTS = 11.1459 MN)

Models	P_{100} responses (base shear) (MN)	Responses ratio (models/MCTS)
ETS	11.5500	1.0363
MFMNS	11.2566	1.0099
eMFMNS	11.4040	1.0232
ETS-Reg$_{SE}$	10.0723	0.9037
ETS-Reg$_{LR}$	11.3220	1.0158

method is very computationally demanding. Thus, the current procedure is introduced by establishing a simple algebraic model in the derivation of the probability distribution of extreme responses. By this simplified approach, the regression models could minimize the computer running time and preserve its accuracy level at the same time, as proven in Table 4.2. Because of its consistency, the ETS-reg$_{LR}$ models are at least equated to the Monte Carlo-based approach employed in deriving the return period of structural responses.

Acknowledgements The paper is financially supported by the Dana Penyelidikan UMT (DP-UMT) 2022-Talent and Publication Enhancement Research Grant (TAPE-RG) (Ref. No.: UMT/TAPE-RG 2022/Vot 55123).

References

1. Li, Y., Kareem, A.: Stochastic response of tension leg platforms to wind and wave fields. J. Wind Eng. Ind. Aerodyn. **36**, 915–926 (1990). https://doi.org/10.1016/0167-6105(90)90088-T
2. Tuah, H., Hudspeth, R.T.: Comparisons of numerical random sea simulations. J. Waterw. Port Coast. Ocean Div. **108**(4), 569–584 (1982). https://doi.org/10.1061/JWPCDX.0000326
3. Morooka, C.K., Yokoo, I.H.: Numerical simulation and spectral analysis of irregular sea waves. Int. J. Offshore Polar Eng. **7**(03), 189 (1997)
4. Abu Husain, M.K., Mohd Zaki, N.I., Johari, M.B., Najafian, G.: Extreme response prediction for fixed offshore structures by Monte Carlo time simulation technique. In: 35th International Conference on Offshore Mechanics and Arctic Engineering, vol. 49941, pp. V003T02A037. American Society of Mechanical Engineers (ASME) Digital Collection (2016)
5. Abu Husain, M.K., Mohd Zaki, N.I., Najafian, G.: Prediction of Extreme Offshore Structural Response: An Efficient Time Simulation Approach. LAP LAMBERT Academic Publishing (2013)
6. Najafian, G.: Local hydrodynamic force coefficients from field data and probabilistic analysis of offshore structures exposed to random wave loading. University of Liverpool (1991)
7. Abu Husain, M.K., Najafian, G.: An efficient Monte Carlo simulation technique for derivation of the probability distribution of the extreme values of offshore structural response. In: 29th International Conference on Ocean, Offshore and Arctic Engineering, vol. 49101, pp. 369–375. ASME (2010)
8. Mukhlas, N.A., Mohd Zaki, N.I., Abu Husain, M.K., Syed Ahmad, S.Z.A., Najafian, G.: Numerical formulation based on ocean wave mechanics for offshore structure analysis—a review. Ships Offshore Struct. **17**(8), 1731–1742 (2022). https://doi.org/10.1080/17445302.2021.1937880
9. Jensen, J.J.: Extreme value predictions using Monte Carlo simulations with artificially increased load spectrum. Probab. Eng. Mech. **26**(2), 399–404 (2011). https://doi.org/10.1016/j.probengmech.2010.09.001
10. Najafian, G.: Application of system identification techniques in efficient modelling of offshore structural response. Part II: Model validation. Appl. Ocean Res. **29**(1–2), 17–36 (2007). https://doi.org/10.1016/j.apor.2007.08.003
11. Mohd Zaki, N.I., Abu Husain M.K., Najafian, G.: Prediction of 100-year responses of fixed offshore structures using the modified version of finite memory nonlinear system models. In: 12th ISOPE Pacific/Asia Offshore Mechanics Symposium, 16, 15. International Society of Offshore and Polar Engineers (2016)
12. Mukhlas, N.A., Mohd Zaki, N.I., Abu Husain, M.K.: Prediction of the 100-year responses by advanced MFMNS. In: 3rd International Conference on Ocean, Mechanical and Aerospace for Scientist and Engineer, pp. 7–8 (2016)

13. Mallahzadeh, H., Wang, Y., Abu Husain, M.K., Mohd Zaki, N.I., Najafian, G.: Efficient derivation of the probability distribution of extreme responses due to random wave loading from the probability distribution of extreme surface elevations. In: 32nd International Conference on Ocean, Offshore and Arctic Engineering, vol. 55331, pp. V02BT02A010. ASME (2013)
14. Syed Ahmad, S.Z.A., Abu Husain M.K., Mohd Zaki, N.I., Mohd M.H., Najafian, G.: Offshore responses using an efficient time simulation regression procedure. In: Trends in the Analysis and Design of Marine Structures: Proceedings of the 7th International Conference on Marine Structures, pp.12. CRC Press (2019)
15. Syed Ahmad, S.Z.A., Abu Husain, M.K., Mohd Zaki, N.I., Mukhlas, N.A., Najafian, G.: Extreme response prediction for fixed offshore structures by efficient time simulation regression procedures. Part 2: Model validation. Ships Offshore Struct. **18**(3), 414–422 (2023). https://doi.org/10.1080/17445302.2022.2059252
16. Syed Ahmad, S.Z.A., Abu Husain, M.K., Mohd Zaki, N.I., Mukhlas, N.A., Najafian, G.: An improved version of ETS-regression models in calculating the fixed offshore platform responses. J. Mar. Sci. Eng. **10**(11), 1727 (2022). https://doi.org/10.3390/jmse10111727

Chapter 5
Mooring Analysis of a Floating Fish Cage in Deep Sea Using Computational Fluid Dynamics

Ahmad Fitriadhy, Muhammad Nazri Othman, and Nur Amira Adam

Abstract In the fisheries industry, meeting market demands through aquaculture at deep sea has become a challenging task. Traditional fish cage farming in shallow waters has limited fish resources. To address this issue while considering the mobility expenses of subsidized traditional fishing vessels, the author proposes designing a floating fish cage for deep-sea operation. Equipped with mooring lines, this floating fish cage aims to counter undesirable rotation and translation caused by environmental forces. The motion analysis, using the quadratic transfer function method in ANSYS-AQWA Workbench, considered various environmental loads such as regular and irregular waves, winds, and currents. Results indicate the highest response amplitude operator (RAO) motion in surge translation (8.079 m) and pitch rotation (3.084°) from 0° wave direction. The actual response shows motion when the mooring system is used, with the highest surge translation at 10.566 m and pitch rotation at 1.866° from split point mooring. Cable tension analysis reveals split point mooring with the highest tension at 952.718 kN on cable 2 and split point (V) mooring with the maximum tension between 38 and 55 kN for each cable.

Keywords Mooring · Floating fish cage · CFD · Regular and irregular wave · RAO

A. Fitriadhy (✉)
Program of Naval Architecture, Faculty of Ocean Engineering Technology, Universiti Malaysia Terengganu, 21030 Kuala Nerus, Terengganu, Malaysia
e-mail: naoe.afit@gmail.com

M. N. Othman
Program of Maritime Technology, Faculty of Ocean Engineering Technology, Universiti Malaysia Terengganu, 21030 Kuala Nerus, Terengganu, Malaysia
e-mail: m.nazriothman@gmail.com

N. A. Adam
School of Port, Logistics and Management, Netherlands Maritime University College, Johor, Malaysia
e-mail: amiraadam.nur@gmail.com

A. Ismail et al. (eds.), *Exploring Diversity in Engineering and Technology for Knowledge and Innovation*, Advanced Structured Materials 215,
https://doi.org/10.1007/978-3-031-64330-9_5

5.1 Introduction

With the growing demand for fisheries, aquaculture has emerged as a crucial method to meet the need for fish supply. Due to the limitations of capture fisheries, aquaculture offers a viable solution to produce a larger quantity of fish. However, operating aquaculture activities near the shore has led to issues such as water pollution from fish excrement and uneaten food as mentioned by Hlordzi et al. [1] and Tom et al. [2]. The fish farms located in near-shore areas, coastal regions, and fjords also tend to generate higher profits compared to land-based farms, primarily because of their investments in water treatment and other operational factors [3, 4]. To address these problems, the proposal is to push aquaculture activities further offshore for fish farming, mitigating the environmental impact and ensuring sustainable fish production.

Farming fish offshore offers a favorable environment, preserving their natural habitat and reducing water pollution as tides and depth carry fish waste away from the cages [5]. To establish floating fish cages offshore, challenges such as tides and waves affecting the structure's position and stability in deep-sea conditions must be addressed.

To secure the floating fish cage and prevent drifting, it must be anchored to the seabed, ensuring stability even during wave actions. The mooring tension line is crucial to withstand wave forces and must be carefully analyzed to avoid breakage. The study aims to identify the correct factors for safe and problem-free operation of the floating fish cage. Mooring analysis is essential to determine the tension in the mooring line under different wave conditions, including amplitude, frequency, and direction.

5.2 Methodology

This section presents the comprehensive methodology employed to achieve the study's aim and objectives. It comprehensively discusses the methods used, including data analysis, ensuring alignment with the problem statement, ultimately leading to valuable insights and solutions (refer Fig. 5.1).

5.2.1 Design Phase

In the design phase, we take the initial model of the floating fish cage, which is built according to the main dimensions outlined (Table 5.1) in the Design Modeler file. This model is then brought into ANSYS-AQWA for a comprehensive mooring analysis, considering its movement in six different degrees of freedom: surge, sway, heave, roll, pitch, and yaw. The resulting optimized design is specifically crafted to excel in deep-sea operations within the South China Sea. To ensure its stability, the design

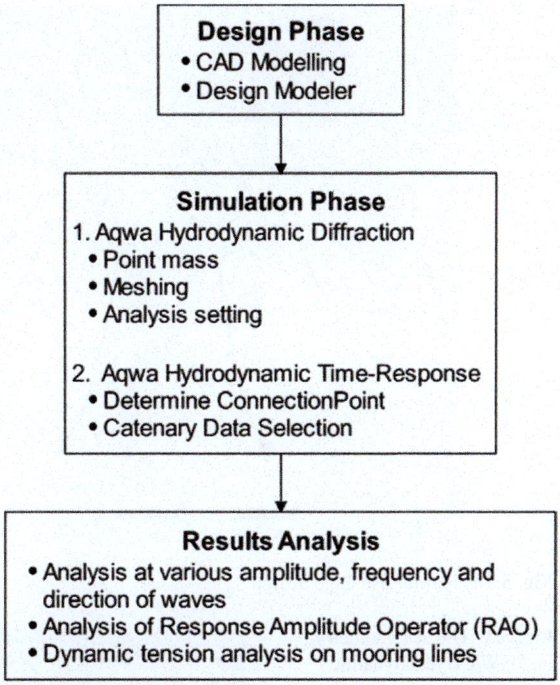

Fig. 5.1 Flow chart of CFD analysis

is meticulously anchored using eight mooring lines. The continuous development of ANSYS introduces further enhancements in modeling techniques, with a focus on the application of specified modeling methods as outlined by Merchant [6, 7] (Fig. 5.2).

In the current phase, precise body alignment along the *XY* plane (− 18 m offset) was achieved through the translation feature. The slice tool in ANSYS Design Modeler is enabling versatile cross sections and sweepable bodies. Utilizing frozen geometry, this phase divided the body into upper and lower sections for thorough analysis as seen in Fig. 5.3. The outline panel facilitated sea parameter adjustments for graphical enhancements, where the sea level alignment defined diffracting and non-diffracting portions, with manual specification options. Additionally, program-controlled point mass elements, derived from structure panel elements, were introduced. However, the program required manual inputs for inertia, radius of gyration, and vertical position (*Z*), determined as shown in Table 5.2.

Table 5.1 Model main dimension

Description	Value (m)
Diameter	80
Depth (*D*)	27
Draft (*T*)	18

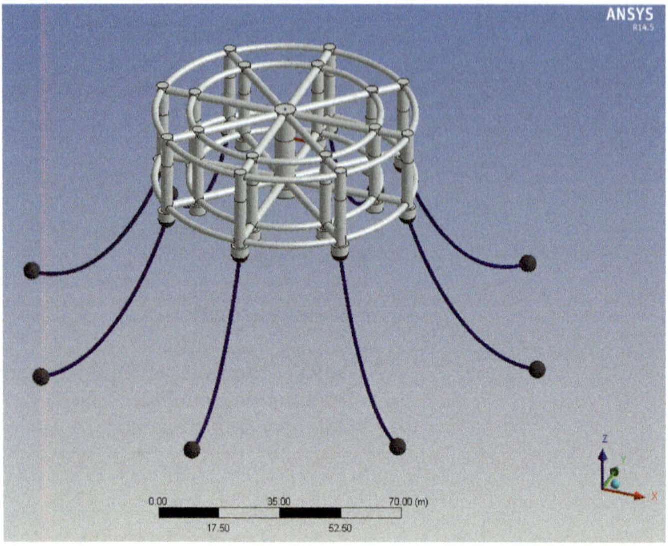

Fig. 5.2 Floating fish cage design

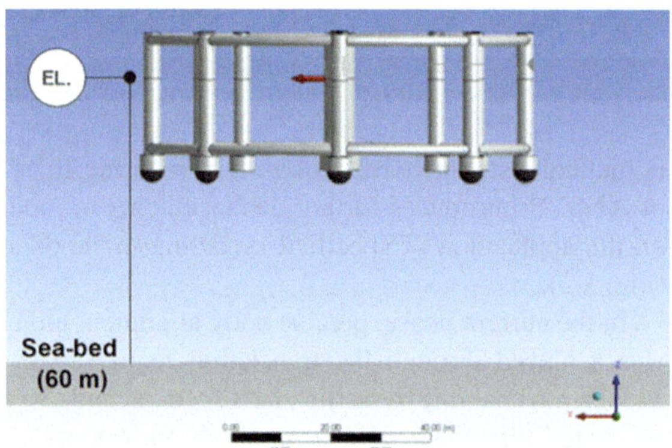

Fig. 5.3 The sea geometry in AQWA Hydrodynamic Diffraction

Table 5.2 Measured inertia of floating fish cage

Description	Value	Unit
I_{xx}	5,031,000,000	kg m^2
I_{yy}	5,031,000,000	kg m^2
I_{zz}	8,297,000,000	kg m^2
Z	– 4.528	M

5.2.2 Simulation Phase

Basically, two simulation scenarios—hydrodynamic diffraction and hydrodynamic time-responses—have been executed as versatile tools to analyze intricate motions and responses in floating structures [8]. The first, hydrodynamic diffraction, undertakes three-dimensional linear and diffraction analyses for multiple bodies, encompassing hydrodynamic interactions. The second, hydrodynamic time response, engages in dynamic analyses for floating structures, factoring in mooring lines and fenders for holistic performance evaluation. These simulations are indispensable in designing reliable, secure, and efficient marine structures across diverse conditions, thereby refining marine operations.

5.2.2.1 Meshing Setup

Mesh generation is an important stage, balancing accuracy and efficiency in ongoing simulations as displayed in Fig. 5.4. Customized for each simulation, it is guided by specific physical preferences. Here, the advanced meshing leads to efficiency and robustness [9], encompassing diverse surface patches [10]. Within mesh generation as summarized in Table 5.3, the defeaturing tolerance adeptly manages fine details; if smaller than this tolerance, a single element spans them; otherwise, mesh size reduces for proper meshing [11]. The maximum element size governs element dimensions, linked to AQWA's diffraction analysis. The meshing type choice defines the algorithm; here, combined meshing suits surface-only geometries in this research.

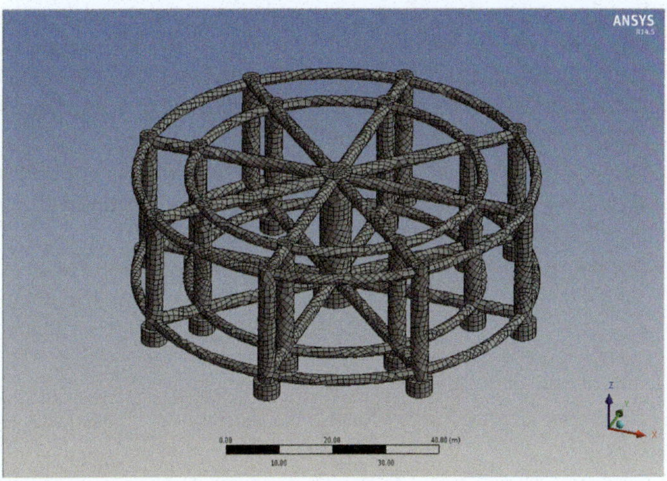

Fig. 5.4 The meshing geometry of the floating fish cage

Table 5.3 Details of mesh

Description	Mesh
Details of mesh	
Defeaturing tolerance	0.90 m
Max element size	1.80 m
Max. allowed frequency	0.38 Hz
Meshing type	Combined meshing
Generated mesh information	
Number of nodes	15,682
Number of elements	15,941
Number of nodes (difference)	9207
Number of element (difference)	9254

In this crucial independent study, the goal is to determine the optimal mesh size and node count for accurate analysis of heaving versus frequency. Through meticulous assessment of various meshing sizes and nodes as presented in Table 5.4, case 2 emerges as the preferred configuration from the graph illustrating heaving-frequency relationships. Its consistent results and reduced time consumption make it the preferred choice for the entire analysis, maintaining accuracy while improving efficiency (see Fig. 5.5).

Figure 5.2 provides a visual representation of the mooring lines of the floating fish cage, showcasing the application of nonlinear catenary cables featuring up to 10 catenary sections and optional joints. These sections and joints are defined within the catenary data object, allowing flexibility in creating diverse cables. Users can conveniently select catenary section types and input unstressed lengths for each section, along with joint options for multi-section cables. For advanced analysis, nonlinear catenary cables can undergo dynamic studies, exploring dynamic forces and structural impacts. By configuring parameters like use dynamics and number of elements, cable dynamics analysis is enabled. This technique is particularly important when connectivity varies, affecting endpoint determination considering seabed slope. This data compilation serves for tension analysis, facilitating a comprehensive understanding of offshore mooring system performance.

Table 5.5 shows the stiffness of the line is defined through stiffness (EA), incorporating Young's modulus (E) and the cross-sectional area (A). The default value is

Table 5.4 Different cases of meshing

Cases	Meshing size (m)	Number of elements	Frequency (Hz)
1	2.0	13,653	0.361
2	1.8	15,973	0.380
3	1.6	17,917	0.393

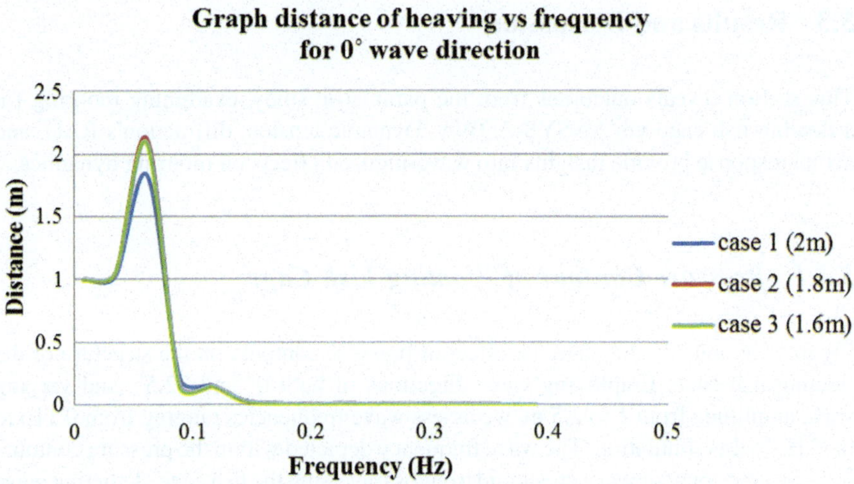

Fig. 5.5 Comparison result for different types of meshing

selected for typical mass per unit length considerations, though this might be inaccurate if buoyancy effects are included. Maximum tension signifies the upper limit for tension values in constructing the mooring-line database, requiring realism to avoid excessive tension ranges or limited coverage. Bending stiffness (EI), determined by Young's modulus (E) and the second moment of area (I), defaults to zero. Equivalent diameter (De) introduces an alternate diameter for drag calculations, defaulting to 0.1 for added mass.

Table 5.5 Catenary cable definition data

Description	Section 1	Section 2	Section 3
Type	Catenary section 1	Catenary section 2	Catenary section 3
Section length (m)	25	25	12
Mass/unit length (m)	50	50	50
Equivalent CSA (m^2)	0.01	0.01	0.01
Stiffness, EA (N)	600,000,000	600,000,000	600,000,000
Maximum tension (N)	7,500,000	7,500,000	7,500,000
Added mass coefficient	1	1	1
Transverse drag coefficient	1	1	1
Equivalent diameter (m)	0.1	0.1	0.1
Longitudinal drag coefficient	0.025	0.025	0.025

5.3 Results and Discussion

This section reveals outcomes from the parametric study, examining mooring for a floating fish cage via ANSYS-AQWA. Dynamic tension, diffraction's RAO, and actual response provide insights into wave-induced effects on mooring dynamics.

5.3.1 *Pressure Contour of Floating Fish Cage*

Figures 5.6 and 5.7 elucidate the effect of pressure contours on the structure of the floating fish cage. Employing wave directions of both 0° and 22.5°, and varying wave amplitude from 1 to 2.5 m, we assess wave frequencies ranging from 0.116 to 0.38 Hz in this simulation. The wave frequency depicted within the pressure contours indicates the count of wave crests and troughs traversing the fish cage, depicting wave lengths across 1–8 oscillations, as shown in Fig. 5.6. At a frequency of 0.225 Hz, three wave crests and troughs traverse the structure, with seven color transitions illustrating wave crests and troughs. The transformation from blue to red–orange vividly illustrates the shift from calm to hazardous conditions as waves interact with the structure, as shown in Fig. 5.10 (Table 5.6).

In Figs. 5.8 and 5.9, the pressure contours are distinctly revealed for different wave directions. Figure 5.8 showcases pressure contours aligned with a 0° wave

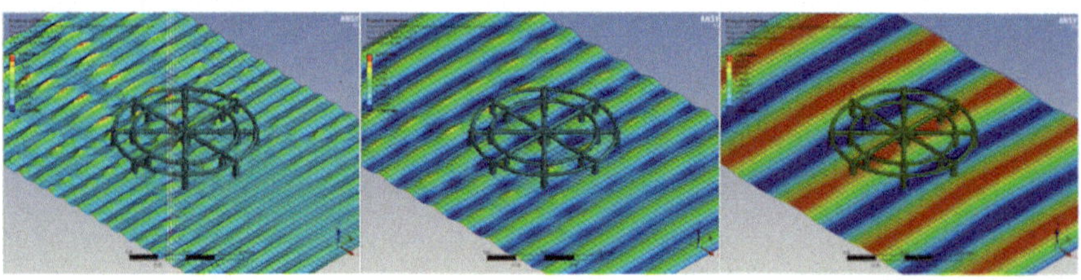

f = 0.380 Hz **f = 0.279 Hz** **f = 0.170 Hz**

Fig. 5.6 Pressure contour with wave amplitude 1 m, and different frequency (wave directions 0°)

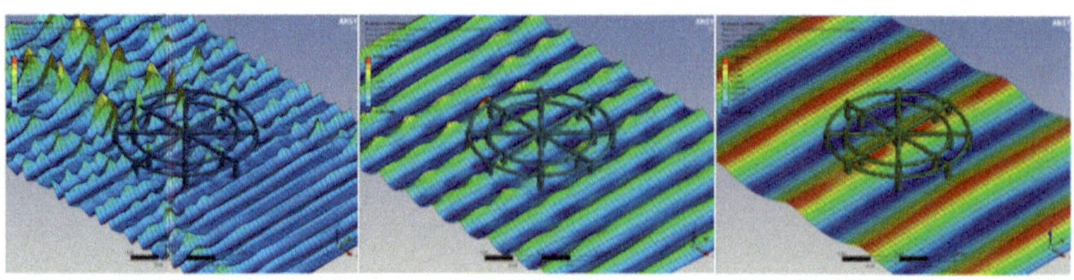

f = 0.380 Hz **f = 0.279 Hz** **f = 0.170 Hz**

Fig. 5.7 Pressure contour with wave amplitude 2.5 m, and different frequency (wave directions 0°)

Table 5.6 Cases of simulation (frequency-domain analysis)

Wave frequency (Hz)	Wave direction (°)	Wave amplitude (m)
0.116	0	1
0.170		
0.225		
0.276	22.5	2.5
0.333		
0.390		

direction, while Fig. 5.8 offers insights into pressure contours corresponding to a 22.5° wave direction. An intriguing aspect emerges as we observe the distinctive patterns in wave amplitude from one figure to another. The wave amplitude varies between a minimum of 1.0 m and a maximum of 2.5 m, a range where the amplitude corresponds to half of the wave height. Specifically, an amplitude of 1 m translates to a wave height of 2.5 m. This wave height serves as a significant indicator of the sea's environmental conditions, as effectively portrayed in Fig. 5.10. Interestingly, while Xiaohui et al. [12] report the South China Sea's highest wave height as 3.5 m, the author extends this parameter to 5 m, thereby revealing more intricate wave patterns within the pressure contours (Table 5.7).

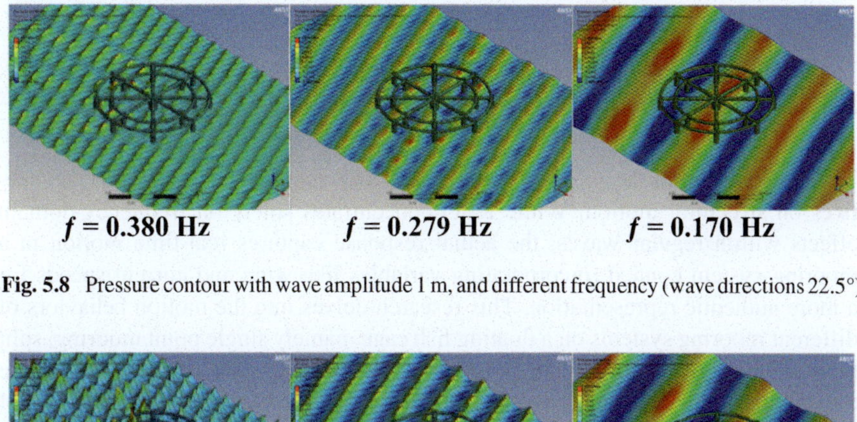

f = 0.380 Hz *f* = 0.279 Hz *f* = 0.170 Hz

Fig. 5.8 Pressure contour with wave amplitude 1 m, and different frequency (wave directions 22.5°)

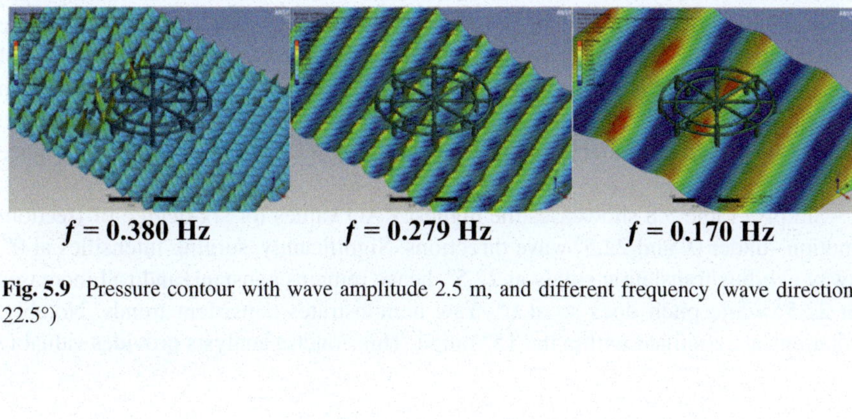

f = 0.380 Hz *f* = 0.279 Hz *f* = 0.170 Hz

Fig. 5.9 Pressure contour with wave amplitude 2.5 m, and different frequency (wave directions 22.5°)

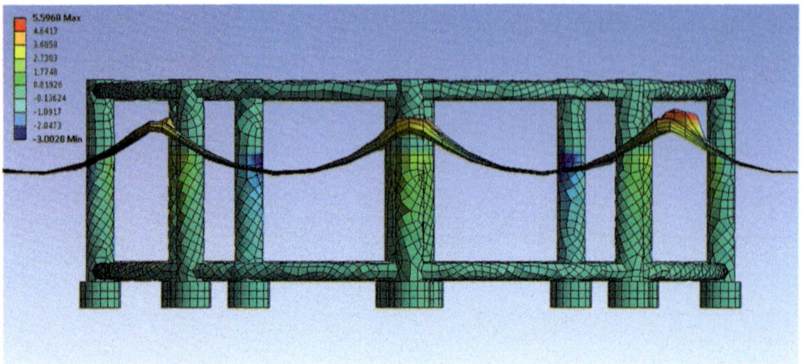

Fig. 5.10 Basic wave pattern

Table 5.7 Environmental loads (time-domain analysis)

Description	Regular wave	Irregular wave
Wave direction (°)	0	0
Wave amplitude (m)	1	1
	2.5	2.5
Wave period (s)	10	10
Current speed (m/s)	–	2.5
Wind speed (m/s)	–	8
		20

5.3.2 Response Amplitude Operator

The response amplitude operator (RAO) and actual response exhibit distinct perspectives on structural motion. While RAO concentrates solely on frequency-domain effects within regular waves, the actual response captures real-time motion in a mooring system context, incorporating variables like wind and current speeds for a more authentic representation. This research delves into the motion behaviors of different mooring systems on a floating fish cage, namely single point mooring, split point mooring, and split point (V) mooring. By analyzing parameters such as wave direction, amplitude, wind, and current speeds, a comprehensive understanding of the six-degree-of-freedom motion is attained. The resulting insights, as depicted in Fig. 5.11, shed light on the varying responses of these mooring systems in deep-sea conditions, offering valuable implications for structural integrity and performance optimization.

Simply, Table 5.8 showcases the highest RAO values for six-degree-of-freedom motions under 0° and 22.5° wave directions. Significantly, surging intensifies at 0°, sway exhibits translation solely at 22.5°, heave remains constant, and roll increases at 22.5° while pitch does so at 0°. Yaw demonstrates consistent trends. Notably, rotations are confined within the 15° range. This concise analysis provides valuable

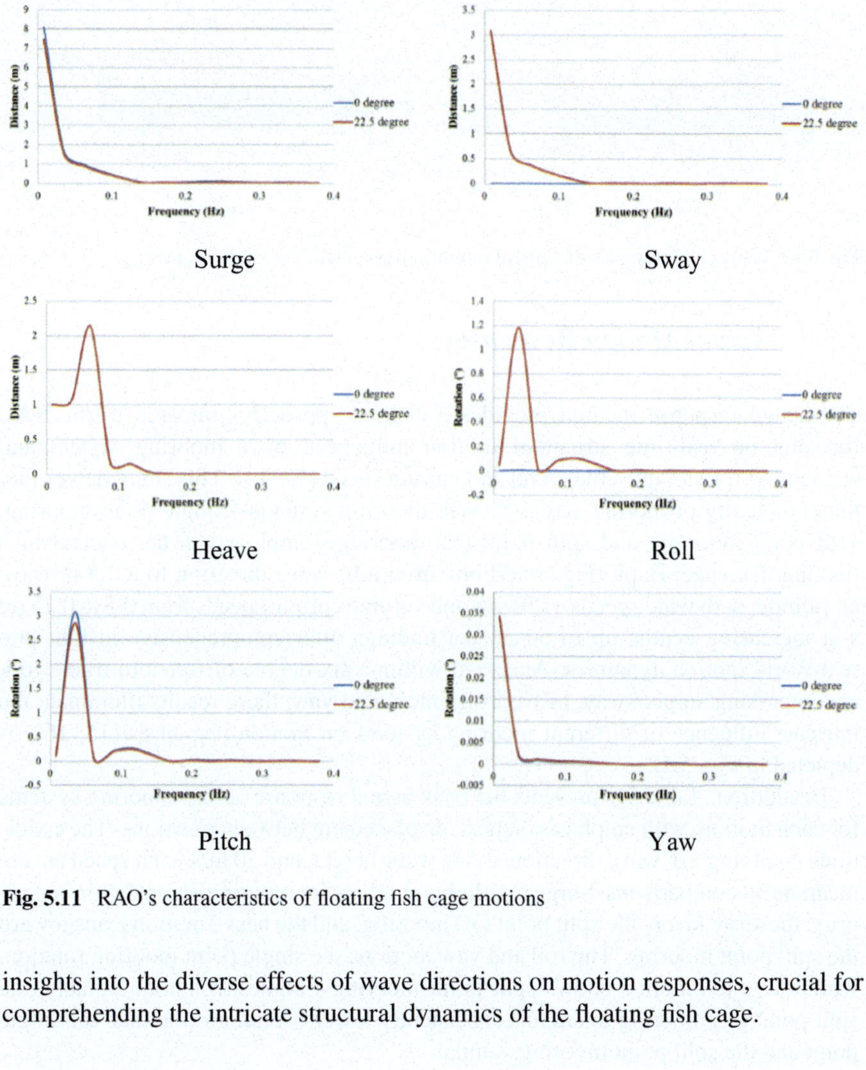

Fig. 5.11 RAO's characteristics of floating fish cage motions

insights into the diverse effects of wave directions on motion responses, crucial for comprehending the intricate structural dynamics of the floating fish cage.

Table 5.8 Maximum RAO for each motion of floating fish cage

Motion	Wave direction	
	0°	22.5°
Surge (m)	0.004–8.079	0.006–7.463
Sway (m)	–	0.002–3.091
Heave (m)	0–2.123	0–2.123
Roll (°)	0–0.007	0.001–1.182
Pitch (°)	0.002–3.084	0.002–2.85
Yaw (°)	0–0.034	0–0.034

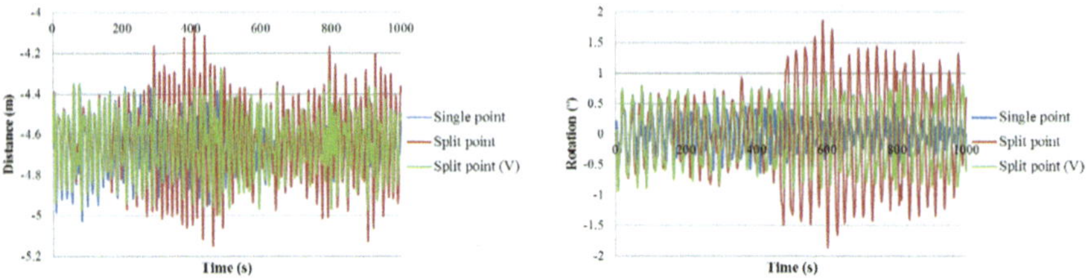

Fig. 5.12 The actual response of vertical motions; heave (left) and pitch (right)

5.3.3 Actual Motion Responses

The actual response method provides a distinct approach compared to the RAO focusing on real-time structural motion influenced by a mooring system and accounting for variables like wind and current speed [13, 14]. This study delves into the complexity of motion across diverse mooring systems—single point mooring, split point mooring, and split point (V) mooring—employed in the context of a floating fish cage. Exploring conditions from a 0° wave direction to a 2.5 m wave amplitude, with wind speeds of 20 m/s and varying current speeds from 0.5 to 0.25 m/s at increasing depths up to 60 m, the findings offer comprehensive insight into real-world motion dynamics. Analyzed within a six-degree-of-freedom framework encompassing surge, sway, heave, roll, pitch, and yaw, these results illuminate the intricate influence of different mooring systems on motion responses [15, 16], as depicted in Fig. 5.12.

In addition, Table 5.9 presents the peak actual response across mooring systems for each motion, with emphasis on peak displacement between positions. The conditions involving a 0° wave direction, 2.5 m wave height, and 20 m/s wind speed ensure meaningful comparisons. Surge highlights single point mooring's translation superiority, the sway favors the split point (V) mooring, and the heave motion leans toward the split point mooring. The roll and yaw motions see single point mooring rotation, while the pitch motion shows split point mooring's highest rotation. Notably, the split point (V) mooring maintains consistently lower overall motion than the single point and the split point mooring, similar.

5.3.4 Mooring Tension

The study delves into the composite catenary mooring line, pivotal for connecting or linking structures. Properties, including section length, maximum tension, stiffness (EA), and mass per unit length, are defined by ECAT cards. System-specific lengths encompass: 20, 20, and 6 m (the single point mooring); 5, 25, and 12 m (split point mooring); 28, 26, and 12 m (split point (V) mooring) (refer Fig. 5.12). For multi-ECAT composite lines, the definition originates from the anchor point.

Table 5.9 Maximum actual response for each motion at different types of mooring configurations

Motion of actual response	Types of mooring system		
	Single point mooring	Split point mooring	Split point (V) mooring
Surge (m)	– 10.405 to 10.566	– 5.511 to 5.508	– 5.493 to 5.586
Sway (m)	– 0.005 to 0.021	0.002 to 0.015	– 0.119 to 0.096
Heave (m)	– 5.026 to – 4.352	– 5.148 to – 4.064	– 4.991 to – 4.27
Roll (°)	– 0.02 to 0.017	– 0.013 to 0.009	– 0.072 to 0.073
Pitch (°)	– 0.675 to 0.71	– 1.856 to 1.866	– 0.93 to 1.03
Yaw (°)	0 to 0.065	– 0.032 to 0.028	– 0.008 to 0.039

Furthermore, Table 5.9 presents the peak tension across cables within diverse mooring systems: the single point, the split point, and split point (v). These tensions are unveiled in Figs. 5.13, 5.14 and 5.15, under specific conditions of a 0° wave, wind, and current direction. In extreme sea conditions—5 m wave height, 20 m/s wind speed, 2 m/s current speed, and frequencies spanning 0.1–0.38 Hz—cable tensions are revealed. For single point mooring, cable 2 displays the highest tension at 109.804 kN, followed by cable 1 at 41.45 kN. Parallel cable orientation preserves their maximum tensions due to force direction change from 0° to 180°. The mooring lines 3–8 are minimally impacted. In split point mooring, cable 2 records the greatest tension at 952.718 kN, with cable 1 following at 726.983 kN, showcasing higher tensions than the single point and the split point (v) mooring. In the split point (V) mooring, cable 4 bears the highest tension at 54.567 kN, and cable 3 at 48.925 kN. The mooring lines's maximum tensions cluster within a range of 38–55 kN, with the split point (V) mooring offering a more balanced force distribution, translating to lower maximum tensions [17]. This analysis informs mooring system design and cable strength considerations across varying force directions (Table 5.10).

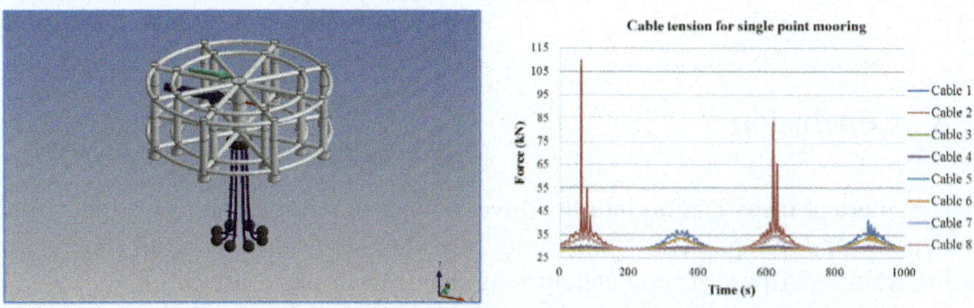

Fig. 5.13 Characteristics mooring-line tensions on the single point mooring system

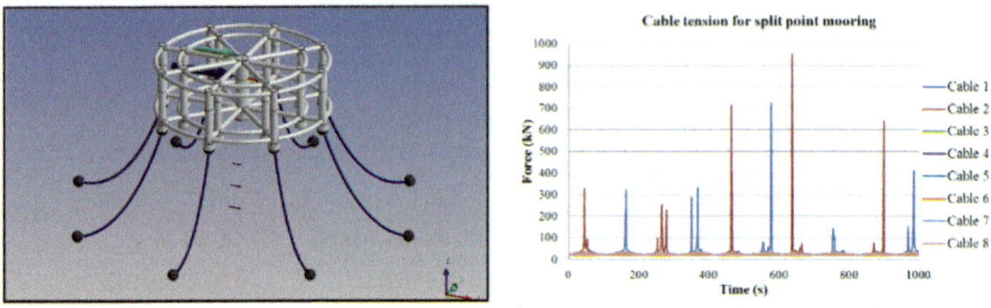

Fig. 5.14 Characteristics mooring-line tensions on the split point mooring system

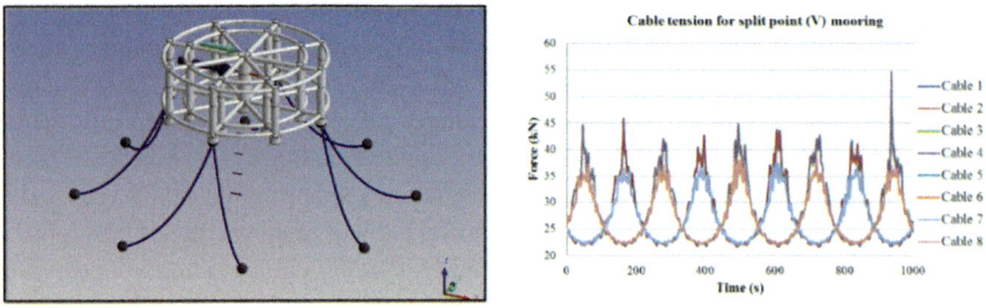

Fig. 5.15 Characteristics mooring-line tensions on the split point (V) mooring system

Table 5.10 Highest tension for each cable on different types of mooring system

Cable	Maximum tension (kN)		
	Single point mooring	Split point mooring	Split point (V) mooring
1	41.45	726.983	45.919
2	109.804	952.718	45.913
3	30.353	24.634	48.925
4	30.352	24.62	54.657
5	35.181	36.362	37.407
6	35.164	36.354	38.266
7	35.656	36.612	37.338
8	35.655	36.653	38.738

5.4 Conclusion

The numerical investigation into the three distinct mooring systems—single point, split point, and split point V—for floating fish cages has been successfully conducted using ANSYS-Aqwa. Several conclusions can be drawn as follows:

- The frequency-domain simulations produced response amplitude operator (RAO) data, unraveling the cage's dynamic behaviors across six degrees of freedom. The cage underwent the surge (8.079 m), the sway (3.091 m), the heave (2.123 m), the roll (1.182 m), the pitch (3.084 m), and exhibited negligible yaw (0.034°) movements.
- Evaluating cage responses across mooring strategies, the single point mooring revealed the broadest surge (− 10.405 to 10.556 m) and yaw (0–0.065°) ranges. Meanwhile, the split point mooring exhibited the widest heave (− 5.148 to − 4.064 m) and pitch (− 1.586° to 1.866°) ranges, and the split point V mooring displayed the highest sway (− 0.119 to 0.096 m) and the roll (− 0.072° to 0.073°) ranges.
- Additionally, dynamic analysis of hydrodynamic time response assessed mooring-line tensions, highlighting variable force distribution among cables across different mooring types. The split point V mooring showed the highest force distribution due to dual line influence, with minimal impact from wave amplitude changes on the line tension.

Acknowledgements This research was not funded by any grant.

References

1. Hlordzi, V., Kuebutornye, F.K., Afriyie, G., Abarike, E.D., Lu, Y., Chi, S., Anokyewaa, M.A.: The use of Bacillus species in maintenance of water quality in aquaculture: a review. Aquac. Rep. **18** (2020). https://doi.org/10.1016/j.aqrep.2020.100503
2. Tom, A.P., Jayakumar, J.S., Biju, M., Somarajan, J., Ibrahim, M.A.: Aquaculture wastewater treatment technologies and their sustainability: a review. Energy Nexus **4**, 100022 (2021). https://doi.org/10.1016/j.nexus.2021.100022
3. El-Thalji, I.: Context analysis of offshore fish farming. IOP Conf. Ser.: Mater. Sci. Eng. **700**(1), 012065 (2019). https://doi.org/10.1088/1757-899X/700/1/012065
4. Bjørndal, T., Tusvik, A.: Economic analysis of land-based farming of salmon. Aquac. Econ. Manag. **23**(4), 449–475 (2019). https://doi.org/10.1080/13657305.2019.1654558
5. Carballeira Braña, C.B., Cerbule, K., Senff, P., Stolz, I.K.: Towards environmental sustainability in marine finfish aquaculture. Front. Mar. Sci. **8**, 343 (2021). https://doi.org/10.3389/fmars.2021.666662
6. ANSYS Inc.: ANSYS-AQWA (2015). http://www.ansys.com/Products/Other+Products/ANSYS+AQWA
7. ANSYS, I.: Patent No. 724.746.3304. United State (2009)
8. Schofield, O., Meredith, M., Newman, L., Sparrow, M., Urban, E.: Implementing a Southern ocean observing system. EOS Trans. Am. Geophys. Union **93**(26), 241–243 (2012). https://doi.org/10.1029/2012EO260002
9. Park, M.A., Loseille, A., Krakos, J., Michal, T.R., Alonso, J.J.: Unstructured grid adaptation: status, potential impacts, and recommended investments towards CFD 2030. In: 46th AIAA Fluid Dynamics Conference, p. 3323 (2016). https://doi.org/10.2514/6.2016-3323
10. Boissonnat, J.D., Chazal, F., Yvinec, M.: Geometric and Topological Inference, vol. 57. Cambridge University Press (2018). https://doi.org/10.1017/9781108297806

11. George, P.L., Borouchaki, H., Alauzet, F., Laug, P., Loseille, A., Marechal, L.: Meshing, Geometric Modeling and Numerical Simulation, vol. 2: Metrics, Meshes and Mesh Adaptation. Wiley (2019). https://doi.org/10.1002/9781119384380
12. Xiaohui, L., Shengjie, Q., Lei, Y., Haibin, Q., Qi, Z.: Analysis of sea condition of Jiaolong deep-sea manned submersible operating in the South China Sea. In: E3S Web of Conferences, vol. 276, p. 01022. EDP Sciences (2021). https://doi.org/10.1051/e3sconf/202127601022
13. Konovessis, D., Chua, K.H., Vassalos, D.: Stability of floating offshore structures. Ships Offshore Struct. **9**(2), 125–133 (2014). https://doi.org/10.1080/17445302.2012.747270
14. Kim, M.: Dynamic Analysis Tol for Moored Tanker Based FPSO's Including Large Yaw Motion (2014). http://www.bsee.gov/TechnologyandResearch/TechnologyAssesmentProgram/Reports
15. Qiao, D., Li, B., Yan, J., Qin, Y., Liang, H., Ning, D.: Transient responses evaluation of FPSO with different failure scenarios of mooring lines. J. Mar. Sci. Eng. **9**(2), 103 (2021). https://doi.org/10.3390/jmse9020103
16. Liu, B., Yu, J.: Effect of mooring parameters on dynamic responses of a semi-submersible floating offshore wind turbine. Sustainability **14**(21), 14012 (2022). https://doi.org/10.3390/su142114012
17. Chrolenko, M.O.: Dynamic analysis and design of mooring lines. Master's thesis, Institutt for marin teknikk (2013)

Chapter 6
Power Plant Planning of Off-Grid Solar Electricity System for Goods Elevator Installation

Woro Agus Nurtiyanto, Wan Mariam Wan Muda, Ahmad Zaki Annuar, and Gunawan

Abstract In planning and estimating the power requirements of goods elevators that are operated through resources from the off-grid solar power plant system to reduce conventional electricity consumption, it is hoped that in the future all electricity use will switch to new renewable energy (EBT). This needs planning is carried out to answer the need for power exchange which originally came from electricity supply through state electricity company to supply new renewable energy (EBT) which will be sourced from solar power plants (PLTS) with an off-grid system, namely the power source is directly connected to the load. In this paper, the author uses a software, namely PVSyst so that it can be seen how much requirements in regard to photovoltaic (PV) installation, batteries, etc., are required. Based on the proposed 4.5 kW power, PT Galva requires 18 solar panel modules with a capacity of 525 Wp with an installation area of 46.5 m^2, power storage of 60 batteries with a capacity of 180 Ah with a voltage of 25.6 V, one unit of solar charge controller (SCC) with a capacity of 100A, and one inverter units with a capacity of 10 kW. This PLTS is capable of producing electrical energy of 11,653 kWh/year and the use of energy that can be used is 9813 kWh/year. The performance ratio in this study was 68.7%, the total energy losses were 176.9 kWh, and the total initial investment cost was Rp. 434,462,000. With a payback period of 8.5 years in a span of 25 years, the return

W. A. Nurtiyanto (✉) · W. M. W. Muda · A. Z. Annuar
Faculty of Ocean Engineering Technology and Informatics, Universiti Malaysia Terengganu, Terengganu, Malaysia
e-mail: p4783@pps.umt.edu.my

W. M. W. Muda
e-mail: w.mariam@umt.edu.my

A. Z. Annuar
e-mail: zannuar@umt.edu.my

W. A. Nurtiyanto · Gunawan
Electrical Engineering Study Program, Faculty of Engineering, Pamulang University, Banten, Indonesia
e-mail: gunawan@gmail.com

A. Ismail et al. (eds.), *Exploring Diversity in Engineering and Technology for Knowledge and Innovation*, Advanced Structured Materials 215,
https://doi.org/10.1007/978-3-031-64330-9_6

of investment (ROI) value is 294.1%. The net present value (NPV) is positive, which is Rp. 1,277,953,837. Hence, this PLTS design research can be said to be feasible.

Keywords Planning · Off-grid PLTS · NPC · ROI · PVSyst

6.1 Introduction

Electricity needs for elevators in Indonesia on average are still supplied by state electricity company, including PT Galva. In order for the electricity needs of the lift/ elevator to have independent electricity, a new and renewable power plant supply is needed as a replacement energy to help provide electricity in the event of a blackout by PLN. Environmentally friendly power plants that are in accordance with Indonesia's geological conditions are solar power in this case PLTS [1].

The ability to advance solar energy generation in Indonesia is very beneficial if it is observed based on the geological area of Indonesia which is located on the earth's equatorial streak. The design of a new renewable energy power plant that functions in order to support the reduction of conventional electricity consumption, the electrical system in the freight elevator that will be designed is an off-grid system, by direct loading on batteries charged by solar panel power [2].

Solar panels or commonly called photovoltaics are one of the energy sources which is very promising from a geographical point of view for Indonesia. In this era, the efficiency of solar cell models is on average between 5 and 20% depending on the constituent objects [3].

Based on information on solar energy radiation collected from 18 regions in Indonesia, solar radiation can be categorized as follows: Western region of Indonesia (KBI) around 4.5 kWh/m^2/day with monthly variables ranging from 10%; and in the Eastern Indonesia (KTI) around 5.1 kWh/m^2/day at a monthly variable of 9%. Therefore, Indonesia's average wind capacity is around 4.8 kWh/m^2/day on a monthly variable of around 9% [4].

PLTS is a power framework that converts energy into electrical energy through photovoltaic cell transformation. The photovoltaic framework converts sunlight-based radiation into energy [5, 6]. The higher the power of solar-based radiation (light) that enters the photovoltaic cell, the higher the electrical power it generates [7]. Since power is in most cases used throughout the day, much of the electrical power generated during the day is stored in the battery at any time for different power devices [8].

Basically, PLTS is a kind of power source that can be used to meet the needs of small to large consumers independently or through hybrid power. The demand for electricity as an energy source is very high [9]. For energy storage media, PLTS on-grid uses the grid as a storage medium and does not use batteries [10]. In the on-grid PLTS, there is an export–import (EXIM) kWh meter which functions to record the amount of energy stored in the PLN network. At the end of the month, it will be calculated to reduce electricity costs for customers according to the policies of the

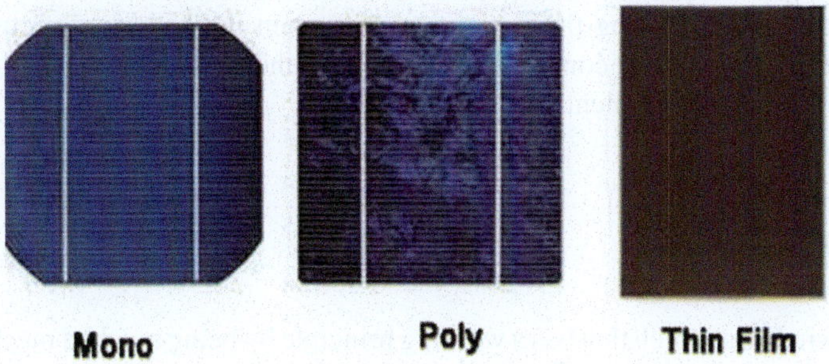

Fig. 6.1 Solar cell

local government/local electricity supplier. Solar panels can be categorized based on the materials that make up the solar cells [11]. In general, there are three types of solar panels, namely monocrystalline silicon (mono-Si), polycrystalline silicon (poly-Si), and thin film solar panels which can be seen in Fig. 6.1 [3].

6.1.1 *Battery/Accumulator*

A battery is a device that functions to store electrical energy using DC voltage and current. On the other hand, the battery also functions to process chemical energy into electrical energy [12–14]. In the installation of solar panels, accumulators are also used as a means of providing electrical energy produced by solar panels. PLTS batteries basically only work during the day or when the sun is there, so for the needs of the night solar panels cannot produce [15]. To deal with this, the electricity produced by solar panels during the day is stored as backup electricity when the sun does not shine on the solar panels, for storage of electricity a battery is needed as an energy charge storage device [16, 17].

6.1.2 *Solar Charge Controller*

A battery controller is a device that functions as a controller of electric current when charging a battery that is supplied through renewable energy such as a wind power plant or a solar power plant.

SCC functions to charge solar batteries in limiting the flow of electrical energy into batteries produced by renewable energy so that it functions as a cross-energy overcharge prevention when charging battery capacity and releasing electric current energy to the load. Thus, the minimum battery load is around 20–30% which has been set, so that the flow of electric current to the load will automatically stop, this principle can reduce the occurrence of empty battery usage which can damage the battery, or by stopping the flow of energy when the battery capacity reaches a specified threshold, the controller helps protect the battery from deep discharge conditions which is detrimental [18].

6.1.3 Limit Switch

A limit switch is a switch that has a working principle by being touched mechanically. This switch has another name, namely a microswitch or basic switch, an electronic device that uses this type of switch is three-dimensional printing. The contact poles are arranged on a flexible plate that can move up or down with a constant position in the middle, this plate has a piece that functions as a spring rotation container. The shape of the spring is shaped like the letter U, this condition causes the contacts to be pressed together on the inside of the switch. The spring in the limit switch can only move in two positions [19].

6.1.4 Elevator

An elevator is a vertical means of transportation used to carry people and goods. Elevators are usually used in high-rise buildings, generally three or more floors in the building. Relatively low buildings usually only use stairs as a medium of transportation from floor A to floor B.

Elevators in this modern era have push buttons that passengers can adjust according to the passenger's goals. There are three types of elevator machines, namely hydraulic, traction which is commonly called a fixed pulley, and hoist or double pulley. The type of hoist can be divided into two directions, namely the push hoist and the pull hoist [20].

6.1.5 Inverters

An inverter is a circuit or electronic device capable of converting direct current into alternating current at the required voltage and frequency in sync with the circuit design. The input current for this power supply is the current input from this inverter, which can be a battery/accumulator or a solar cell [21].

6.1.6 PVSystNet Present Cost Software

PVSyst is a software that is used for the stages of learning, measuring (sizing), and analyzing photovoltaic system data accurately. This application was developed by the Geneva University which has an interconnected system with a network (grid connected) or commonly called an on-grid, a stand-alone system or commonly called an off-grid, and a pump system (pumping) or commonly called a hybrid system, and the power supply network for public transport (DC-grid) [22].

6.2 Methodology

The block diagram in Fig. 6.2 consists of the process of capturing solar radiation by PV which will charge the battery and is controlled by the SCC which will then be converted by an inverter from direct current (DC) to alternating current (AC) to turn on the load, namely a LIFT.

The first flowchart stage starts from determining the voltage and current generated by the solar panel to then get the amount of power generated; see Fig. 6.3. The next step is to measure the amount of sunlight with a lux meter and convert it into solar radiation by calculating the intensity solar radiation (W/m^2) $\approx$ Lux $\times$ conversion factor [23]. After getting the amount of solar radiation, the incoming power can be determined with calculations that are in accordance with the flowchart and then one can get the efficiency value of the PV.

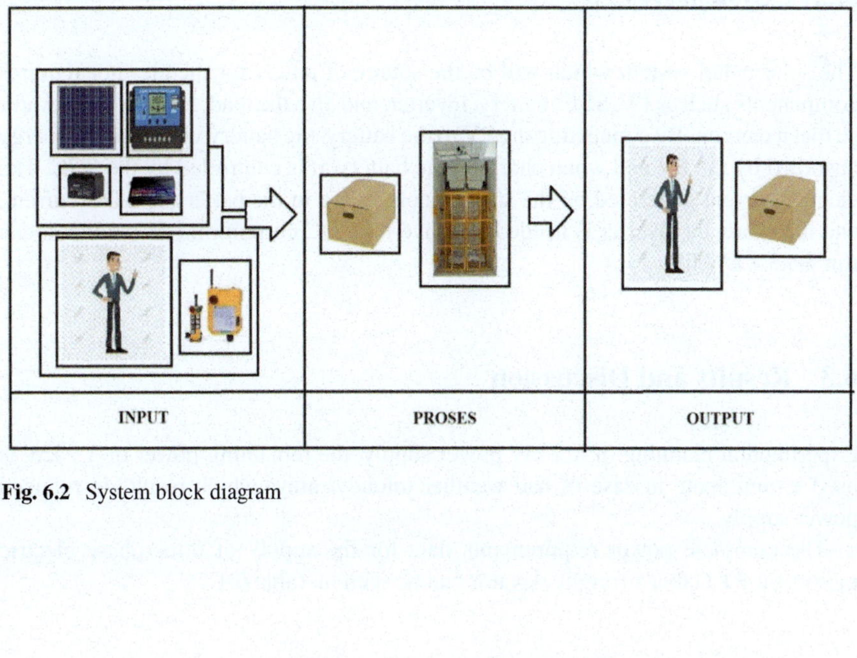

Fig. 6.2 System block diagram

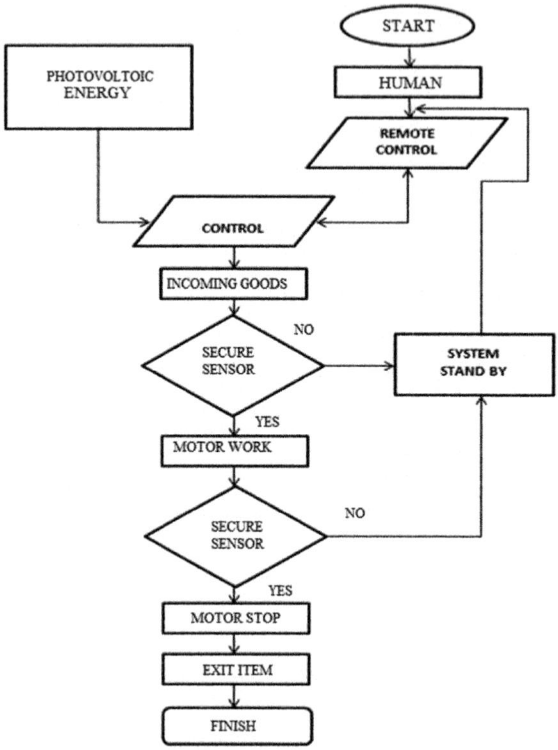

Fig. 6.3 Flowchart

6.2.1 System Design

The solar panel system which will be the source of power for the elevator requires components such as PV, SCC, battery, inverter, and also the load, namely the elevator so that it can run. By connecting the PV to the battery, the battery will store the energy absorbed by the PV and when charging the battery, it is controlled by the SCC. The electric current generated by the solar panels stored in the battery is a DC current, and therefore, an inverter is needed to convert the DC current into AC so that it can run the 3.7 kW lift.

6.3 Results and Discussion

Experimental planning of 4.5 kW power supply, the remaining power of 3.7 kW is used to anticipate in case of bad weather (cloudy/rainy) which results in reduced power supply.

The complete power requirements data for the supply of three-phase electric motors on PT Galva's freight elevator can be seen in Table 6.1.

Table 6.1 Specification of PT Galva goods elevator motor

Hoist model	Kamuichi FC/R/KC Type Hoist
Hoist capacity	3 Ton
Winging-up motor	3.7 kW 6P
Power supply	200 V 50 Hz
Acceptable of 3 core cab tire cable	5.5 × 33 mm^2
Remark	60 A

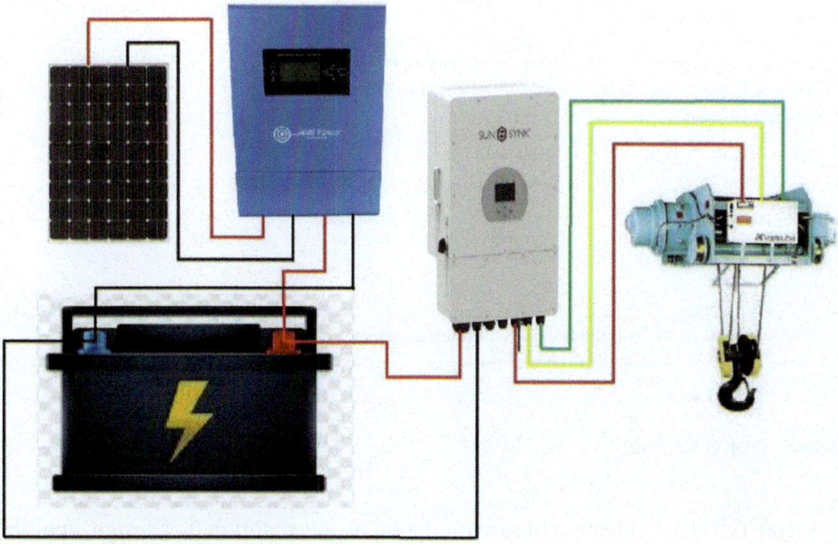

Fig. 6.4 PLTS composition

6.3.1 PLTS Composition

The solar power plant that will be used in this research is the PLTS off-grid AC tree phase coupling. This system will supply power by continuously covering the power requirements of the freight elevator; see Fig. 6.4.

6.3.2 Simulation Report

The following is a report on the results of the simulation of a solar power plant using the PVSyst software with the inputted factors. PT Galva is at − 6.1617962 latitude and 106.8198350 longitude.

Project: PT Galva

Variant: New simulation variant

PVsyst V7.2.14
VC0, Simulation date:
13/05/22 15:52
with v7.2.14

Project summary

Geographical Site	Situation		Project settings	
Jakarta	Latitude	-6.21 °S	Albedo	0.20
Indonesia	Longitude	106.85 °E		
	Altitude	15 m		
	Time zone	UTC+7		

Meteo data
Jakarta
Meteonorm 8.0 (2010-2014), Sat=100% - Synthetic

System summary

Stand alone system — **Stand alone system with batteries**

PV Field Orientation		User's needs	
Fixed plane		Daily household consumers	
Tilt/Azimuth	30 / 180 °	Constant over the year	
		Average	27.4 kWh/Day

System information

PV Array		Battery pack	
Nb. of modules	18 units	Technology	Lithium-ion, LFP
Pnom total	9.45 kWp	Nb. of units	60 units
		Voltage	102 V
		Capacity	2700 Ah

Results summary

Available Energy	11653 kWh/year	Specific production	1233 kWh/kWp/year	Perf. Ratio PR	68.72 %
Used Energy	9813 kWh/year			Solar Fraction SF	98.23 %

Fig. 6.5 Project summary

Figure 6.5 is a description of the simulation results report in this design, applying PV with tilt and azimuth 30° and 180° facing north. In accordance with the proposed load, it requires an average energy of 27.4 kWh/day, so it takes 18 units of modules that will produce a power of 9.45 kWp. This solar power plant is capable of producing electrical energy of 11,653 kWh/year with the consumption of energy that can be used by consumers as much as 9813 kWh/year.

6.3.3 Performance Ratio

The performance ratio is the most important part of a solar power plant planning, where this PR is used as a benchmark for the quality of PLTS planning which is a comparison of the output of photovoltaic energy between actual and theory. The PR in this study was 0.687 or in percent, meaning 68.7%; see Fig. 6.6.

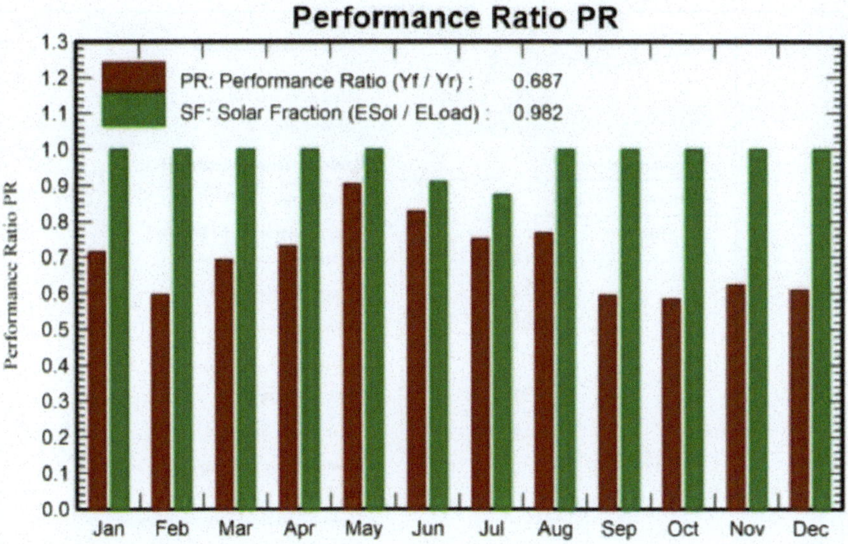

Fig. 6.6 Performance ratio graphic

6.3.4 *Energy Loss*

Figure 6.7 shows the lost energy, the total global horizontal irradiation (GHI) is 1757 kWh/m^2 while the effective solar irradiation is 1461 kWh/m^2, the PV array energy at STC is 13,809 kWh after experiencing some losses. The energy obtained is 10747 kWh of effective energy at the output of the PV array. The total energy that reaches PLTS users is 9913 kWh with an energy load of 9990 kWh.

In planning this solar power plant, the author will analyze the feasibility of PLTS investment which will be carried out using the PVSyst software and will calculate the overall LCOE, NPV, payback period, and ROI by inputting the results of the calculation of the research components as described above.

6.3.5 *Cost Calculation*

This cost calculation is part of the required investment costs to be incurred for the construction of an off-grid solar power plant installation on PT Galva's freight elevator. Below is a breakdown of the component costs that will be used in this plan; see Table 6.2.

Civil works (estimated cost of installation, etc.) 0.1$/kWh (Tetra Tech ES, 2018). According to the detailed table of component costs which corresponds to component prices, it can be estimated that the initial investment in this study was Rp. 434,462,000.

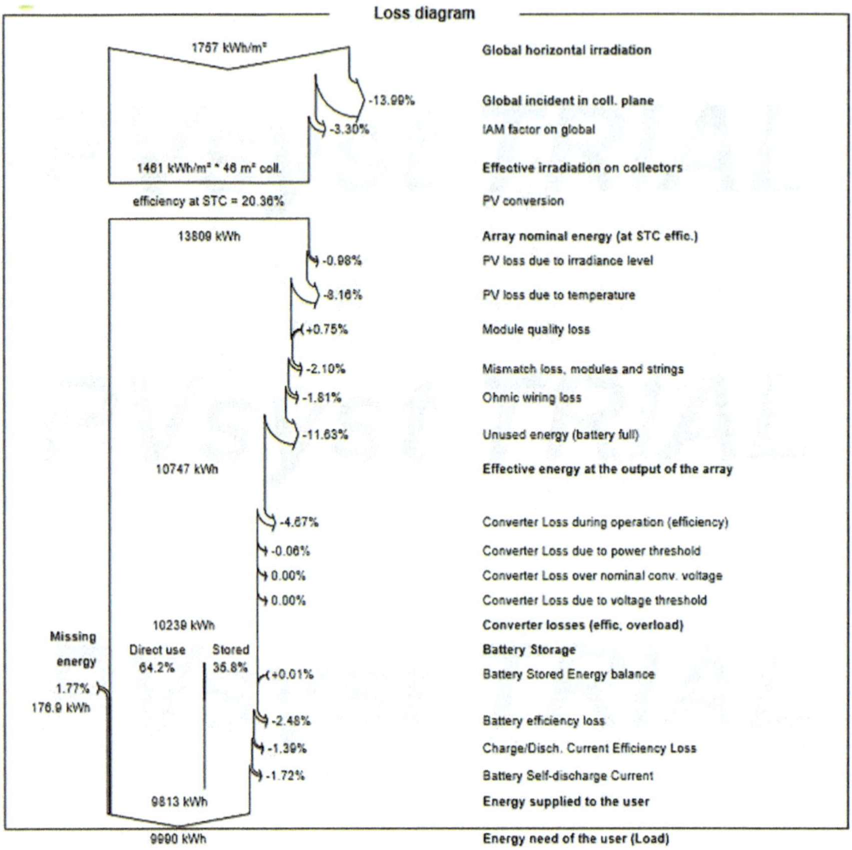

Fig. 6.7 Energy losses

Table 6.2 Cost breakdown

No	Equipment	Amount	Price (Rp.)	Total (Rp.)
1	Solar module	18	8,925,000	160,650,000
2	Battery	60	3,947,000	236,820,000
3	SCC	1	10,180,000	10,180,000
4	Inverters	1	13,132,000	13,132,000
5	Installation fee	0.1$/watt = 0.1 × 9.5 kW		13,680,000
6	Total			434,462,000

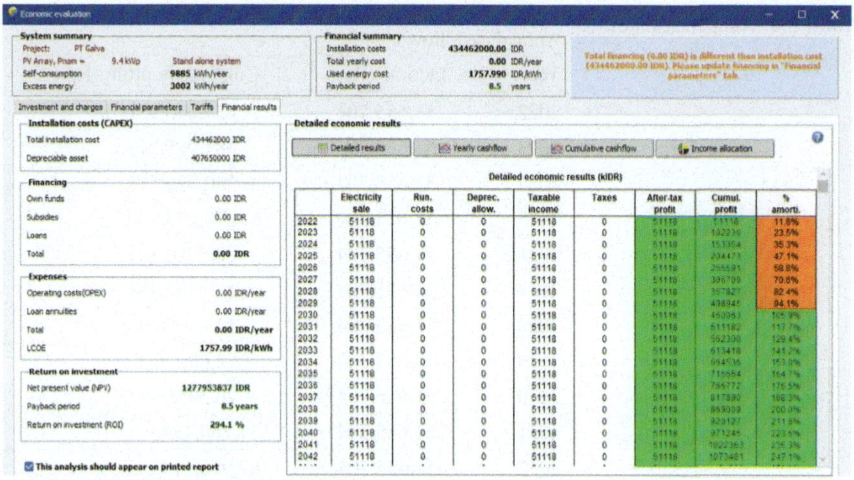

Fig. 6.8 Financial result

6.3.6 *Feasibility Analysis Generator Investment*

In planning this solar power plant, the author will analyze the feasibility of the PLTS investment which will be carried out using the PVSyst software and will calculate the overall LCOE, NPV, payback period, and ROI by inputting the results of the research component calculations as described in Fig. 6.8.

The results of the PVSyst software simulation for investment feasibility where the initial investment cost is Rp. 434,462,000 and the price of electrical energy is Rp. 5.171/kWh. The payback period in this study was 8.5 years; see Tables 6.3, 6.4, and 6.5.

In this study, the cumulative profit in the 13th year was Rp. 134,571,623, the value of return of investment (ROI) is 294.1%, and the net present value (NPV) is positive, which is Rp. 1,277,953,837. So this PLTS design research can be said to be feasible.

Table 6.3 Investment details

Total investment	
Initial cost	Rp. 434,462,000
O&M cost	Rp. 4,344,625
Amount	Rp. 438,806,625

Table 6.4 Annual cash flow

Annual cash flow		
Year	Income (Rp.)	Cumulative profit (Rp.)
2022	51,655,704	– 433,640,921
2023	51,655,704	– 381,985,217
2024	51,655,704	– 330,329,513
2025	51,655,704	– 278,673,809
2026	51,655,704	– 227,018,105
2027	51,655,704	– 175,362,401
2028	51,655,704	– 123,706,697
2029	51,655,704	– 72,050,993
2030	51,655,704	– 20,395,289
2031	51,655,704	31,260,415
2032	51,655,704	82,916,119
2033	51,655,704	134,571,623
Total	619,868,448	134,571,623

Table 6.5 Feasibility analysis

Investment feasibility analysis	
Net present value (NPV)	Rp. 1,277,953,837
Payback period	8.5 Year
Returns of investment (ROI)	294.1%

6.4 Conclusion

Based on the proposed 4.5 kW power, PT Galva requires 18 solar panel modules with a capacity of 525 wP with an installation area of 46.5 m^2, power storage of 60 batteries with a capacity of 180 Ah with a voltage of 25.6 V, one unit of SCC with a capacity of 100 A, and one inverter unit with a capacity of 10 kW. This PLTS is capable of producing electrical energy of 11,653 kWh/year with a usable energy consumption of 9813 kWh/year. The performance ratio in this study is 0.687 or in percent, which means 68.7% and the total energy losses are 176.9 kWh, which occurred in June as many as 72.0 kWh and in June as many as 104.8 kWh.

The total initial investment cost is Rp. 434,462,000. With a payback period of 8.5 years in a span of 25 years. The return of investment (ROI) value is 294.1%. The result of net present value (NPV) is positive, which is Rp. 1,277,953,837. So this PLTS design research can be said to be feasible.

Acknowledgements This research was supported by the Ministry of Higher Education Malaysia, through the Fundamental Research Grant Scheme (Ref: FRGS/1/2018/TK07/UMT/02/1). The authors would like to express their gratitude to the Faculty of Ocean Engineering Technology and Informatics, Universiti Malaysia Terengganu, for supporting the facilities during the course of the research.

References

1. Matahari Untuk PLTS di Indonesia. Selasa, 19 Juni 2012. https://www.esdm.go.id/id/media-center/arsip-berita/matahari-untuk-plts-di-indonesia. Accessed 15 Mar 2022
2. Agung, P.: Implementasi Peraturan Menteri ESDM tentang PLTS Atap. Kementrian Energi dan Sumber Daya Mineral, Siaran Pers Nomor: 35.Pers/04/SJI/2022 (2022). https://ebtke.esdm.go.id/post/2022/01/21/3058/implementasi.peraturan.menteri.esdm.tentang.plts.atap. Accessed 16 Mar 2022
3. Romario, H.: Studi Perencanaan Pembangkit Listrik Tenaga Surya pada Komplek Perumahan Royal Gardenia Medan. Skripsi. Universitas Sumatera Utara, Medan (2018). https://repositori.usu.ac.id/handle/123456789/9631
4. Sri, A.N.: Sumber Energi Listrik Alternatif, Kementrian Pendidikan dan Kebudayaan, Energi listrik yang dapat dikembangkan di Indonesia (2019)
5. Slamet, H., Gatot, S., Subandi, dan Arifin, N.: Perencanaan Pembangkit Listrik Tenaga Surya (PLTS) On-Grid Dengan System DC Coupling Berkapasitas 17kWp Pada Gedung Graha Cendekia. Jurnal **5**, 2502–8782. Univeristas Muhammadiyah Prof. DR. Hamka. https://journal.uhamka.ac.id/index.php/teknoka/article/view/10271. https://doi.org/10.22236/teknoka.v5i.300
6. Dwi, M.C.R., Pradipta, A.R.S., Afrida, N.U., Rizal, Z.A., dan Jimmy, T.P.: Perancangan Pembangkit Listrik Tenaga Surya (PLTS) pada Kolam Budidaya di Daerah Sentono Menggunakan Software PVsyst. Jurnal Pendidikan Teknik Elektro. **6**(2). Universitas PGRI Madiun (2020). https://doi.org/10.25273/jupiter.v6i2.10519
7. Alvin M.R., Bambang, W., dan Agung, N.: Analisis Potensi Dan Unjuk Kerja Perencanaan Pembangkit Listrik Tenaga Surya (PLTS) Di Departemen Teknik Elektro Universitas Diponegoro Menggunakan Software PVSyst 6.43. Transient **7**(4). Universitas Diponegoro (2019). https://doi.org/10.14710/transient.v7i4.883-890
8. Usman, M.K.: Analisis Intensitas Cahaya Terhadap Energi Listrik Yang Dihasilkan Panel Surya. J. Power Elektron. **9**(2). Politeknik Harapan Bersama (2020). https://doi.org/10.30591/polektro.v9i2.2047
9. Bentar, S.: Usaha Menanggulangi Pemanasan Global. Kementrian Pendidikan dan Kebudayaan, Menggunakan Energi Terbarukan (2019)
10. Tomi, E.: Perencanaan Pembangkit Listrik Tenaga Surya Untuk Catu Daya Tambahan Pada Hotel Kini Kota Pontianak. Skripsi, Universitas Tanjungpura, Pontianak (2017)
11. Angryani, N.S.: Studi Analisis Perubahan Putaran Motor Induksi Satu Fasa Akibat Output PLTS Aplikasi Kipas Angin. Skripsi, Universitas HKBP Nommensen, Medan (2018). http://repository.uhn.ac.id/handle/123456789/5152
12. Dian, F.A., Juara, M.T.: Pengaturan Tegangan Pembangkit Listrik Tenaga Surya (PLTS) 1000 WATT. Jurnal. Kajian Teknik Elektro. **1**(1). STT PLN Jakarta. https://media.neliti.com/media/publications/259756-pengaturanteganganpembangkitlistriktenag-2a5e5696.pdf
13. Yuli, S.: Pengukuran Efisiensi Panel Surya Tipe Monokristalin dan Karakterisasi Struktur Material Penyusunnya. Skirpsi, Institut Teknologi Sepuluh, Surabaya (2017). https://repository.its.ac.id/3165/1/1113100031-Undergraduate_Theses.pdf
14. Mengenal Sistem Panel Surya, 15 Juni 2021. https://sunenergy.id/blog/sistem-panel-surya/. Accessed 20 Mar 2022
15. Muslih, N.: Karakteristik Baterai Sebagai Penyimpan Energi Listrik Secara Spesifik. Jurnal. **6**(1). Universitas Islam Sumatera Utara, Medan (2021). https://jurnal.uisu.ac.id/index.php/jet/article/view/3797

16. Achmad, F.H.: Rancang Bangun Kendali Hybrid Power System (PLTS-PLN). Skripsi, Institut Teknologi Nasional Malang, Malang (2018). http://eprints.itn.ac.id/7345/1/1552001%20%282018%29.pdf
17. Kurnia, R., Mirza, Z.: Manajemen Baterai Pada Pembangkit Listrik Tenaga Surya 920 Wp Dengan Beban Alat Pengering Ikan Otomatis 600 Watt. Jurnal. **18**(21). Universitas Bung Hatta, Padang (2021)
18. Haris, I., Priyan, G., Budiyanto: Analisa Simulasi Kinerja Sel Surya 10 Wp dengan Energi Terbarukan Sumber Energi Cahaya Buatan Sebagai Pengganti Sinar Matahari. Jurnal. e-ISSN : 2460-8416. Universitas Muhammadiyah Jakarta, Jakarta (2016)
19. Aristyo, A.: Rancang Bangun Mesin Pemotong Balok Kayu Serbaguna Dengan Sistem Kontrol Otomatis. Skripsi, Universitas Islam Majapahit, Mojokerto (2019)
20. Kosmas, P.: Sistem Pengendalian Motor DC Aplikasi Lift Dengan Pengendali Digital Berbasis Arduino. Skripsi, Universitas HKBP Nommensen, Medan (2018)
21. Tim Fakultas Teknik Universitas Negeri Yogyakarta. Teknik Dasar Rectifier dan Inverter. Yogyakarta. Modul. TU15. Sekolah Menengah Kejuruan (2003)
22. Ketut, I.S., Wayan, R., Made, S. I.: Studi Analisis Pengaruh Perubahan Posisi Terhadap Efisiensi Panel Surya LPJU By Pass Ngurah Rai. E-Journal SPEKTRUM **5**(1). Universitas Udayana Denpasar, Denpasar (2018)
23. Ahmad, M.: Analisis Rangkaian Inverter 12VDC-220VAC Dengan Sumber Panel Surya Pada Beban Motor Listrik Satu Fasa. Skripsi, Universitas Muhammadiyah Sumatera Utara, Medan (2018)

Chapter 7
Study of Coastal Morphology Changes Due to Tsunami in the Sunda Megathrust—A Case Study at Pelabuhan Ratu Bay

Alamsyah Kurniawan, Ardi Iman Malakani, Munawir Bintang Pratama, and Ahmad Fitriadhy

Abstract Pelabuhan Ratu Bay is located in an area facing the Sunda Megathrust which has the potential for a Tsunami to occur. The danger from the Tsunami is not only in the form of run-up and inundation but also the danger of changing the morphology of the area. This study aims to investigate the morphological changes of the area in the form of changes in erosion and sedimentation due to the Tsunami generated by the earthquake in the Sunda Megathrust with various seismicity scenarios and sediment grain size without being influenced by other external effects such as tides and wind. This study was performed using two modelling tools, namely COMCOT for modelling Tsunami propagation and Delft3D for modelling the sediment transport. The result in instantaneous water level of the tsunami, wave propagation, and morphological changes is discussed. The geometry and slopes of Pelabuhan Ratu Bay allow the wave to experience a shoaling which escalates the wave height to at least tripled compared to the offshore value. The incoming waves induced sedimentation at the offshore region and mixed erosion and sedimentation at the nearshore region. Findings in each scenario are compared, and the impact of seismicity and sediment size is evaluated.

A. Kurniawan (✉) · A. I. Malakani
Ocean Engineering Program, Faculty of Civil and Environmental Engineering, Institut Teknologi Bandung, Jl. Ganesha No. 10, Bandung, Jawa Barat, Indonesia
e-mail: alamsyahk@itb.ac.id

A. I. Malakani
e-mail: ardimalakani@gmail.com

M. B. Pratama
School of Engineering, The University of Edinburgh, Edinburgh EH9 3JU, UK
e-mail: munawirbintang@gmail.com

A. Fitriadhy
Program of Naval Architecture, Faculty of Ocean Engineering Technology and Informatics, Universiti Malaysia Terengganu, 21300 Kuala Terengganu, Terengganu, Malaysia
e-mail: a.fitriadhy@umt.edu.my

A. Ismail et al. (eds.), *Exploring Diversity in Engineering and Technology for Knowledge and Innovation*, Advanced Structured Materials 215,
https://doi.org/10.1007/978-3-031-64330-9_7

Keywords Coastal morphology · Tsunami · Sediment transport · Numerical modelling · Pelabuhan Ratu Bay

7.1 Introduction

Geographically, Indonesia is an archipelagic country located at the confluence of four tectonic plates; therefore, a large potential for an earthquake that can generate a tsunami exists [1, 2]. Previous studies had documented the location of the epicentres of earthquakes that occurred based on the earthquake relocation recorded by the Meteorology, Climatology, and Geophysics Agency (BMKG) in 2009 to 2018, including the location of the earthquake epicentre that caused a tsunami at the south of Java Island [3]. In addition, a previous study explained that there is a seismic gap that allows for a large earthquake in the future, where in this study it is explained that there is a possibility of a tsunami reaching a tsunami height of up to 20 m in the worst-case scenario, in a case where the two megathrust segments spanning Java break at the same time.

The region of Teluk Pelabuhan Ratu (Pelabuhan Ratu Bay) is one of the areas that are highly vulnerable in the event of a tsunami in the Sunda Subduction Zone (Fig. 7.1). These hazards are not only in the form of the inundation height to the distance of the tsunami inundation that enters the Pelabuhan Ratu Bay area but also in the changing of morphology at the Pelabuhan Ratu Bay. The tsunamis can cause dramatic morphological changes, such as erosion of barrier island, sand dunes, and other coastal features [4]. These are induced by the interaction of the hydrodynamic force of the tsunami wave, slope of the bathymetry and topography, wave overtopping, and the influence of the roughness of the seabed causing the tsunami sediment transport process [5, 6].

This study aims to investigate the morphological changes (erosion or sedimentation) of the Pelabuhan Ratu Bay due to the tsunami generated by the earthquake in the Sunda Megathrust Zone (Sunda Megathrust) within various seismicity scenarios and sediment grain sizes. This modelling is performed using two modelling tools, namely Cornell Multi-grid Coupled Tsunami (COMCOT) for numerical modelling of tsunamis and Delft3D-FLOW for numerical modelling of sediment transport.

7.2 Materials and Methods

7.2.1 Model Description

Two modelling tools are used in this study, COMCOT for modelling tsunami wave propagation and Delft3D-FLOW for the morphology model. COMCOT description will be presented in this study. For Delft3D-FLOW description, it is referred to the

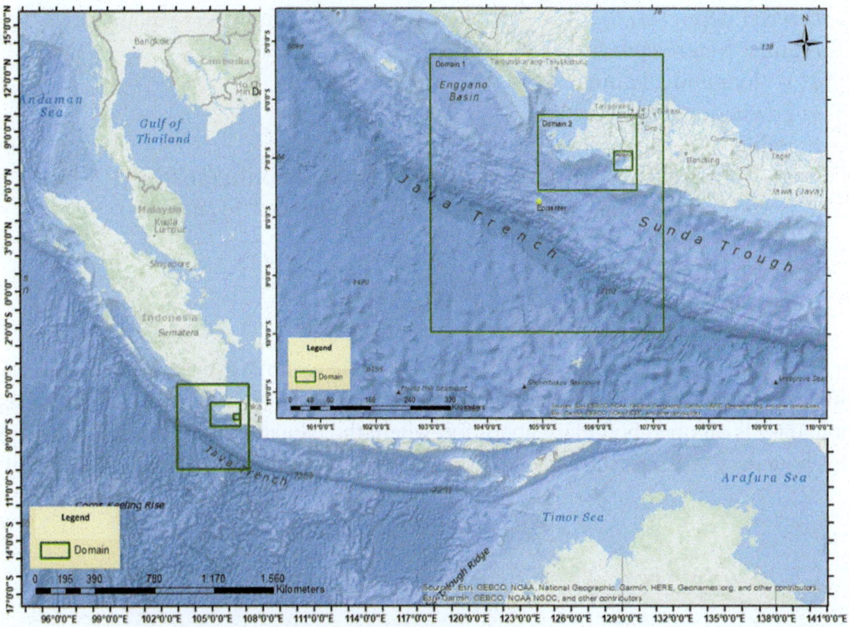

Fig. 7.1 Study area, Southern Java Sea to Pelabuhan Ratu. Green boxes mark the modelling domain (domain 1; large, domain 2; medium, domain 3; small)

software manual [7]. The COMCOT model is a numerical model based on depth-averaged-based models with shallow water equations (SWE) developed by Cornell University. The governing equations in the SWE model are the continuity equation (conservation of mass) and the momentum equation (motion) which are dispersive equations because the propagation velocity/wave group is a function of depth and does not depend on the period/frequency [8].

The model is expressed in spherical coordinates due to a large coverage of domain and benefitted with bathymetric input in degrees. The governing equations in spherical coordinates are given in Eqs. (7.1–7.3).

$$\frac{\partial \eta}{\partial t} + \frac{1}{R\cos\varphi}\left\{\frac{\partial P}{\partial \psi} + \frac{\partial}{\partial \varphi}(\cos\varphi Q)\right\} = -\frac{\partial h}{\partial t} \tag{7.1}$$

$$\frac{\partial P}{\partial t} + \frac{1}{R\cos\varphi}\frac{\partial}{\partial \psi}\left\{\frac{P^2}{H}\right\} + \frac{1}{R}\frac{\partial}{\partial \varphi}\left\{\frac{PQ}{H}\right\} + \frac{gH}{R\cos\varphi}\frac{\partial \eta}{\partial \psi} - fQ + F_x = 0 \tag{7.2}$$

$$\frac{\partial Q}{\partial t} + \frac{1}{R\cos\varphi}\frac{\partial}{\partial \psi}\left\{\frac{PQ}{H}\right\} + \frac{1}{R}\frac{\partial}{\partial \varphi}\left\{\frac{Q^2}{H}\right\} + \frac{gH}{R}\frac{\partial \eta}{\partial \varphi} + fP + F_y = 0 \tag{7.3}$$

where

η: water surface level (m)

h: water depth (m)
H: total water depth (m)
φ, ψ: longitude, latitude
g: gravitational acceleration
R: radius of earth
P, Q: volume flux in—x (west-east) and—y (south-north) (m^3/s)
u, v: current speed in—x (west-east) and—y (south-north) (m/s)
f: Coriolis effect
Ω: Earth's rotation rate
Fx, Fy: bottom friction in X and Y directions
n: manning roughness ($s/m^{1/3}$)

The land boundary is a closed boundary where no wave and energy formation occur because the boundary absorbs the incoming wave energy, while the sea boundary is an open boundary condition where the tsunami waves that pass through this boundary will pass through the model domain without reflecting into the domain.

7.2.2 Study Area

The study area is in Pelabuhan Ratu Bay, located in West Java Province as shown in Fig. 7.1. The tsunami waves propagate from the ocean (where the model can use a coarse resolution) to the coastal and low-land area (in which a much finer resolution is needed). Therefore, a nesting technique is implemented. The nesting technique will allow a smooth transition between the ocean and coastal-scale domain [2, 9–11]. Further, the research is constructed into three domains, using a two-stage nested grid system, as depicted in Fig. 7.1. Each transition applies a refinement factor of 6. The mesh resolution (length-characteristic) is 2160 m at domain 1, the largest domain. Thus, the resolution at domain 2 (medium) and domain 3 (small) is 360 m and 60 m, respectively. Domain 1 covers a deep-water region where the effect of bottom friction can be neglected. Therefore, domain 1 uses a more simplified approach, the linear SWE. Meanwhile, domains 2 and 3 cover a shallow water and coastal region; thus, the nonlinear SWE is applied.

7.2.3 Bathymetry and Topography

The model used Batimetri Nasional or the National Bathymetry (BATNAS) to generate bathymetry in the three domains. The National Bathymetry is issued by the Geospatial Information Agency (BIG). This dataset is the result of the inversion of gravity anomaly data resulting from altimetric data processing by adding sounding data carried out by various government-owned agencies with single and multibeam surveys [12]. The spatial resolution of the BATNAS data is 6 arc-seconds using the MSL datum.

Elevation for land area or the topographic data is derived from the National DEM or DEMNAS [13]. It was constructed from several data sources including IFSAR data (5 m resolution), TERRASAR-X (5 m resolution), and ALOS PALSAR (11.25 m resolution), by adding the stereo-plotting mass point data [14]. The spatial resolution of DEMNAS is 0.27 arc-second, using the EGM2008 vertical datum [13].

7.2.4 Modelling Scenarios

The seismic parameters in this study are adapted from Setiyono et al. [16] and Tanioka et al. [15] who had conducted research related to the tsunami in the Sunda Strait region [15, 16]. The parameters from these two studies were used to formulate four modelling scenarios. Each scenario contained a different value of strike, dip, rake, slip, fault length, and sediment type (median sediment particle size = d50).

The details of each scenario are given in Table 7.1. Scenario 1 implemented seismic parameters from Tanioka et al. [15]. Scenario 2 modified the fault length of scenario 1. Scenario 3 applied the conditions in Setiyono et al. [16]. The d50 in the scenarios 1–3 are 0.5 mm, based on Perkasa [17], which according to the Wentworth Scale is a type of coarse sand sediment [18]. In scenario 4, a larger value is used, which is 2.0 mm, equivalent to gravel-type sediment. A constant manning coefficient of 0.1 is used during the simulations.

Table 7.1 Seismic and sediment input for each scenario

Parameters	Scenario 1	Scenario 2	Scenario 3	Scenario 4
Mag (MW)	8.5	8.5	8.5	8.5
Depth (km)	10	10	10	10
Strike	300	300	297	300
Dip	8	8	45	8
Rake	83	83	90	83
Slip (°)	10	10	14	10
Fault-length (km)	300	300	195	300
Fault-wide (km)	80	50	65	80
d50 (mm)	0.5	0.5	0.5	2.0
Reference	[14]	[14]	[15]	[14]

7.3 Results and Discussion

As mentioned in Table 7.1, two simulation scenarios are adapted from previous studies [15, 16]. Further, model calibration in this study is conducted by comparing model output and the finding in two referred works. In practice, the obtained inundation at Pelabuhan Ratu in the model is compared with the value in those works. Tanioka et al. [15] with parameters in scenario 1 and Setiyono et al. [16] with parameters in scenario 3 resulted in an inundation of 7 and 5 m at Pelabuhan Ratu, respectively. In comparison, the model findings are shown in Fig. 7.2. Based on the figures, the inundation height at Pelabuhan Ratu for scenarios 1 and 3 is around 7.1 m and 5 m, respectively. It indicates that the model can replicate the findings in the referred works [15, 16].

7.3.1 *The Initial Condition*

In tsunami modelling, the initial condition is more influential than the boundary condition, in governing the physics of the tsunami waves propagation. Initial condition is defined as an instantaneous water level fluctuation generated by the earthquake. This section describes the varying initial conditions at the four scenarios, due to the difference seismic parameters, as presented in Table 7.1.

Figure 7.3 shows the initial condition for scenarios 1, 2, and 3 (For this section, scenario 1 equals to scenario 4). The significant difference observed is the orientation

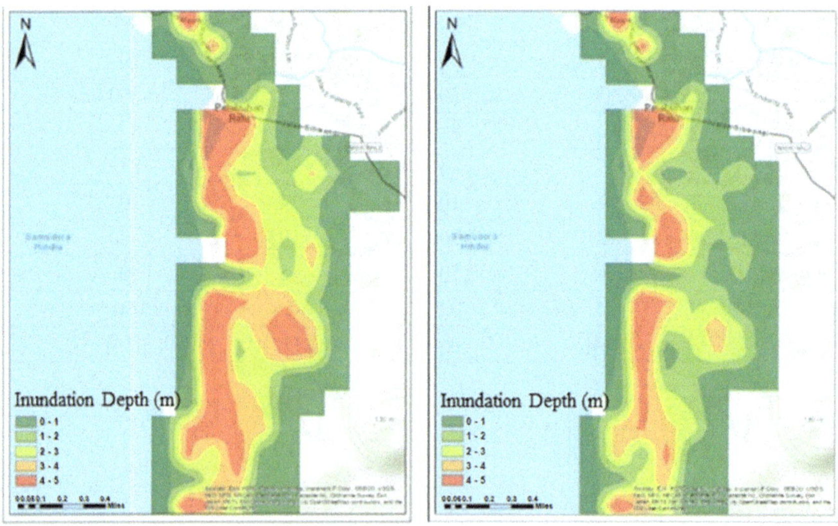

Fig. 7.2 Inundation depth at Pelabuhan Ratu resulted from (left) scenario 1 and (right) scenario 3

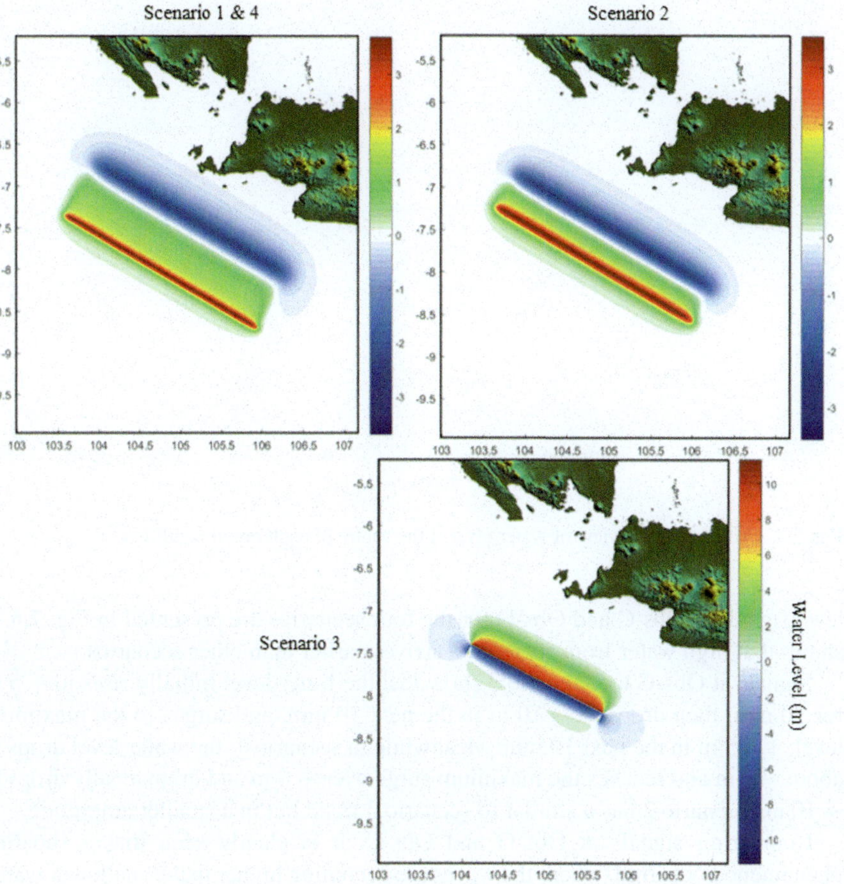

Fig. 7.3 Instantaneous water levels/initial conditions due to seismic input for each scenario

of the lower wave. Scenario 3 had its lower wave at the southern side, unlike the other scenarios. This is influenced by the slip and rake parameters.

Additionally, Fig. 7.4 shows the longitudinal view of water level at a transect inside the active region of initial condition. It is seen that the wavelength of scenario 3 is smaller than the others. Scenario 2 has a smaller fault-wide than scenario 1 and 4; therefore, the distance between wave crest-trough is shorter.

7.3.2 Tsunami Wave Propagation

Two observation points (one each for the deeper (Obs-O) and shallower depth (Obs-C)) are introduced in Fig. 7.5 which is used to observe how the wave propagates in each scenario (for this section, scenario 1 equals to scenario 4). The time-series water

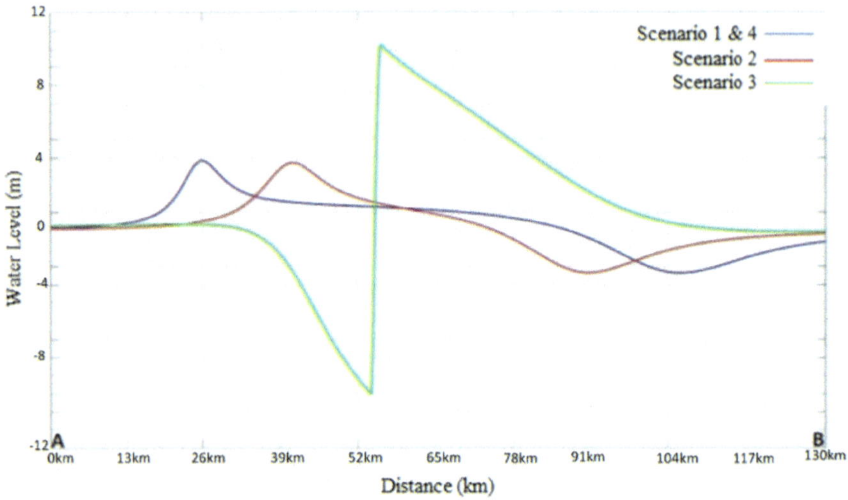

Fig. 7.4 Cross-sectional view of water level at the centre of each initial condition

level signals at Obs-C and Obs-O for the four scenarios are presented in Fig. 7.5. It shows that high water from scenario 3 arrives earlier than other scenarios.

Signals at Obs-C for scenario 3 show that the water level initially rises to + 5 m for 30 min, then drops to − 10 m in the next 10 min, and surges to the maximum level, + 6.5 m in the next 10 min. Meanwhile in scenario 1, the water level drops to about − 2 m and reaches the maximum surge over + 8 m and dramatically drops to − 10 m. Scenario 2 has a similar to scenario 3 trend but in a smaller amplitude.

Comparing signals at Obs-O and Obs-C, it is clearly seen that a shoaling phenomenon occurred, where the waves are becoming higher in the shallower water. In the offshore, Obs-O, the wave amplitude is between 3 to 4 m. Once the wave reaches the shallow area, Obs-C, the amplitudes tripled at the least. The maximum wave amplitude is 18 m, observed in scenario 4.

7.3.3 Tsunami-Induced Morphology Changes

All scenarios are run separately to evaluate how the tsunami waves influence the ocean floor, especially at Pelabuhan Ratu Bay. Figure 7.6 shows the final erosion and sedimentation of scenarios 1, 2, 3, and 4. Overall, all scenarios produce the same pattern. Sedimentation occurs at the offshore water. While, near the coastline, a noticeable erosion and sedimentation pattern are consistent in each scenario. Nonetheless, there are differences between scenario 1 to the other three scenarios which are depicted in Fig. 7.7. Positive and negative values indicate that scenario 1 induces a larger sedimentation and erosion compared to the other scenario.

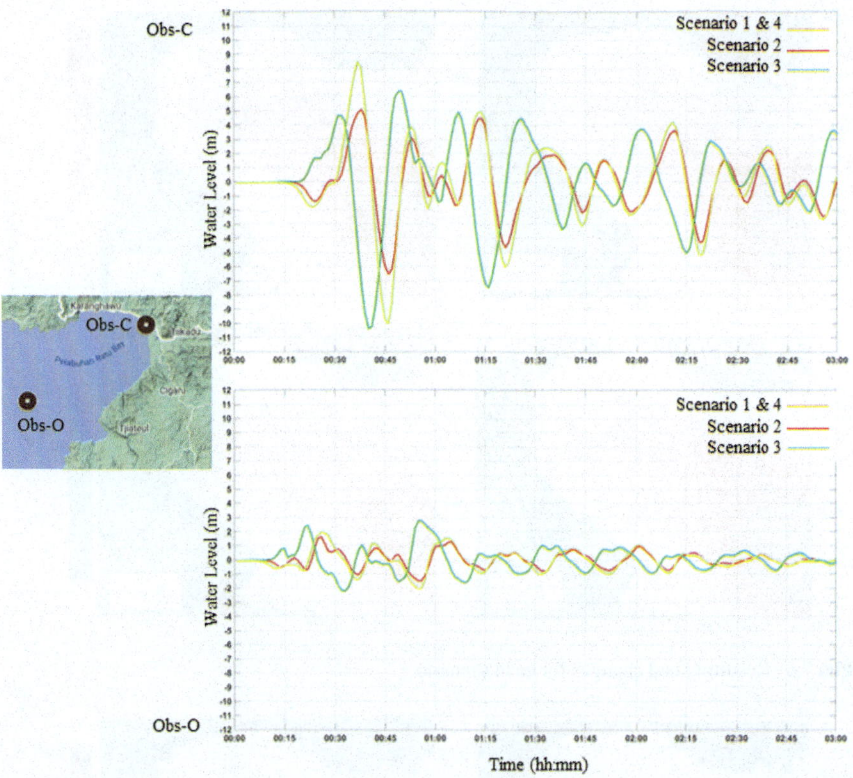

Fig. 7.5 Time series of water level at two stations (Obs-O: Offshore; Obs-C: Coastline) for each scenario

Figure 7.7 presents the difference between bed changes in scenarios 1 and 2. It shows that scenario 1 with a wider fault produces a higher wave height (see Fig. 7.5) thus resulting in higher sedimentation at the shallower region. Especially near the coastline, the seabed experiences a higher erosion and sedimentation in scenario 1 (also see Fig. 7.6). However, the offshore region saw higher sedimentation in scenario 2.

Scenarios 1 and 3 which are produced by a very different of seismic input (see Table 7.1) are seen producing a comparable erosion and sedimentation intensity at near the coastline, as shown in Fig. 7.6. The differences are displayed in Fig. 7.7. Scenario 1 induces more sedimentation in shallower regions. At the offshore, the bed changes are similar.

Figure 7.7 also displays a noticeable difference between scenarios 1 and 4. Scenario 4 is run with an identical seismic input as scenario 1; however, it uses a larger d50, see Table 7.1. With a coarser sediment, bed changes in scenario 4 are less dynamic than scenario 1. It can be observed at offshore and nearshore, either erosion or sedimentation in scenario 1 is greater than scenario 4.

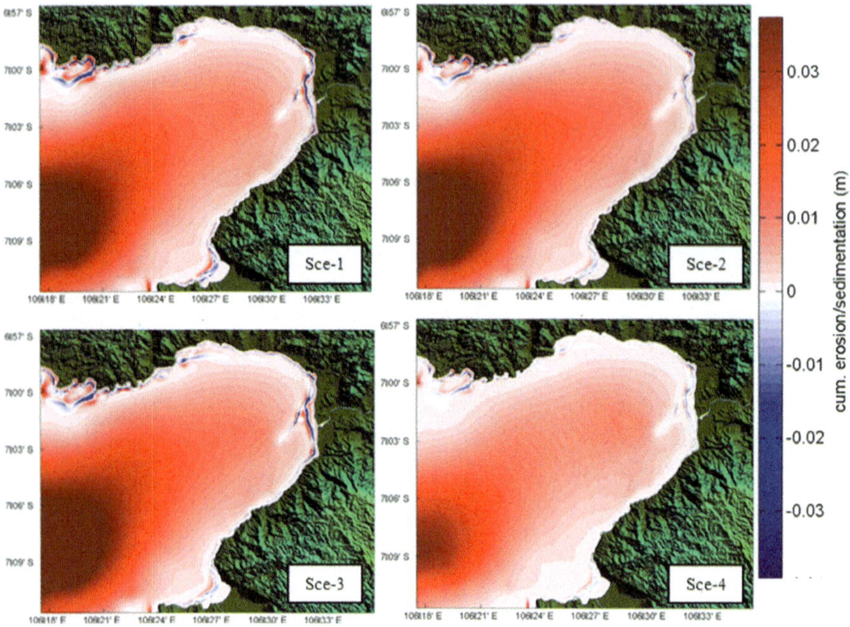

Fig. 7.6 Resulting bed changes for each scenario

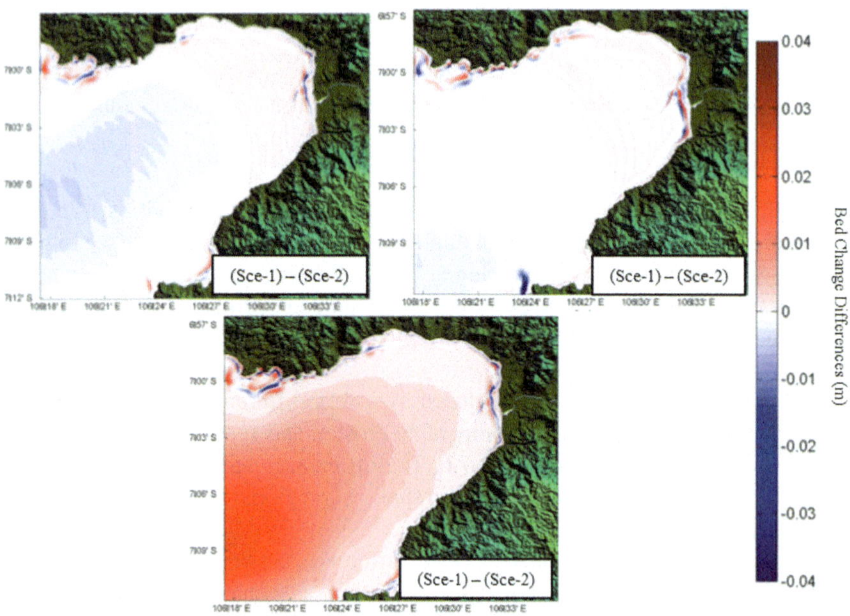

Fig. 7.7 Differences in bed changes for scenarios 1, 2, and 3 relative to scenario 1

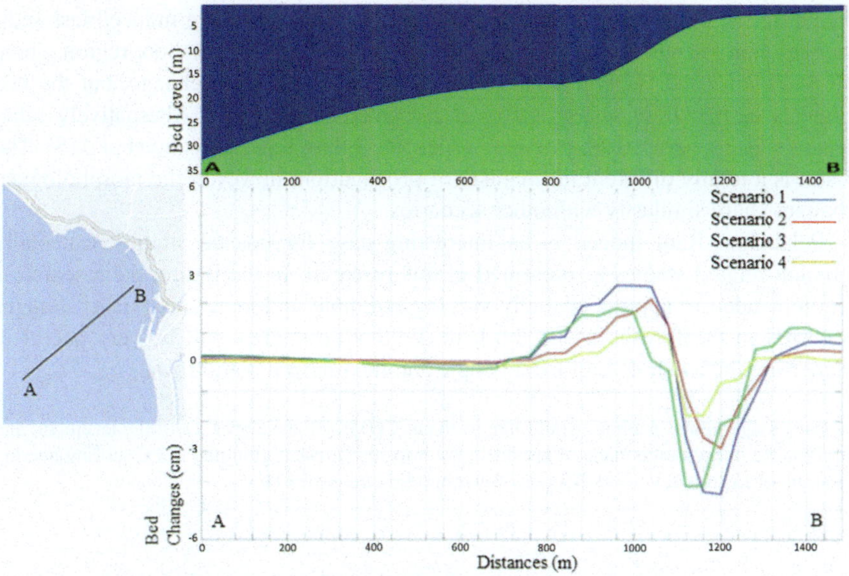

Fig. 7.8 Cross-sectional view of bed level and bed changes at Pelabuhan Ratu

A cross section at Pelabuhan Ratu, as seen in Fig. 7.8, is used to better visualize bed changes for each scenario. The profile displays that the depth went from 33 to 3 m in a span of 1500 m. At two segments marked with yellow and red boxes, a noticeable bed change occurs as displayed in Fig. 7.8. Scenario 1 saw the highest sedimentation and erosion, about 3 and 4 cm. Scenarios 2, 3, and 4 have similar trends but in a lower value. Especially in scenario 4 due to a larger sediment, the bed become less dynamic.

7.4 Conclusion

Four scenarios of tsunami model have been run and evaluated in this study where each model has a different seismic or sediment input. A semi-enclosed geometry and bed slopes at Pelabuhan Ratu Bay allow the waves to develop amplitude up to three to four times at the nearshore region. Scenario 1 which is based on Tanioka et al. [16] produced an 18 m wave height, the largest among other scenarios. The waves arrive at the coast initially low water and in 10–15 min later surging to the maximum amplitude. Bed changes impacted by the waves in this scenario are dominated by sedimentation at the offshore region. Meanwhile, at the nearshore region, mixed erosion and sedimentation are occurred.

Where the fault-wide is reduced as tested in scenario 2, the waves height arriving at the coast will be lower. However, it is still destructive and possesses a paramount danger to the people and infrastructure. The less energetic wave induced less sedimentation at the shallower water but produced more at more offshore region where the sediment used is bigger. The wave propagation does not change, but the bed changes are becoming less dynamic, as presented in scenario 4. Alternatively, other sets of seismic parameters are used in scenario 3, based on Setiyono et al. [15]. The result is a totally different mechanism of propagation; however, the morphological changes share similarity with other scenarios.

Pelabuhan Ratu indeed is an interesting case for coastal studies, especially tsunami-related studies as presented in this research. In the future, the research is recommended to apply a spatially varying sediment and more analysis to changes occurred in the foreshore and the land. Follow-up studies will be very useful in increasing the awareness of local people and the relevant authorities.

Acknowledgements Authors would like to thank COMCOT developer, Delft3D developer, and BIG for the open-source material provided. Authors also present gratitude to Ocean Engineering Program of Institut Teknologi Bandung for supporting this research.

References

1. Pranantyo, I.R., Cummins, P.R.: Multi-data-type source estimation for the 1992 Flores earthquake and tsunami. Pure Appl. Geophys. **176**, 2969–2983 (2019). https://doi.org/10.1007/s00024-018-2078-4
2. Kurniawan, A., Satria, R.A., Pratama, M.B.: Analyzing tsunami hazard using numerical modelling: study case Palu, Sulawesi Tengah, Indonesia. IOP Conf. Series Mater. Sci. Eng. **982**, 012036 (2020). https://doi.org/10.1088/1757-899X/982/1/012036
3. Widiyantoro, S., Gunawan, E., Muhari, A., Rawlinson, H., Mori, J., Hanifa, N.R., Susilo, S., Supendi, P., Shiddiqi, H.A., Nugraha, A.D., Putra, H.E.: Implications for megathrust earthquakes and tsunamis from seismic gaps south of Java Indonesia. Sci. Rep. **10**, 15274 (2020). https://doi.org/10.1038/s41598-020-72142-z
4. Tehranirad, B., Kirby, J.T., Shi, F.: A numerical model for tsunami-induced morphology change. Pure Appl. Geophys. **178**, 5031–5059 (2021). https://doi.org/10.1007/s00024-020-02614-w
5. Tanaka, N., Sato, M.: Scoured depth and length of pools and ditches generated by overtopping flow from embankments during the 2011 Great East Japan tsunami. Ocean Eng. **109**, 72–82 (2015). https://doi.org/10.1016/j.oceaneng.2015.08.053
6. Ramalho, I., Omira, R., Moussaoui, S.E., Baptista, M.A., Zaghloul, M.N.: Tsunami-induced morphological change—a model-based impact assessment of the 1755 tsunami in NE Atlantic from the Morocco coast. Geomorphol. **319**, 78–79 (2018). https://doi.org/10.1016/j.geomorph.2018.07.013
7. Deltares: Delft3D-FLOW user manual. Netherlands (2022)
8. Wang, X.: COMCOT user manual guide. Cornell University, Ithaca, USA (2009)
9. Pratama, M.B., Venugopal, V., Ajiwibowo, H., Ginting, J., Novico, F.: Modelling tidal flow hydrodynamics of Sunda Strait, Indonesia. Indones. J. Mar. Sci. **25**(4), 165–172 (2020). https://doi.org/10.14710/ik.ijms.25.4.165-172
10. Kurniawan, A., Barli, P.P.A., Pratama, M.B., Fitriadhy, A.: Potential study of tidal stream turbine farm at Toyapakeh Strait, Bali. ILMU KELAUTAN Indones. J. Mar. Sci. **26**(3), 155–162 (2021). https://doi.org/10.14710/ik.ijms.26.3.155-162

11. Kurniawan, A., Tay, S.H.X., Ooi, S.K.: Numerical investigation of tide and sea level anomalies propagation in Malacca Strait. Terr. Atmos. Ocean. Sci. **32**, 571–582 (2021). https://doi.org/10.3319/TAO.2021.10.12.01
12. BATNAS: Batimetri Nasional. https://tanahair.indonesia.go.id/demnas/. Accessed 22 Jan 2022
13. DEMNAS: Seamless digital elevation model (DEM) Nasional. https://tanahair.indonesia.go.id/demnas/. Accessed 22 Jan 2022
14. Julzarika, A., Harintaka: Indonesian DEMNAS: DSM or DTM? In: 2019 IEEE Asia-Pacific conference on geoscience, electronics and remote sensing technology (AGERS), pp 31–36 (2019). https://doi.org/10.1109/AGERS48446.2019.9034351
15. Setiyono, U., Gusman, A.R., Satake, K., Fujii, Y.: Pre-computed tsunami inundation database and forecast simulation in Pelabuhan Ratu, Indonesia. Pure Appl. Geophys. **174**, 3219–3235 (2017). https://doi.org/10.1007/s00024-017-1633-8
16. Tanioka, Y., Latief, H., Sunendar, H., Gusman, A.R., Koshimura, S.: Tsunami Hazard Mitigation at Pelabuhan Ratu Indonesia. J. Disaster Res. **7**(1), 19–25. https://doi.org/10.20965/jdr.2012.p0019
17. Perkasa, R.T.: Dinamika Perubahan Garis Pantai Antara Sungai Cimandiri dan PPN Palabuhanratu Kabupaten Sukabumi, Jawa Barat. Final Project, Institut Pertanian Bogor (2016)
18. Wentworth, C.K.: A scale of grade and class terms for clastic sediments. J. Geol. **30**, 377–392 (1922). https://doi.org/10.1086/622910

Chapter 8
Analysis of Solar Energy Potential Based on Slope Angle Using PV-SYST 7.2.6 at the University of Pamulang—South Tangerang

Aripin Triyanto, Natalita Maulani Nursam, Nora'Aini Ali, Hasiah Salleh, Rizky Maulana, and Ariyawan Sunardi

Abstract The need for electrical energy always increases every year. Sources of electrical energy can be generated from various types of generators. Electrical energy can also be generated from renewable energy, one of which is sunlight. Indonesia is a tropical country located on the equator; the potential for solar energy produced in an area is influenced by the movement of the sun and depends on atmospheric conditions in that year. The Pamulang University Viktor building has a large rooftop area and can be used as a source of renewable electrical energy, namely the sunlight-powered plant (PLTS) with the availability of a large rooftop area. Energy from the sun is contained in Pamulang University Viktor using the PV-Syst 7.2.6 software through simulations using various variations of tilt angle experiments. Simulation

A. Triyanto (✉) · R. Maulana · A. Sunardi
Electrical Engineering Department, Faculty of Engineering, Pamulang University, Jl. Raya Puspiptek, Buaran, Kec. Pamulang, 15310 Kota Tangerang Selatan, Banten, Indonesia
e-mail: dosen01315@unpam.ac.id

R. Maulana
e-mail: maulanarizky835@gmail.com

A. Sunardi
e-mail: dosen00322@unpam.ac.id

N. M. Nursam
Research Center for Electronics, National Research and Innovation Agency (BRIN), P2ET LIPI Campus LIPI Building 20 Floor 4, Jl. Sangkuriang, Cisitu 40135, Bandung, Indonesia
e-mail: natalita.mn@gmail.com

N. Ali · H. Salleh
Renewable Energy and Power Research Interest Group (REPRIG), Faculty of Ocean Engineering Technology and Informatics, Universiti Malaysia Terengganu, 21300, Kuala Nerus, Kuala Terengganu, Malaysia
e-mail: noraaini@umt.edu.my

H. Salleh
e-mail: hasiah@umt.edu.my

A. Ismail et al. (eds.), *Exploring Diversity in Engineering and Technology for Knowledge and Innovation*, Advanced Structured Materials 215,
https://doi.org/10.1007/978-3-031-64330-9_8

experiments were carried out with angles of 6°, 12°, 18°, 24°, 30°, and 36° facing north, azimuth = 0°; this is adjusted to the latitude of the sun's position at that location. Based on the simulation results that have been carried out using the slope angle, it can be said that all of them are good because the average results of the energy produced are sufficient for the required energy. As for the six simulations of slope angle variations that have been carried out, the best simulation results are obtained by setting a slope angle of 12° facing north, azimuth = 0°, with the simulation results obtained the total amount of solar energy is 146,974 kWh in a year.

Keywords Solar panels · Energy · Irradiation · PV-Syst software · Tilt angle

8.1 Introduction

The development of technology in this modern era causes the need for electrical energy to increase every year. This is due to the large use of electrical energy in industry and the increasing population [1]. Energy sources that may produce electricity include hydroelectric power plants (PLTA), steam power centers (PLTU), nuclear power centers (PLTN), gas power centers (PLTG), and diesel power centers (PLTD). In addition, electrical energy can also be generated from renewable energy and one of them is sunlight, namely sunlight-powered plants (PLTS). Indonesia is a tropical country located on the equator which has a large enough potential for solar energy. Solar energy has the potential to be used directly as renewable energy [2]. The potential of solar energy produced in an area is influenced by the sun's movement, so this potential of solar energy will change every time, depending on atmospheric conditions, place, and time of year. Based on presidential regulation No. 79 of 2014, there is an increase in renewable energy sources commonly known as innovative renewable energy (EBT) in the national energy mix, which is 23% in 2025 [3]. Utilization of using the PLTS system is currently experiencing an increase in research. Because the source of electrical energy from sunlight will not run out, it can be used as a source of electrical energy [4]. Efforts are being made to increase the utilization of innovative renewable energy (EBT) originating from solar radiation sources, so an analysis of the potential of solar energy at Pamulang University is carried out, precisely in the rooftop area utilizing PV-Syst software. The experiment was carried out by considering the angle of inclination on the layout of the solar panels, namely 6°, 12°, 18°, 24°, 30°, and 36° facing north, azimuth = 0°, this is adjusted to the latitude of the sun's position at that location. The analysis and simulation are carried out to know the layout based on the angle of inclination so that the maximum absorption of solar energy is obtained within a year.

Study Literature

Utilization of photovoltaic using a using a solar emulator with the output of polycrystalline solar cells with a maximum output power of 10 W peak. The results of the solar panel test are 10, the light intensity is 6900 lx, the voltage is 17.7 V, and the current is 0.02 amperes in obtained. At a light intensity of 121.100, a voltage of 20.2 V, and a current of 0.53 amperes [5], the tilt angle adjustment for monthly solar panels in the Meulaboh area is influenced by the elevation angle. January and March, the optimum angle value starts from 35.54° to 12.39° May to December, the panel orientation is facing south of the panel surface facing down which is indicated by the negative notation [6]. Photovoltaic development planning with both on- and off-grid systems in south Sumatra province. The capacity was 5308 GWh in 2020. With the literature study method, PLTS that has been installed until 2021 has only reached 12.441 MW or around 4.12% of the South Sumatra RUEN target of 296.6 MW in 2025 [7]. The simulation results using the PV-Syst software for one year are 40,400 kWh, option two is 4300 kWh, and option three is 8300 kWh [8].

System Design Theory

The potential of solar energy in a solar panel can be optimized, one of which is by adjusting the direction of the solar panel according to the position of the sun's rays in the area. The amount of potential solar energy generated depends on the arrangement of the solar modules. This solar module is the main component of a solar panel. If the solar modules are arranged in series, the resulting voltage will be greater, and if the solar modules are arranged in parallel, the current obtained is greater. This figure (see Fig. 8.1) shows an example of a typical solar panel type [5, 6].

A solar power plant (PLTS) or commonly known as a solar cell or solar energy is a tool for generating electrical energy whose main source is utilizing the sun's rays available on earth. Photovoltaic energy conversion found in PLTS to produce DC electrical energy that is arranged in series and parallel and can be combined in a photovoltaic module totaling approximately 40 PV in the form of an array. The

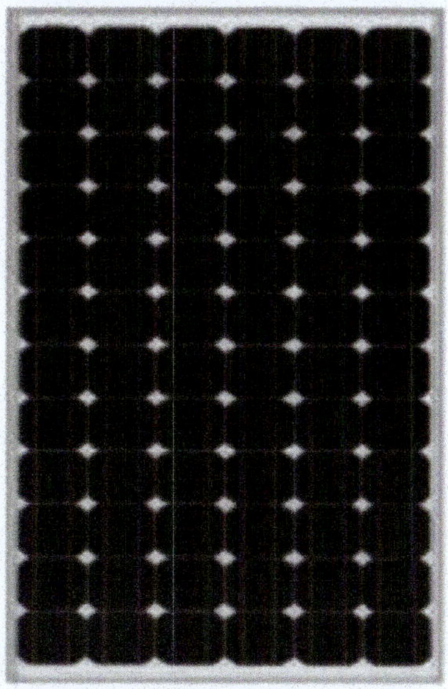

Fig. 8.1 Monocrystalline solar module [5, 6]

generated DC electrical energy can be converted into AC electrical energy by using inverter auxiliary equipment [9]. PV mini-grid systems are categorized based on the configuration of each component. Three varieties exist solar panels: on grid/grid connected (linked to the PLN electrical network), standalone/off-grid (not linked to PLN electricity, standalone), and hybrid (a blend of on grid and off-grid).

PLTS Grid Connected

PLTS grid connected is a type of PLTS device whose network connection is combined with the network from PLN, as shown in Fig. 8.2. There are two types of PLTS grid-connected systems, some of which work using a backup battery and some that do not use a backup battery or it can be called PLTS grid connected without a battery [9].

PLTS Standalone/Off-Grid

A solar power plant known as PLTS standalone/off-grid is not a part of the PLN electrical grid. This PLTS stands alone in this type of PLTS equipped with a battery as a storage device for the electrical energy produced, according to Fig. 8.3. In its use, the battery is expected to store energy when the weather conditions are cloudy or the level of solar radiation produced is not optimal [10].

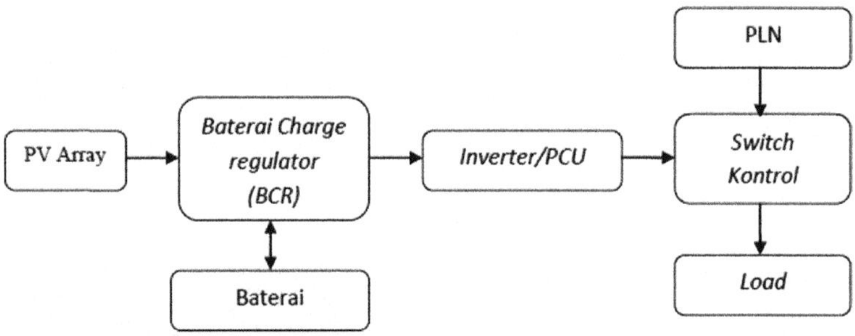

Fig. 8.2 PLTS grid-connected system diagram [9]

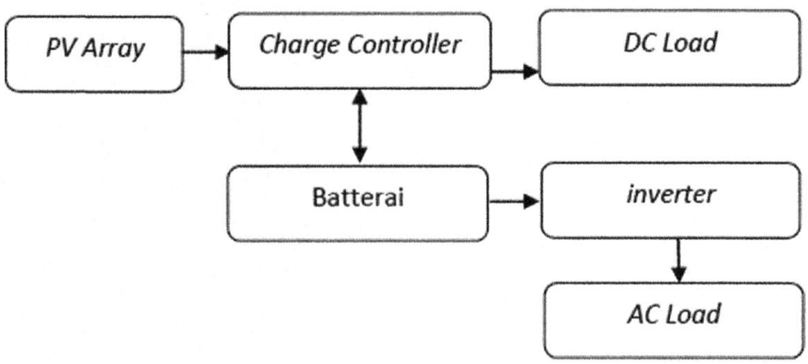

Fig. 8.3 PLTS system diagram standalone [10]

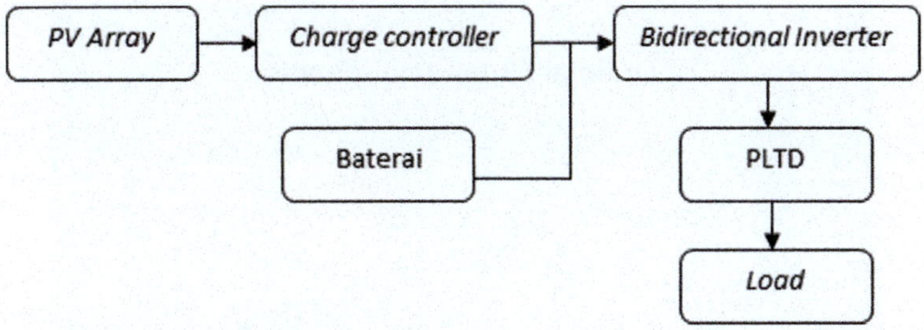

Fig. 8.4 Hybrid PLTS system diagram [11]

PLTS hybrid

A hybrid power take-off system (PLTS) is one that is linked to either PLTD or other current power plants. The purpose of this PLTS hybrid system (see Fig. 8.4) is to decrease fuel consumption and extend the running hours and services of the current system.

The factors that can affect the maximum value obtained from the absorption of solar energy potential depend on [11, 12]:

1. Solar radiation

Solar radiation on earth that falls on earth will have different variables at each location depending on the state of the sun's spectrum on earth. Solar insolation has little effect on voltage (V) and greatly affects current (I).

2. Air temperature

The temperature of a solar cell can help maximize the performance of the solar cell if the cell temperature is in normal condition, which is 25 °C. The voltage (V °C) in the solar cell will decrease if the temperature is greater than its normal state. The total energy produced will be doubled for every increase in cell temperature per 10 °C so when the solar cell temperature has an increase of 25 °C it will decrease by about 0.4%.

3. The state of the earth's atmosphere

The state of the earth's atmosphere is very influential on the maximum amount of electric current obtained from solar cells. Several atmospheric conditions can affect them, such as sunny, cloudy, cloudy weather, water vapor levels, air pollution levels, smoke, and so on.

4. The position of the solar cell to the sun

Maintaining the stability of sunlight so that it falls on the surface of the solar module directly can obtain maximum energy. Therefore, to maintain stability between sunlight on the solar module, the best arrangement of the location of the solar module is carried out, because the sun's latitude changes every time.

Tilt Angle and Direction of PLTS

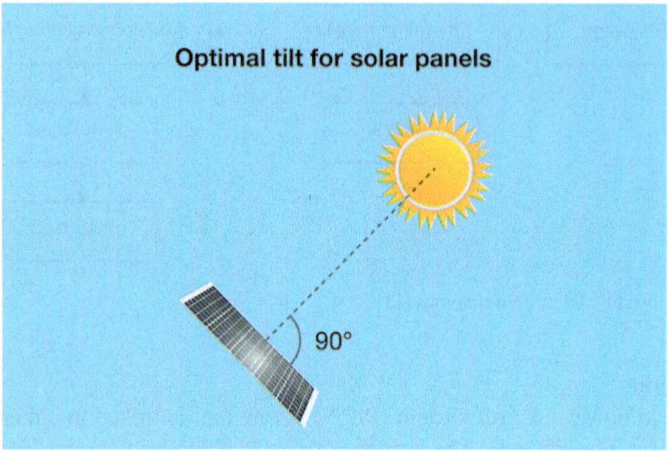

Fig. 8.5 Module in vertical state or tilt angle 90°

In maximizing the potential of solar energy absorbed by solar modules to be converted into electrical energy in a solar power plant, two angle parameters must be considered so that the absorption of solar radiation produced can be optimal, namely:

1. Tilt Angle

The tilt angle is the arrangement of the solar module angles measured from the horizontal plane. The tilt angle setting is set based on the latitude coordinates of the location. This tilt angle must be oriented perpendicular to the panel surface to the direction of the sunlight falling as seen in Fig. 8.5, so that this tilt angle can be estimated at 90 degrees, or the module is in a vertical position.

Determination of the maximum tilt angle on the solar module is the basis of the effectiveness of the operation in the solar panel because if the tilt angle setting is not appropriate, it will lose the potential of solar energy that should be obtained optimally. This formula may be used to get the angle of inclination [13].

2. Azimuth Angle

The azimuth angle is the module facing angle, which is set to two types of directions, namely facing north (azimuth 0°) or facing south (azimuth 180°). In order not to be covered by shadows, it is better if the location of the center of the sun is north of the equator, and then the solar module is positioned facing south. Likewise, if the location of the center of the sun is south of the equator, the solar module is positioned facing north opposite the center of the sun. Loss of inclination will occur, and the resulting potential will not be ideal if the two directions confronting this module are not parallel to the direction of the sun's center. Figure 8.6 that follows is an illustration of the placement of the solar module, and the azimuth angle is adjusted to the position of the sun .

3. Solar Module Stand Construction

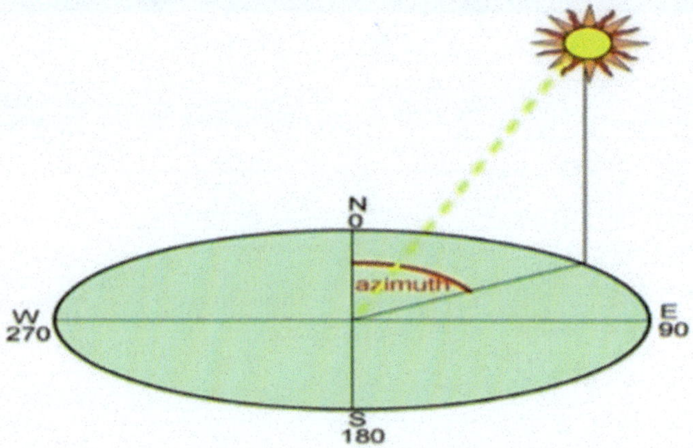

Fig. 8.6 The azimuth direction of the solar module concerning the sun

A design or design is required for the construction of a support during the installation of a solar panel in order to serve as a solar module holder. It is also necessary for the design of the solar module holder to pay attention to the material used and the tilt angle to obtain maximum solar radiation energy. In the design of the building on the rooftop of Universitas Pamulang, the solar module holder will be made of hollow aluminum metal material with the angle of the solar module facing north. The solar module supports holder's construction (Fig. 8.7).

PV-Syst

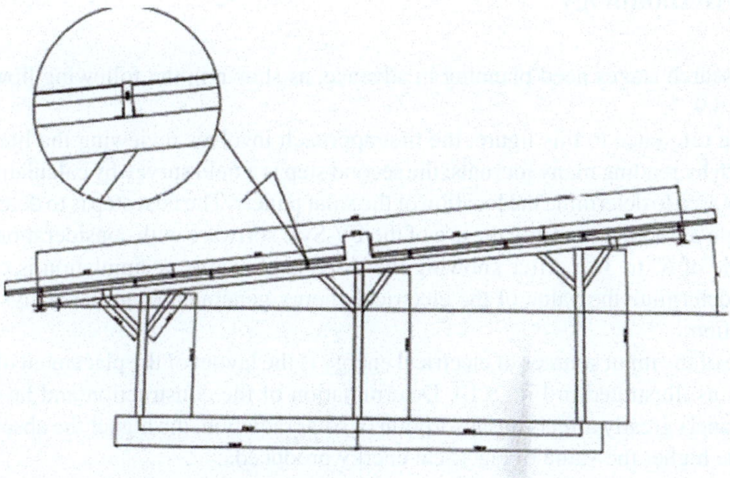

Fig. 8.7 Solar module mount construction design

Fig. 8.8 Menu databases PV-system

PV-Syst which stands for photovoltaic system is a software used in sunlight-powered plants (PLTS) systems to simulate, create designs, measure things, and go at data pertaining to PV systems. We are unable to cover every issue that may come up when using the PV-Syst. In Fig. 8.8, the PV-Syst sign is visible.

8.2 Methodology

The research stages need planning in advance, as shown in the following flowchart in Fig. 8.9.

With reference to this figure, the first approach involves reviewing the literature research by reading many journals; the second step is a field survey by calculating the rooftop area to determine the location of the solar panels. The next step is to determine the angle of inclination from the use of the PV-Syst software with consideration from an angle of 6° to 36°. After knowing the slope angle value, a simulation is carried out to determine the value of the electrical energy generated from the analysis and simulation.

One of the input sources of electrical energy is the layout of the placement of solar panels, as illustrated in Fig. 8.10. Determination of the construction and layout of solar panels greatly affects the absorption of solar radiation, the higher the absorption rate, the higher the value of electrical energy produced.

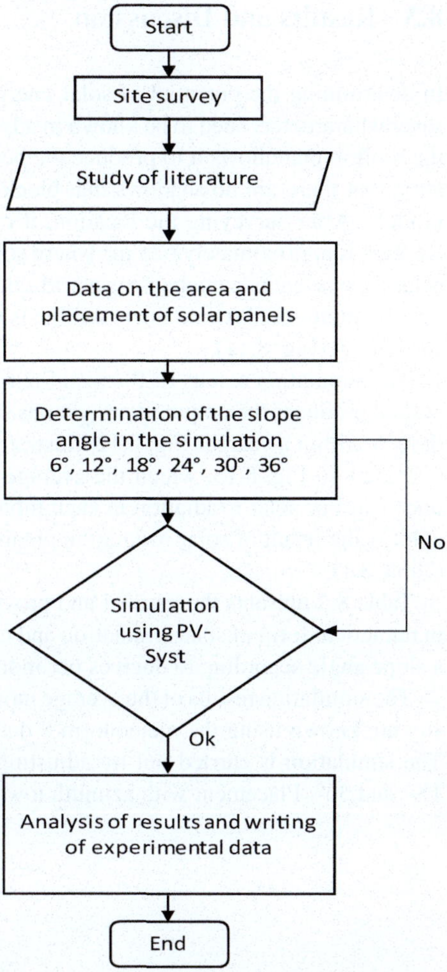

Fig. 8.9 Research flowchart

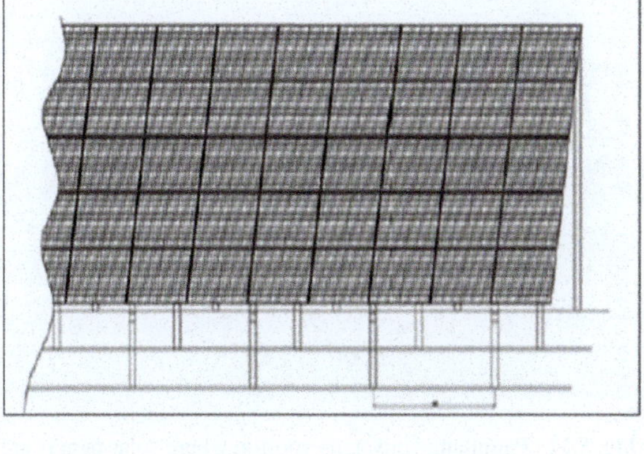

Fig. 8.10 Solar panel arrangement according to construction

8.3 Results and Discussion

In determining the potential of solar energy on the Universitas Pamulang rooftop, several parameters need to be known in advance, including knowing the zone around the rooftop that allows it to produce the best solar energy absorption with consideration that there are no shadows that block the fall of sunlight directly on the solar module. After surveying the location, it can be seen that the selection of the best location is approximately 625 m^2 where solar panels will be installed with a suitable solar panel system, namely using a grid-connected/on-grid system with the selection of solar modules. Type monocrystalline is based on literature studies that have been carried out (Fig. 8.11).

The simulation results of PV-Syst 7.2.6 diagram of solar radiation show that the system produces the output in one year as shown; the system output produces (kWh/day) or within a year, as Fig. 8.12 illustrates.

Based on Fig. 8.13, which the average solar irradiation produced per month in one year. For solar irradiation in each month, it is shown in dark red and the green color is the result of solar irradiation obtained every month for a year within a year (Table 8.1).

Table 8.2 presents the amount and provides a detailed breakdown of the amount in numerical form of solar irradiation and energy output in the resulting system with a slope angle according to degrees per month in one year.

The simulation results of the average monthly radiation value at Pamulang University are known using the meteonorm 8 database available in the PV-Syst software. The simulation is carried out by adjusting the slope angle variations between 6°, 18°, and 36°. Placement with azimuth toward the north = 0°. The following are the

Fig. 8.11 Pamulang University rooftop where solar panels are placed

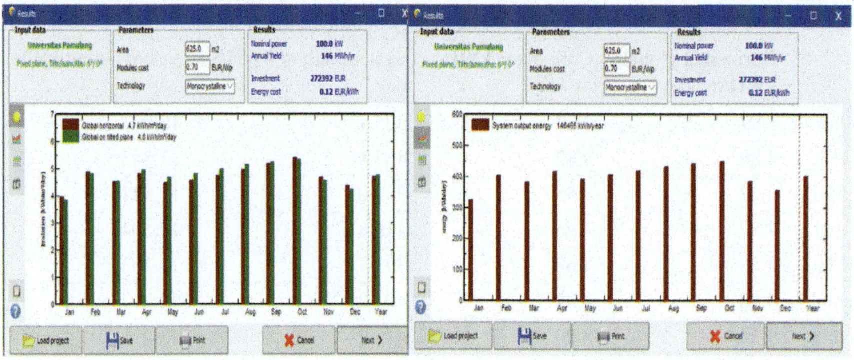

Fig. 8.12 PV-Syst 7.2.6 simulation results diagram

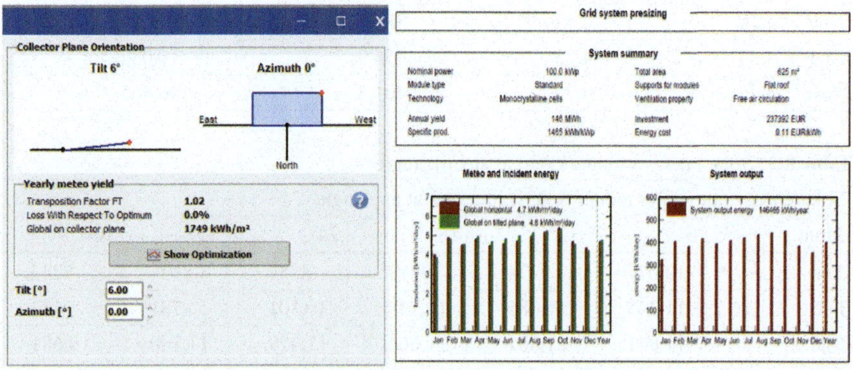

Fig. 8.13 Result chart with tilt angle setting 6°

Table 8.1 Slope angle formula [13]

Latitude	Slope
Latitude value < 25°	Latitude value × 0.87
25° ≤ Latitude value ≤ 50°	(Latitude value × 0.76) + 3.1°
Latitude value > 50°	Special rules

results of the simulation that has been carried out on the rooftop area of Pamulang University (Table 8.3).

The simulation results in Fig. 8.13 show that the highest level of solar irradiation with a 6° tilt angle setting was 13,974 kWh in October indicated by the green column and the lowest level of solar irradiation was 10,077 kWh occurring in January, represented, by the column that is yellow. Another observation from the graphic is that, on average, 146,465 kWh of energy are produced annually.

The simulation results in Fig. 8.14 indicate that the maximum degree of solar irradiation with a 12° tilt angle setting was 13,796 kWh in October indicated by the

Table 8.2 Result table of solar radiation and energy output on the system

	Horizontal global kWH/m²/day	Coll. plane kWH/m²/day	System output kWH/day	System output kWH
Jan	3.97	3.87	325.1	10,077
Feb	4.89	4.82	405.1	11,343
Mar	4.52	4.55	382.0	11,841
Apr	4.82	4.96	416.8	12,504
May	4.48	4.68	393.7	12,203
June	4.58	4.85	407.2	12,217
July	4.74	5.00	419.9	13,017
Aug	4.98	5.16	433.8	13,447
Sep	5.18	5.26	442.0	13,259
Oct	5.41	5.36	450.8	13,974
Nov	4.70	4.58	385.1	11,552
Dec	4.37	4.23	355.9	11,032
Year	4.72	4.77	401.3	146,465

Table 8.3 Solar energy output (kWh) at any tilt angle

Solar energy output per month (kWH) in a year at any angle						
Mont	6°	12°	18°	24°	30°	36°
Jan	10,077	9734	9319	8851	8324	7742
Feb	11,343	11,086	10,738	10,301	9781	9183
Mar	11,841	11,801	11,664	11,429	11,100	10,681
Apr	12,504	12,741	12,869	12,887	12,793	12,590
May	12,203	12,623	12,937	13,141	13,232	13,211
June	12,217	12,774	13,220	13,552	13,766	13,858
July	13,017	13,568	15,001	14,312	14,498	14,555
Aug	13,447	13,796	14,028	14,140	14,132	14,002
Sep	13,259	13,345	13,316	13,173	12,916	12,550
Oct	13,974	13,750	13,416	12,977	12,438	11,800
Nov	11,552	11,174	10,709	10,163	9549	8880
Dec	11,032	10,582	10,052	9468	8816	8109
Total energy (kWH) in a year	146,465	146,974	146,269	144,393	141,343	137,162

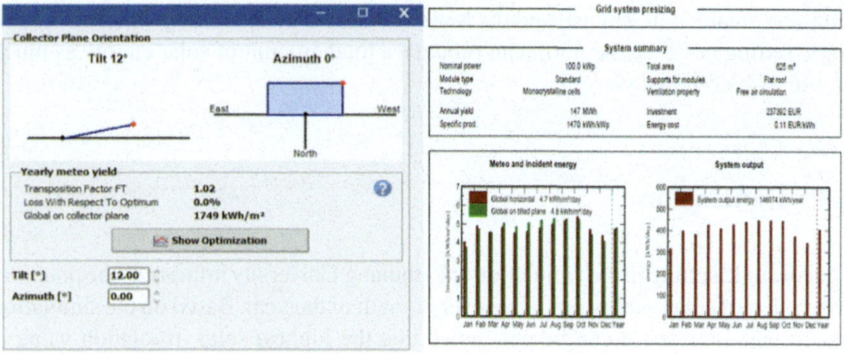

Fig. 8.14 Result chart with tilt angle setting 12°

green column and the lowest level of solar irradiation was 9734 kWh occurring in January, indicated by the yellow column; from the diagram, it can also be seen that the average amount of energy obtained is 146,974 kWh in a year.

Figure 8.15 simulation results show that the highest level of solar irradiation with a 36° tilt angle setting was 7742 kWh in October indicated by the green column and the lowest level of solar irradiation was 14,555 kWh occurring in January, as shown by the column that is yellow, it is also evident from the graphic that the average annual energy output is 137,162 kWh.

To obtain the best solar energy in a solar power plant, several simulation experiments were carried out with tilt angle settings. As for the overall simulation results, the best slope angle that has been carried out is at an angle of 12 °C facing north because it is adjusted to the latitude of the sun at the location of the PLTS installation. The following can be seen in the graph of the overall simulation results for variations in the angle of inclination carried out. The highest value of solar energy conversion

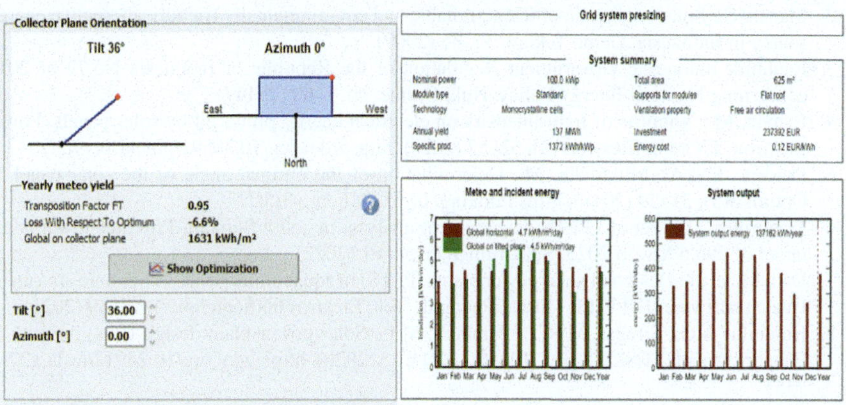

Fig. 8.15 Result chart with tilt angle setting 36°

data was obtained in August, and the lowest solar energy in January with the best tilt angle setting is 12° to the north can produce a total amount of solar energy as much as 146,974 kWh in a year.

8.4 Conclusion

The tilt angle setting in the simulation at Pamulang University influences the potential absorption of solar energy produced every month of the year. Based on the simulation results and analysis, it can be concluded that the highest solar irradiation value is from April to October with the solar panel facing north, and the best tilt angle setting is 12° and azimuth equal to 0°. Simulation experiments with slope angles of 6°, 12°, 18°, 24°, 30°, and 36° can all be said to be good because based on the simulation results that have been carried out, the energy generated reaches more than 137 MWh in one year. The results of the analysis of the solar energy potential of the Surya panel mini-grid at Pamulang University through the PV-Syst software simulation can be said to be good because the calculation of the total amount of solar energy produced is 146,974 kWh in a year.

Acknowledgements Critical comments to Professor Ahmad Fitriadhy and the IUP2022 Organizing Committee on the manuscript were greatly appreciated by the writer.

References

1. Azhar, M., Satriawan, D.A.: Implementation of new energy and renewable energy policies in the context of national energy security. Adm. Law Gov. J. **1**(4), 398–412 (2018). https://doi.org/10.14710/alj.v1i4.398-412
2. Manan, S.: Solar energy, an efficient, reliable and environmentally friendly alternative energy source in Indonesia. Gema Teknol. 31–35 (2009)
3. Republik Indonesia: Government Regulation of the Republic of Indonesia No.79 of 2014 concerning National Energy Policy. Huk. Online, pp. 1–60 (2014)
4. Usman, M.: Analysis of light intensity on electrical energy produced by solar panels. Power Elektron. J. Orang Elektro **9**(2), 52–57 (2020). https://doi.org/10.30591/polektro.v9i2.2047
5. Pandria, T.M.A., Mukhlizar, M.: Determination of the optimal angle of the solar panel. J. Optim. **3**(5), 21–29 (2018). https://doi.org/10.35308/jopt.v3i5.277
6. Manurung, D.Y., et al.: PLTS development analysis in south Sumatra Province towards the target of 296.6 Mw in 2025. Spektrum **9**(1), 54–61 (2022)
7. Dwi, M., et al.: Design of solar power plants (PLTS) in aquaculture ponds in Sentono area using PVsyst software. JUPITER (Jurnal Pendidik. Tek. Elektro) **06**(September), 18–30 (2021)
8. Hutajulu, A.G., Siregar, M.R.T., Pambudi, M.P.: Solar power plant design (Plts) on grid Di Ecopark Ancol. TESLA J. Tek. Elektro **22**(1), 23 (2020). https://doi.org/10.24912/tesla.v22i1.7333
9. Halim, L., Oetomo.: Early design and implementation of solar inverter for off grid solar power plant.J. Teknol. **12**(1), 31–38 (2020). https://jurnal.umj.ac.id/index.php/jurtek/article/view/4105

10. Saodah, S., Hariyanto, N.: Hybrid power generation system design with 3 kVA Capacity. Pros. Semin. Nas. Penelit. Pengabdi. Pada Masy, 187–190 (2019)
11. Kencana, B., et al.: Solar power plant feasibility study guide (PLTS). Indones. Clean Energy Dev. **II**(November), 68 (2018)
12. Kananda, K.: Preliminary study of solar energy potential in lampung region: a case study of the Sumatran Institute of Technology (ITERA) campus towards a smart campus. J. Sci. Appl. Technol. **1**(2), 75–81 (2017)
13. Rachmi, A., Prakoso, B., Berchmans, H., Devi Sara, I., Winne.: Guide to planning and utilization of rooftop PV mini-grid in Indonesia. PLTS Atap, 94 (2020)

Chapter 9
A Study on Factors Affecting the Stability of Fishing Vessels in Hazardous Conditions at Sea

Qasrina Mohamed and Mohammed Ismail Russtam Suhrab

Abstract Fishing vessels are the most likely to capsize due to instability caused by hazardous conditions at sea. About 24,000 fatalities at sea among fishers occur each year. Many accidents and major catastrophes occur despite advancements in naval architecture, weather forecasting, communication, and other areas. This research was conducted to identify and rank the factors affecting the stability of fishing vessels located at Kuala Kedah and then to rule out possible solutions to overcome this issue. Based on the literature review, questionnaires are developed for research purposes. This questionnaire used the Likert scale to measure and rank the data. After the full survey was conducted, the results were plotted into a series of charts to give an overview of the study's outcome. By gathering responses from the target group, this study shows that the stability of fishing vessels is associated with various factors, with the highest agreement by the respondents on the effects of wind and waves at sea. This suggests that the environmental hazard is most likely to occur compared to the other factors. Therefore, some recommendations are made to lay out the solutions to reduce the effect of fishing vessels' instability.

Keywords Fishing vessels · Stability · Hazardous conditions · Kuala Kedah

9.1 Introduction

Fishing activities unfold in challenging environments, subjecting workers to hazardous conditions driven by economic pressures, thereby jeopardizing the safety of both the fleet and crew [1]. This challenge is notably acute in Galicia, where over half of the national fishing fleet operates, contributing to a disproportionately high

Q. Mohamed · M. I. R. Suhrab (✉)
Faculty of Maritime Studies, Universiti Malaysia Terengganu, 21030 Kuala Terengganu, Malaysia
e-mail: m.ismail@umt.edu.my

Q. Mohamed
e-mail: s56166@ocean.umt.edu.my

A. Ismail et al. (eds.), *Exploring Diversity in Engineering and Technology for Knowledge and Innovation*, Advanced Structured Materials 215,
https://doi.org/10.1007/978-3-031-64330-9_9

number of fatal accidents during work hours, second only to the construction sector [2–4]. Similarly, in Malaysia, fishing is a common occupation in rural areas, particularly in villages, with Kedah standing out for hosting the highest number of fishermen on licensed vessels in the Peninsular Malaysia [5, 6]. Our research focuses on the Kedah River, traversing the Kedah countryside and impacting Alor Setar, before emptying into the Straits of Malacca at Kuala Kedah. Notably, Kuala Kedah serves as a fishing port and a departure point for ferries to Langkawi, a renowned tourist destination.

Accidents concerning stability pose the greatest threat, characterized by swift occurrences that leave insufficient time for the crew to secure safety or employ life-saving equipment, often resulting in vessel loss [7]. Most stability casualties manifest in rough seas, with capsizing typically involving forces beyond waves and wind. The IMO SLF sub-committee asserts that stormy seas heighten the likelihood of stability accidents, particularly in a beam sea with gusty wind, the following sea, or a quartering sea with broaching [8, 9]. Capsizing is more prevalent in small vessels, with a boat US survey revealing that 15–19 footers, constituting 41% of all capsizes, were common fishing boats with challenging-to-drain cockpits, occasionally venturing into adverse weather conditions and overloading [10].

This study centers on the fishing activities within the Kedah River in Kedah, a state boasting the highest number of fishermen in the Peninsular Malaysia. The vulnerability of fishing vessels to capsizing due to instability in hazardous sea conditions, compounded by their compact size, poses a severe threat to the livelihoods and economic growth of local fishermen. A fish market adjacent to a Kuala Kedah pier exemplifies the integral role of these activities. Despite the longstanding operation of fishing vessels, they persist as a significant threat to crews, vessels, and the environment. Advancements in naval architecture, weather forecasting, and communication have not curbed the occurrence of accidents and major catastrophes. To unravel the intricate link between capsizing and vessel stability, a comprehensive understanding of the factors influencing stability is imperative.

The objectives of this research, derived from the underlying research questions, encompass identifying factors influencing fishing vessel stability, ranking these variables based on their frequencies of occurrence, and proposing recommendations to enhance the stability of fishing vessels. This holistic approach aims to address the complex dynamics of fishing vessel stability in hazardous conditions at sea.

9.2 Methodology

In the methodology employed for this study, data collection encompassed a systematic approach, involving a literature review to identify core research issues. The Likert scale was utilized to gauge respondents' agreement on factors influencing fishing vessels' stability, with survey questions covering demographic details and respondents' knowledge of the study domain. A random sampling procedure selected survey respondents, and subsequent data analysis included computing mean values and generating descriptive statistics. Data analytics, a crucial step, involves examining molded data for informed interpretation and decision-making. Likert scale values were instrumental in producing and showcasing respondents' knowledge and agreement. Graphical representation facilitated the determination of central tendency measures, while nominal and ordinal data enriched demographic insights. The culmination of these steps informs a comprehensive research report, enabling discussions and the formulation of recommendations to address critical issues identified in the study.

9.3 Results and Discussion

In the first section, the survey questions were administered to ten fishermen operating around Kuala Kedah, either in deep or shallow waters. The demographic data included age, years of experience, and educational background. Respondents' ages varied from 18 to 69 years, with the majority aged over 50. The target group exhibited diverse experience, ranging from one year to over 30 years, and varied educational backgrounds, spanning primary, secondary, and post-secondary or pre-university education. Figures 9.1, 9.2, and 9.3 visually present the demographic distribution in terms of age, years of experience, and educational background, respectively.

The second section focused on evaluating respondents' knowledge or experience in various aspects related to fishing vessel stability. This included understanding vessel structures, experiences of accidents during fishing operations, agreements on the impact of vessel size on stability, and familiarity with maritime regulations and agencies. The Likert scale (values 1–5) measured respondents' knowledge, with the mean values calculated and presented in Table 9.1. Notably, respondents demonstrated the highest level of knowledge regarding relevant maritime agencies. A bar chart (Fig. 9.4) further illustrates these findings.

The third section explored factors affecting fishing vessel stability, identified through a literature review. Twelve factors, including wind and waves, improper ballasting, overloading, and various operational activities, were assessed based on

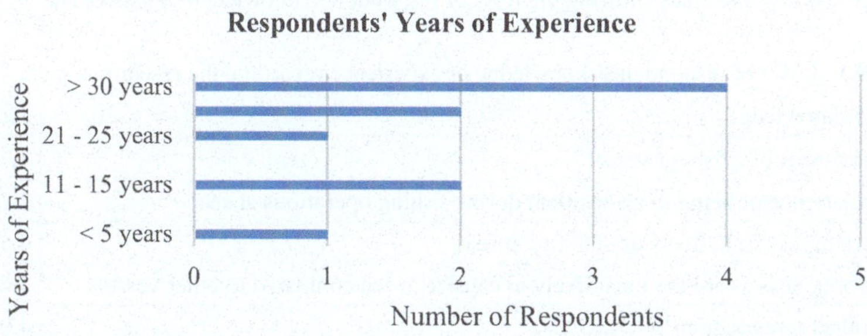

Fig. 9.1 Horizontal bar chart showing the respondents' years of knowledge or experience

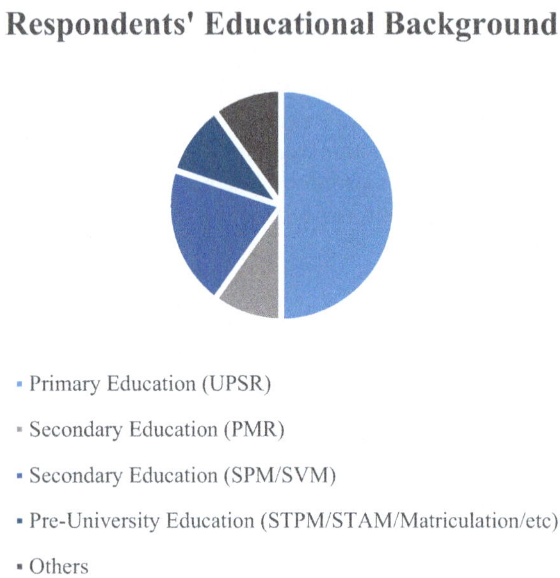

Fig. 9.2 Pie chart showing the respondent's educational background

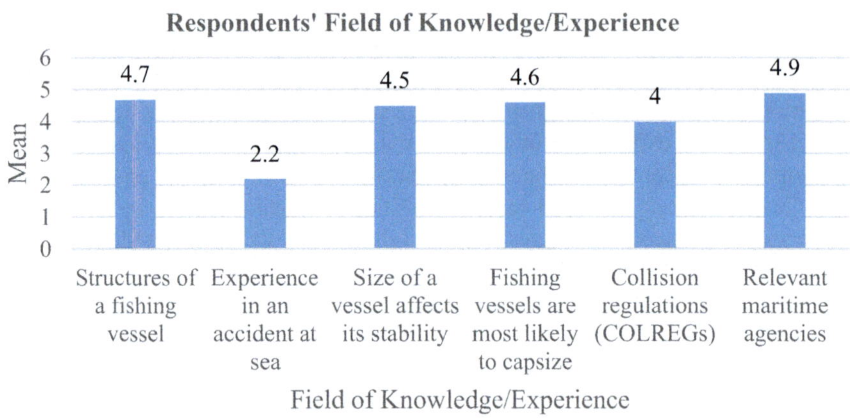

Fig. 9.3 Vertical bar chart showing the level of respondents' knowledge or experience

Table 9.1 Level of respondents' knowledge or experience regarding the research topic

Field of knowledge	Mean
The structures of a fishing vessel	4.7
The experience of being in an accident during fishing operations at sea	2.2
The size of a vessel affects its stability at sea	4.5
The fishing vessels are the most likely to capsize at sea compared to other vessels	4.6
The collision regulations (COLREGs)	4.0
The relevant maritime agencies	4.9

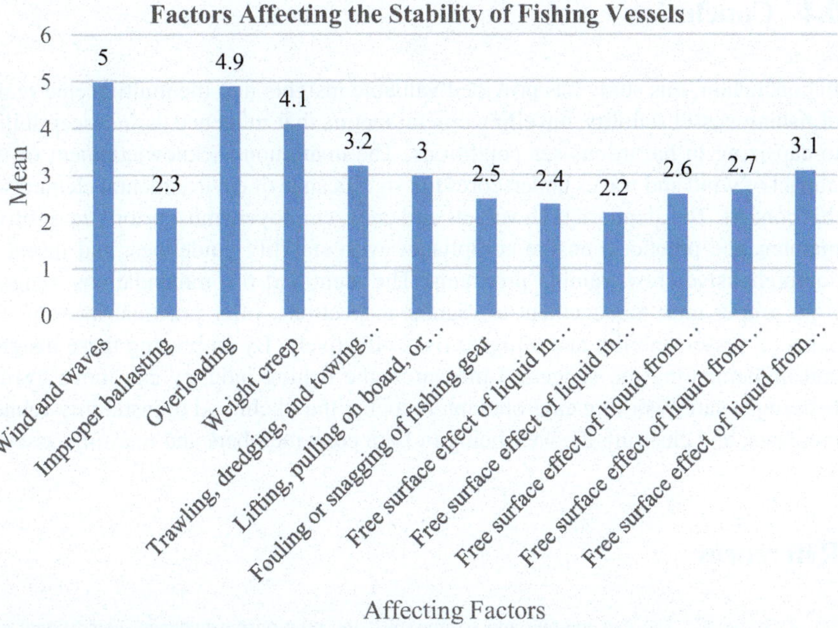

Fig. 9.4 Vertical bar chart showing the factors affecting the stability of fishing vessels

respondents' agreement using the Likert scale. The results, presented in Table 9.2, revealed that respondents considered wind and waves as the most impactful factors on vessel stability. A bar chart visually depicts the ranked factors and their mean values, providing valuable insights into the perceived significance of each factor.

Table 9.2 Ranking of factors affecting the stability of fishing vessels

Variables	Mean	Rank
Wind and waves	5.0	1
Improper ballasting	2.3	11
Overloading	4.9	2
Weight creep	4.1	3
Trawling, dredging, and towing	3.2	4
Lifting, pulling on board, or splitting the catch	3.0	6
Fouling or snagging of fishing gear	2.5	9
Free surface effect of liquid in slack tanks	2.4	10
Free surface effect of liquid in cross-connected port/starboard tanks	2.2	12
Free surface effect of liquid from flooding of deckhouses	2.6	8
Free surface effect of liquid from water on deck	2.7	7
Free surface effect of liquid from progressive downloading	3.1	5

9.4 Conclusion

In conclusion, this study has provided valuable insights into the multifaceted realm of fishing vessel stability, unveiling crucial factors that influence their susceptibility to capsizing in hazardous sea conditions. The unanimous acknowledgment of the impact of wind and waves underscores the significance of environmental elements in this context. To safeguard both vessels and crew, employers must prioritize rigorous maintenance practices, ensure compliance with stability guidelines, and invest in comprehensive crew training programs. The identified recommendations, ranging from proper tank management to regular inspections, offer practical measures to enhance vessel stability and mitigate risks effectively. By embracing these insights and implementing the suggested measures, the fishing industry can move toward fostering a safer working environment, reducing the likelihood of instability-related incidents, and ensuring the well-being of both crew members and maritime assets.

References

1. Zickgraf, C.: The fish migrate and so must we: the relationship between international and internal environmental mobility in a Senegalese fishing community. J. Int. Relat. **16** (2018)
2. Pita, P., Villasante, S.: The building of a management system for marine recreational fisheries in Galicia (NW Spain). Ocean Coast Manag. **169** (2019). https://doi.org/10.1016/j.ocecoaman.2018.12.027
3. Fernández-González, R., Pérez-Pérez, M.I., Pérez-Vas, R.: Impact of the COVID-19 crisis: analysis of the fishing and shellfishing sectors performance in Galicia (Spain). Mar Pollut Bull. **169** (2021). https://doi.org/10.1016/j.marpolbul.2021.112463
4. Pita, P., Fernández-Márquez, D., Antelo, M., Macho, G., Villasante, S.: Socioecological changes in data-poor S-fisheries: a hidden shellfisheries crisis in Galicia (NW Spain). Mar. Policy **101** (2019). https://doi.org/10.1016/j.marpol.2018.09.018
5. Mohamed Shaffril, H.A., Hamzah, A., D'Silva, J.L., Abu Samah, B., Abu Samah, A.: Individual adaptive capacity of small-scale fishermen living in vulnerable areas towards the climate change in Malaysia. Clim. Dev. **9** (2017). https://doi.org/10.1080/17565529.2016.1145100
6. Teh, L.S.L., Teh, L.C.L., Sumaila, U.R.: Quantifying the overlooked socio-economic contribution of small-scale fisheries in Sabah, Malaysia. Fish Res. **110**, (2011). https://doi.org/10.1016/j.fishres.2011.06.001
7. Woodward, J.L., Pitblado, R.M.: LNG Risk Based Safety: Modeling and Consequence Analysis (2010)
8. Kobyliński, L.: New generation stability norms-how to approach the task. Polish Marit. Res. **25**, (2018). https://doi.org/10.2478/pomr-2018-0023
9. Vassalos, D., Paterson, D.: Reconfiguring passenger ship internal environment for damage stability enhancement. J. Mar. Sci. Eng. **8**, (2020). https://doi.org/10.3390/JMSE8090693
10. Sundar, V., Sannasiraj, S.A., Murali, K., Sundaravadivelu, R.: Runup and inundation along the Indian Peninsula, including the Andaman Islands, due to Great Indian Ocean Tsunami. J. Waterw. Port. Coast Ocean Eng. **133**, (2007). https://doi.org/10.1061/(asce)0733-950x(2007)133:6(401)

Chapter 10
Failure Analysis of a Connecting Rod Bearing of a Diesel Engine Generator Set

Wilarso, Che Wan Mohd Noor, Ahmad Faisal Mohamad Ayob, and Wan Nurdiyana Wan Mansor

Abstract The performance of the generator set is significantly impacted by damage to the rod bearing components. On cylinder number 10, which was discolored in the shank area, is where the damage to the connecting rod of the type 3516 diesel engine first manifested itself. The condition is normal after inspection of the pin-end and crank-end regions. The rod bearing's state or the road's metal, meanwhile, received a minor scratch. The research's suggested remedy is to examine the diesel engine's oil. Based on engine oil testing, the goal of this study is to identify the primary factor that led to the connecting rod bearing's failure. Engine oil testing is the procedure used in the research on rod bearing analysis. The quality of the oil used in the 3516 B series diesel engine degraded, according to the analysis of the lubrication system data from this investigation. Increased levels of Fe (30 ppm), Pb (13 ppm), oxidation, and sulfation in engine oil are signs of this. The examination of engine oil has led researchers to the conclusion that deteriorated oil performs worse. Given that the indicated viscosity value limit is $+ 3$ cSt, there may be signs that an engine issue is present. At 100 C, the viscosity fluctuated by ± 2 and ± 3 cSt, respectively.

Wilarso (✉) · C. W. M. Noor · A. F. M. Ayob · W. N. W. Mansor
Faculty of Ocean Engineering Technology, Universiti Malaysia Terengganu, 21030 Kuala Nerus, Malaysia
e-mail: p4759@pps.umt.edu.my

C. W. M. Noor
e-mail: che.wan@umt.edu.my

A. F. M. Ayob
e-mail: ahmad.faisal@umt.edu.my

W. N. W. Mansor
e-mail: nurdiyana@umt.edu.my

Wilarso
Sekolah Tinggi Teknologi Muhammadiyah Cileungsi, Jl. Anggrek No. 25, Perum PT. SC, Cileungsi, Bogor, Jawa Barat, Indonesia 16820

C. W. M. Noor
Centre for Automotive Engineering, Universiti Malaysia Pahang, 26600 Pekan, Pahang, Malaysia

A. Ismail et al. (eds.), *Exploring Diversity in Engineering and Technology for Knowledge and Innovation*, Advanced Structured Materials 215,
https://doi.org/10.1007/978-3-031-64330-9_10

Keyword Low engine oil pressure · Diesel engine lubrication system · Lubrication system

10.1 Introduction

The efficiency of the power supply to state electricity company, the client is impacted if a diesel engine malfunctions, which is frequently caused by insufficient lubricant maintenance. Adequate lubrication is necessary to increase the effectiveness and lifespan of diesel engine components [1]. Finding the root of the problem is crucial because there are many causes for component failure. In the most recent technology, specialized diagnostics with engine oil testing can be performed to identify the failed component. Advanced diagnostic methods, however, are unable to identify all component faults [2], particularly the failure of the rod bearing components [3].

Numerous reports have suggested that the failure of engine crankshaft bearings is caused by a breakdown of hydraulic lubrication [4, 5]. In this regard, routine lubricating oil testing may assist in detecting rod bearing problems by displaying the pattern of wear on these components [6]. For instance, Fig. 10.1 depicts the state of the worn-out rod bearing surface that is damaged.

A generator from the type 3516 B series has revealed a rod bearing abnormality in this work [7]. To ascertain whether the rod bearing is impacted, a historical review of lubrication testing is necessary [8]. Such a preliminary examination is required to look into the origin of failure and the symptoms that are connected to it. This report will examine lubricating oil to look for any damage indicator, and to further explore the reason for failure, the lubricating oil test results are compared to the manufacturer's standard [9].

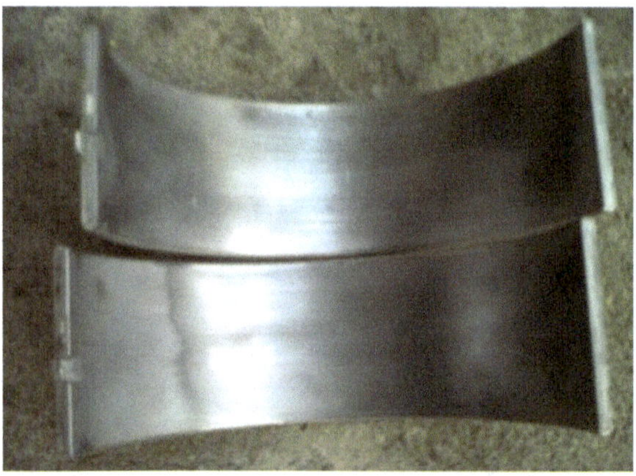

Fig. 10.1 Rod bearing

10.2 Methodology

The failure analysis of the connecting rod bearing on the 3516 B series diesel engine went through several stages, as illustrated in Fig. 10.2, which depicts the process flow.

10.2.1 Work Procedures

A set of consecutive processes known as work procedures must be followed for the work to be completed. They also clearly show the direction or flow that must be taken from the starting point of the work to the finishing point—Identification of rod bearing damage in 3516 B series diesel engines, data collection on the findings of engine oil testing, and verification of the health of the damaged bearings.

10.2.2 Visual Assessment

While a visual inspection of the damaged component is being done, the goal of the visual inspection is to determine the type of damage impacting the connecting rod [10]. Testing the engine oil is required to determine the root reason. To test the engine oil, the manufacturer of the damaged engine's recommendations must be followed. This includes knowing the material composition that is worn on the bearing, starting with Fe, Cu, and Pb, which is the basic material of the bearing. Before the oil is drained, the engine oil test method is collected and placed in a suitable location utilizing HP-427 under the requirements of the oil to be tested. The required oil sample is taken, and then it is forwarded to the lab for testing. Without proper sample techniques, time and resources are lost, and incorrect conclusions might be drawn from flawed data. It is crucial to identify, comprehend, and practice the procedures necessary to obtain a representative engine oil sample to enhance trust in the diesel engine oil analysis program [11].

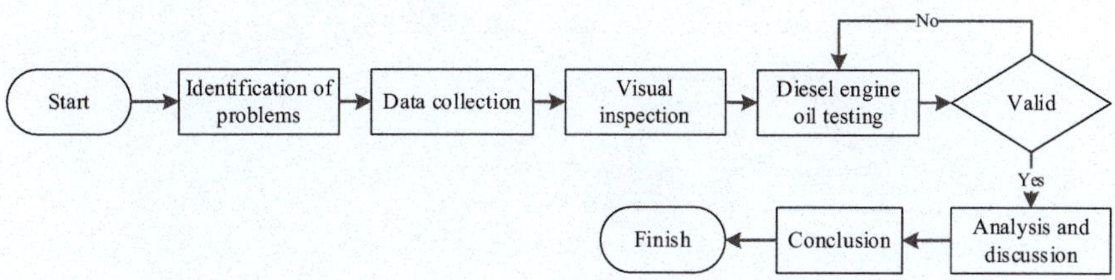

Fig. 10.2 Flowchart for research

10.2.3 Lab Assessment

By contrasting the test findings with industry norms, the goal of this test, also known as an oil test using Fourier Transform Infrared Analysis (FTIR), is to determine which characteristics have changed as the oil level has risen. Services are practical management solutions that increase the functionality and lifespan of your equipment. Both downtime and operating costs can be reduced at the same time. Oil analysis includes four main types of testing for engine, hydraulic, and drive train oils: oil identification, oil condition, oil contamination, and component wear rate [12].

10.3 Results and Discussion

The operation and maintenance manual (OMM) states that low engine oil pressure, high engine oil pressure, excessive oil consumption, and bearing attrition, particularly in heat-damaged piston rods, are signs of lubrication system problems in diesel engines. The first nine elements in Fig. 10.3 are: (1) oil pan, (2) suction bell/inlet screen, (3) and (4) oil pump and relief valve, (5) oil filter and bypass valve, (6) oil cooler, (7) oil gallery, (8) oil flow, and (9) oil return line.

Figure 10.3 presents a schematic representation of the lubrication system for a 3516 B diesel engine. The lubrication system's job is to circulate oil inside the engine. In addition to other associated engineering jobs including engine power upgrading, reverse engineering, and bearing failure analysis, the analysis of engine bearings is a critical component of the engine design process and keeps it cool.

Figure 10.4 displays the graph of the circulating engine oil pressure. In the same manner that high oil flow increases viscosity and decreases oil temperature, it is stated in the operation and maintenance manual that excessive engine oil pressure harms the engine due to overheating. However, the actual viscosity value of the oil must be

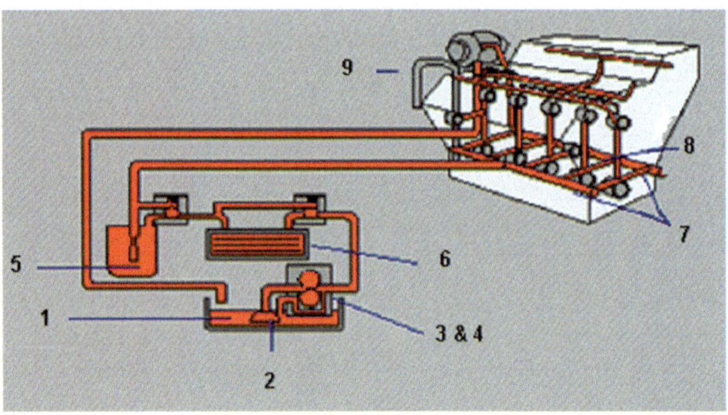

Fig. 10.3 Schematic for the lubrication system of the 3516 B series diesel engine is shown

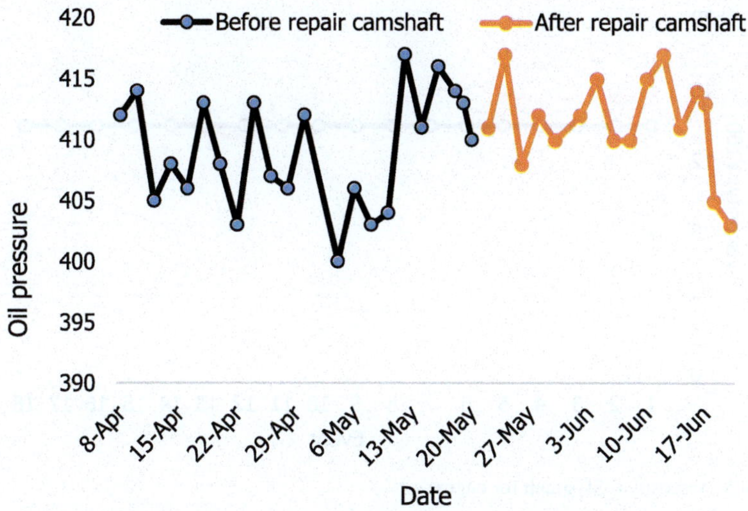

Fig. 10.4 Engine oil pressure (kPa) before and after a camshaft modification is shown

compared to ascertain whether the high oil pressure condition is within permissible bounds or not. A lack of oil supply to the diesel engine components will have much more severe effects on how quickly those components wear than in situations when the oil pressure is too low. Because of component wear, it leads to overheating.

According to the operation and maintenance manual, a diesel engine's minimum permissible oil pressure is 287 kPa (Test Spec 3516 B FDN, 2006). From the data on the diesel engine, the oil pressure of 3516 B series, the analysis above calculates an average oil pressure of 409 kPa, showing that the oil pressure is still operating over the minimum value.

Additionally, this diesel engine runs on SAE 15W-40 Meditran oil. By contrasting the results of the engine oil pressure research in Fig. 10.5, it can be seen that the oil's viscosity value has increased. The oil viscosity value during the first periodic service after replacing the left camshaft is 17.56 cSt [13].

Considering that the average viscosity value limit is + 3 cSt, this suggests that there may be indications of an engine problem. At 100 °C, the viscosity changed (monitor) by ± 2 cSt, and it changed (action needed) by ± 3 cSt.

Figure 10.6 shows an increase in viscosity as well as values for oxidation and sulfation in the engine oil. The level of metal wear (wear element), which was estimated using an analysis of an oil sample to determine the number of metal particles per part million, is shown in Figs. 10.7 and 10.8 (PPM). The analysis is divided into two categories: metal wear related to the wear of diesel engine components and the presence of oil additive deterioration.

The other metal elements, such as Al, Cu, Cr, or Sn, are the main components, while only the Fe element is high in terms of component wear, especially that attached to the piston rod, as shown in Fig. 10.7 (maximum limit of 14). The piston and

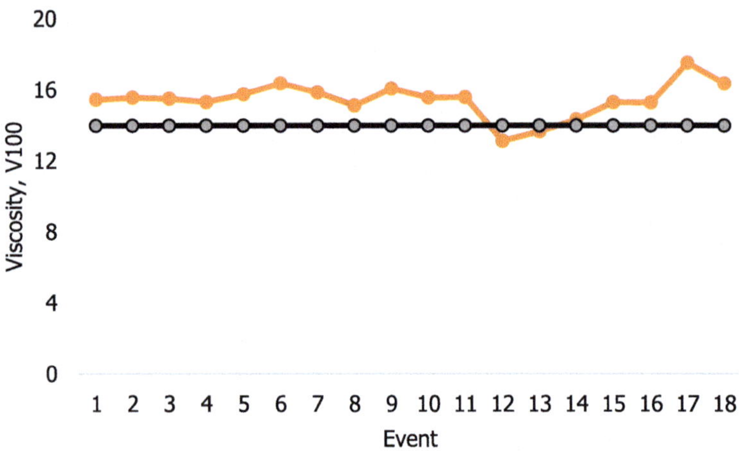

Fig. 10.5 Viscosity (cSt) graph for engine oil

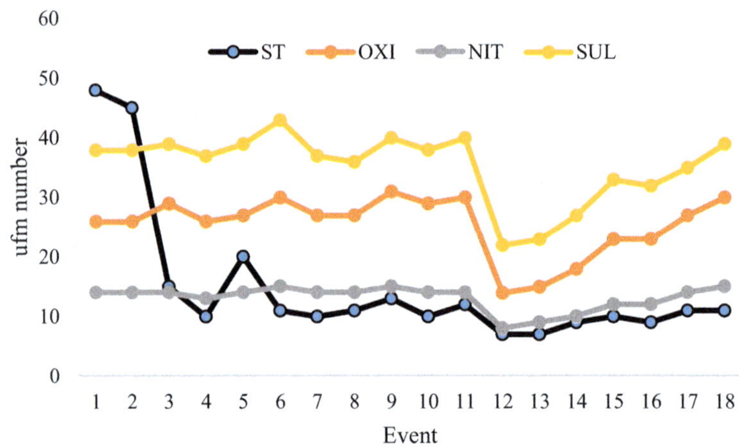

Fig. 10.6 Uses the UFM (subtracted FTIR approach) method to display the soot, oxidation, nitration, and sulfation data

crankshaft bores both have fewer than four bearings. According to the standard Caterpillar S.O.S. SM Services Wear, the wear level on the machine's bearing components is thus still within acceptable limits [14]. Similar findings were made with the investigation of oil additives in engines, where it was discovered that only Ca had exceeded the legal limit, while the concentrations of Mg, B, Na, Zn, P, and Mo particles were still within the permissible range.

Multiple evaluations of Fig. 10.8 lubrication system make it abundantly evident that the oil used in the 3516 B diesel engine has deteriorated in quality. These include an increase in the oil's oxidation and sulfation levels as well as an increase in the

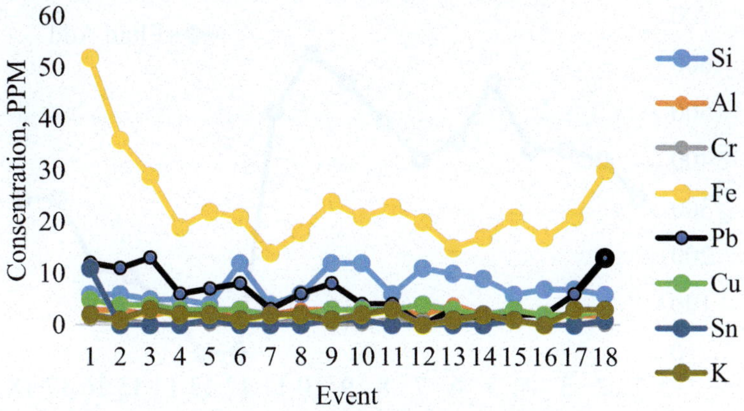

Fig. 10.7 Graph of metal wear elements and engine oil additive values

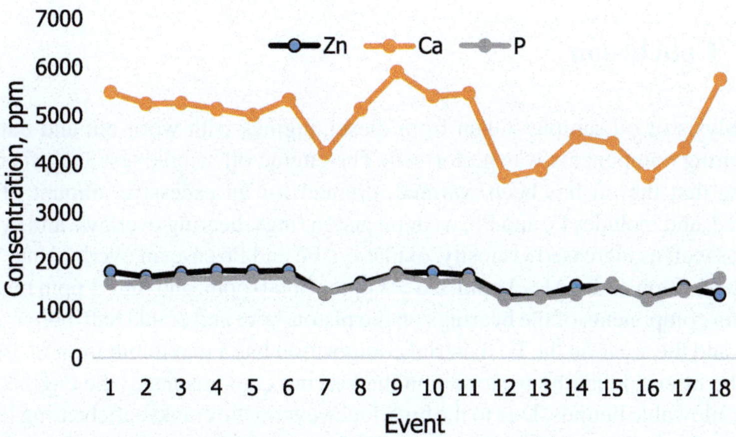

Fig. 10.8 Engine oil additives element graph

elements Fe, Pb, and Ca. The relationship between test results and guidelines for S.O.S. services from Caterpillar. Reduced viscosity and Pb are signs of diluted fuel, wear on the bearing overlays, temperature-related oil degradation, and fuel–oil mixing while the engine is running [15].

Similar patterns may be seen in Fig. 10.9, a graph of the engine's oil consumption during operation, which shows an increasing trend in oil addition and a standard limit of 85 L/500 h. It is thus significantly connected with the growth in engine operating temperature based on the increasing trend [16]. One of the reasons why piston rods become discolored is an inefficient maintenance method [11]. As a result of these assessments, the trend data for the difference in exhaust gas temperature exhibits indications of engine damage. Before a machine malfunctions, it takes some time for data trends to become apparent in terms of maintenance.

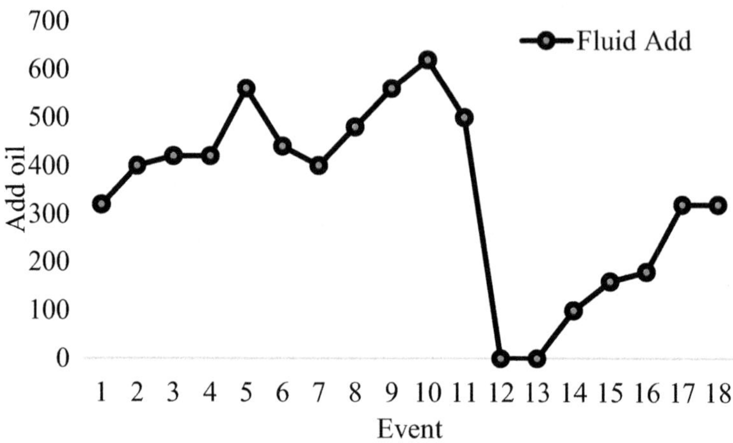

Fig. 10.9 Changes in engine oil consumption from March 2022 to June 2022

10.4 Conclusion

An analysis of oil samples taken from diesel engines with worn-out and damaged rod bearing components was performed. The engine oil sample test analysis results indicate that the oil has been warmed, drained for an excessive amount of time, oxidized, and includes Fe and Pb, wear on piston rings, bearing overlays, and cylinder liners as well as increased viscosity, oxidation, Fe and Pb case oil overheating. While the composition of the Al > 3 ppm, Cu > 3 ppm, Cr > 0 ppm, and Sn > 1 ppm materials the main components of the bearings on the piston-bore and crankshaft-bore is below 4 ppm and the wear on the Fe material composition has a maximum limit of 14 ppm, it can be inferred that the wear rate on the bearing components of the engine is still within allowable bounds. Due to the high fuel wear on the crankshaft, bearing linings, and cylinder liners from March to June 2022, replenishing oil is typical.

Acknowledgements I want to thank my coworkers who participated in the engine oil testing research as well as company PT. ABCD for allowing me to conduct the research.

References

1. Cesur, I., et al.: Investigation of the Effects of 0W20 and 10W40 Lubricating Oils on Cylinder Liner Wear, vol. 1, no. 1, pp. 88–92 (2023). https://doi.org/10.59287/icaens.970
2. Gao, Z., Huo, B., Tong, Y., Jiang, Z.: Fault Diagnosis of the Connecting Rod Bush of Internal Combustion Engines Based on Wireless Temperature Monitoring, vol. 46, no. 4 (2019). https://doi.org/10.13543/j.bhxbzr.2019.04.012
3. Arso et al.: Analysis of Damage to Main Bearings in Unit of BMGS Mitsubishi Generator, vol. 5, no. 2, pp. 151–160 (2021). https://doi.org/10.31289/jmemme.v5i2.4926

4. Kulakov, A.T., Barylnikova, E.P., Talipova, I.P.: Lubrication conditions and development of pre-failure state of crankshaft bearings. Lecture Notes in Mechanical Engineering (2020). https://doi.org/10.1007/978-3-030-22041-9_108
5. Witek, L., Zelek, P.: Stress and failure analysis of the connecting rod of diesel engine. Eng. Fail. Anal. **97** (2019). https://doi.org/10.1016/j.engfailanal.2019.01.004
6. Wilarso, Domodite, A., Sholih, H., Mujiarto: Analysis of SOS results for engine lubricants contaminated by the fuel in the 3516 TA diesel engine generator set. J. Phys. Conf. Ser. **1764**(1) (2021). https://doi.org/10.1088/1742-6596/1764/1/012172
7. Wilarso, W., Noor, C.W.M., Ayob, A.F.M., Mansor, W.N.W.: Investigation and failure analysis of a diesel generator connecting rod. Mech. Eng. Soc. Ind. **2**(2), 63–70 (2022). https://doi.org/10.31603/mesi.6624
8. Wan, M.N.C., et al.: An investigation of using palm biodiesel blend on the performance of a stationary marine diesel engine. Def. S T Tech. Bull. **9**(2), 86–96 (2016) [online]. Available https://www.researchgate.net/publication/311985911_An_investigation_of_using_palm_biodiesel_blend_on_the_performance_of_a_stationary_marine_diesel_engine
9. Nadkarni, R.: Guide to ASTM Test Methods for the Analysis of Petroleum Products and Lubricants, 2nd edn. (2007). https://doi.org/10.1520/mnl44-2nd-eb
10. Wasilczuk, M., Wasilczuk, F.: Combined thrust radial bearing of a submarine main shaft—design and analysis of failure. Eng. Fail. Anal. **115** (2020). https://doi.org/10.1016/j.engfailanal.2020.104651
11. Raposo, H., Farinha, J.T., Fonseca, I., Ferreira, L.A.: Condition monitoring with prediction based on diesel engine oil analysis: a case study for urban buses. Actuators **8**(1) (2019). https://doi.org/10.3390/act8010014
12. Berau, V., et al.: Basic of Oil Analysis, p. 32 (2020)
13. Hermawan, A., Rahardja, I.B., Syam, M.Y., Sukismo, H., Patah, N.: Analysis of viscosity of lubricating oil on generator machine working hours at KP. Macan Tutul 4203 **1**(3), 69–74 [online]. Available https://jurnal.umj.ac.id/index.php/JASAT/article/view/4312/3086
14. Hasannuddin, A.K., et al.: Performance, Emissions and Lubricant Oil Analysis of Diesel Engine Running on Emulsion Fuel, vol. 117 (2016). https://doi.org/10.1016/j.enconman.2016.03.057
15. Schwendemann, S., Amjad, Z., Sikora, A.: A survey of machine-learning techniques for condition monitoring and predictive maintenance of bearings in grinding machines. Comput. Industr. **125** (2021). https://doi.org/10.1016/j.compind.2020.103380
16. Noor, C.W.M., Mamat, R., Najafi, G., Bakar, A.A., Samo, K.: Determination of bio-diesel engine combustion pressure using neural network based model. J. Eng. Sci. Technol. **14**(2), 909–921 (2019)

Chapter 11
Real-Time Capability of Megathrust Tsunami Detection by WERA Ocean Radar System

Budianto Ontowirjo, Andri Ramdhani, Semeidi Husrin, Achmad Witjaksono, and Leonardus Muryono

Abstract Radar technology has been proven to be able to capture trans-oceanic tsunami signatures. With this advancement, an ocean radar system has been set up for real-time tsunami monitoring along the Sunda Strait, extended from the Tanjung Lesung Observation Station (TOS), Tanjung Lesung, Banten province, to Kahai Observation Station (KOS), Kahai, Lampung province, Indonesia. The 6.7 MW earthquake occurred at 09.05 UTC and generated small tsunami waves minutes after and activated the alert mode of InaTOC SUN buoy. The anomalies on the water surface generated by the earthquake were detected by the radar in a form of strong changes in measured combined currents at distances of 30 km West of the Tanjung Lesung observation station. The radar systems are able to track the very strong current pattern in real-time following the wave propagation toward the coastlines. The real-time detection on strong current has shown potential application of ocean radar technology for offshore monitoring and navigation safety.

Keywords Earthquake · Tsunami · Current · Phased-array · Ocean radar

---

B. Ontowirjo (✉)
Bakrie University, Jl. H. R. Rasuna Said Kav. C-22, Jakarta 12940, Indonesia
e-mail: budianto.ontowirjo@bakrie.ac.id

A. Ramdhani
Indonesian Meteorological, Climatology and Geophysical Agency BMKG, Jakarta, Indonesia
e-mail: andri.dhani@gmail.com

S. Husrin
Ministry of Marine and Fisheries KKP, Jakarta, Indonesia
e-mail: semeidi@gmail.com

A. Witjaksono
National Agency for Research and Innovation BRIN, Jakarta, Indonesia
e-mail: achm005@brin.go.id

L. Muryono
PT Terindo Bumi Raya, Jakarta, Indonesia
e-mail: muryono@gmail.com

A. Ismail et al. (eds.), *Exploring Diversity in Engineering and Technology for Knowledge and Innovation*, Advanced Structured Materials 215,
https://doi.org/10.1007/978-3-031-64330-9_11

11.1 Introduction

High-frequency ocean radars like the WERA® system are used for real-time monitoring of ocean surface currents and sea state parameters. They operate by using electromagnetic waves to measure ocean conditions and have a wide coverage area, extending up to 300 km offshore [1]. These radar systems have various applications, including ocean current and wave mapping [2], search and rescue, and vessel traffic services [3].

Based on the theoretical approach described in [4], in this context, the WERA system was optimized for wave and tide-induced strong current observation along the coast of the Sunda Strait. Tsunami waves can alter sea surface current velocities, leading to specific patterns that can be observed through radar measurements. These changes are influenced by the height of the tsunami waves and the transition from deep ocean to shallow water. The continental shelf acts as a boundary where the behavior of the tsunami wave and its impact on current change significantly. These observations are consistent with existing understanding of tsunami current and their relationship to water depth [4]. These observations were not purely theoretical but were validated through real-world measurements during an actual 2011 Japan Tsunami event [5].

11.2 Methodology

The primary focus of this work is the use of a high-frequency (HF) ocean radar systems like WERA for the early detection and monitoring of tsunamis in the Sunda Strait, using radar technology to track changes in water surface velocities induced by waves and tides. These strong current detection warning systems can help to improve navigation safety and generate alerts in the event of an earthquake-generated tsunami [6].

The Sunda Strait Tsunami Early Warning Project operates two WERA ocean observatories at Tanjung Lesung, Banten province, and Kahai, Lampung province, Indonesia. The system coverage area is shown in Fig. 11.1.

The radar system operates at a frequency of 16.15 MHz and uses a linear array of 12 antenna elements along the coast. The estimation of strong current is based on sea surface current spectrum analysis, filtering out residual currents [7].

The Sunda Strait radar system is connected to a web-enabled monitoring system (WEMS) for rapid detection and notifications are sent to subscribed users via various methods.

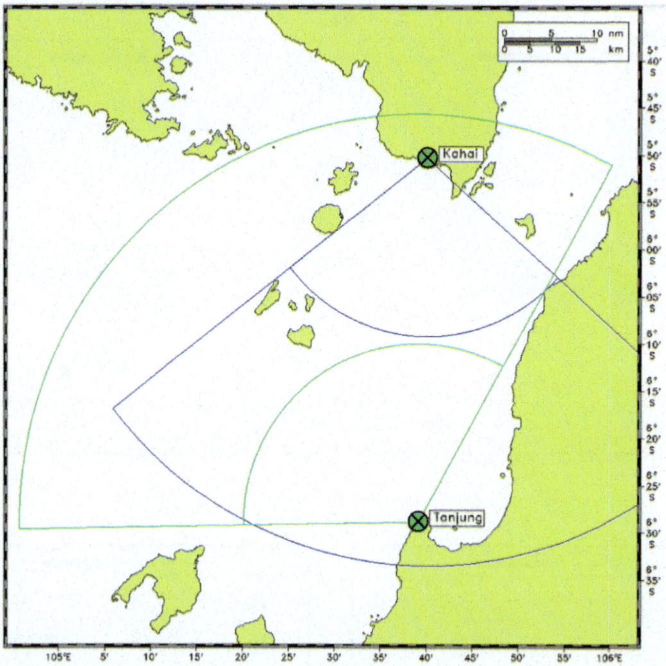

Fig. 11.1 Sunda Strait ocean observation system: WERA system (blue green circular sector)

11.3 Results and Discussion

The WERA radar system recorded unusual changes in surface current velocities on January 14, 2022. These changes were observed in the form of fast intensification in incoming surface currents.

While the changes in current velocity were significant, they might be solely due to the natural tidal effects.

To separate potential tsunami waves, the radar system has to evaluate residual current velocities, filtering out tidal current. This process took only a few seconds and provided visualizations of tsunami alerts for decision-making.

Figure 11.2 shows a series of January 14, 2022 water surface current anomalies measured by the radar system. The images show a dark blue color area which indicates combined strong currents. The observed strong current event took about 1.0 h.

Figure 11.3 shows a series of January 14, 2022 water surface current anomalies measured by the radar system. These images show a dark blue color area which indicates combined strong currents.

Figures 11.4, 11.5 and 11.6 exhibited strong currents generated by combined high tide and earthquake events. Further analysis of InaTOC SUN ocean bottom pressure (Fig. 11.7) and Sebesi island IDSL tide gauge data (Fig. 11.8) support an earthquake generated tsunami.

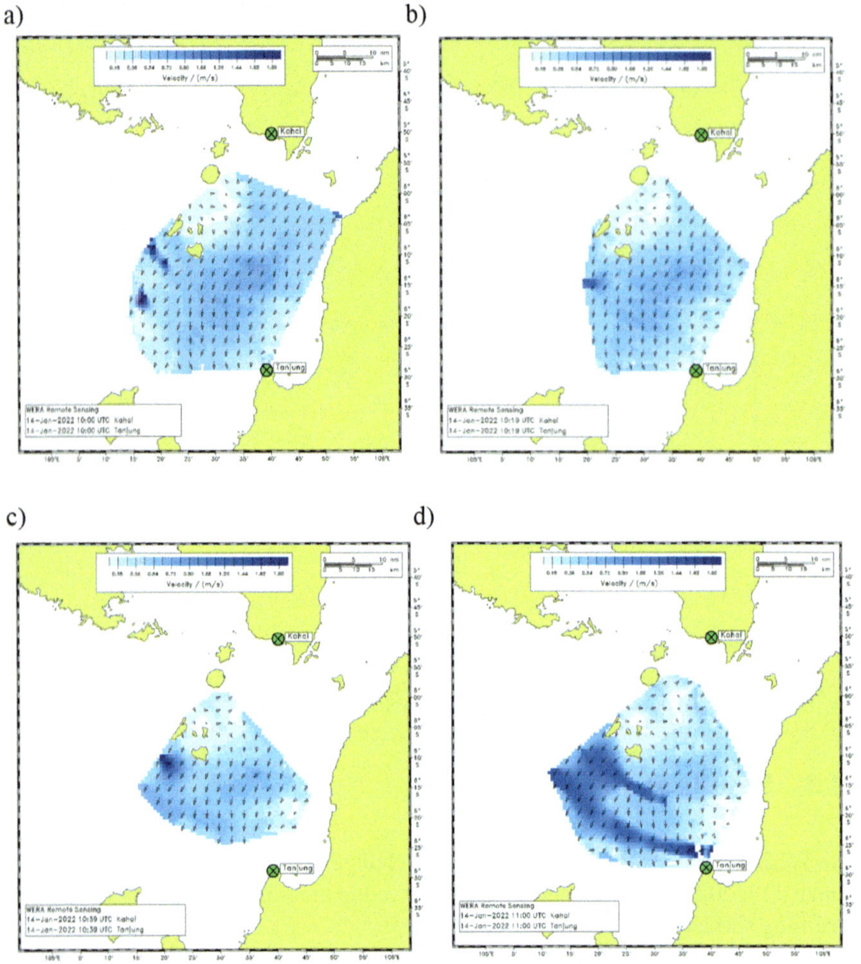

Fig. 11.2 Velocity maps **a** 10:00 UTC, **b** 10:19 UTC, **c** 10:39 UTC, **d** 11:00 UTC January 14, 2022

It was suggested that the observed strong current event off the coast of Kahai, Lampung province, was likely caused by natural tidal conditions (Fig. 11.9).

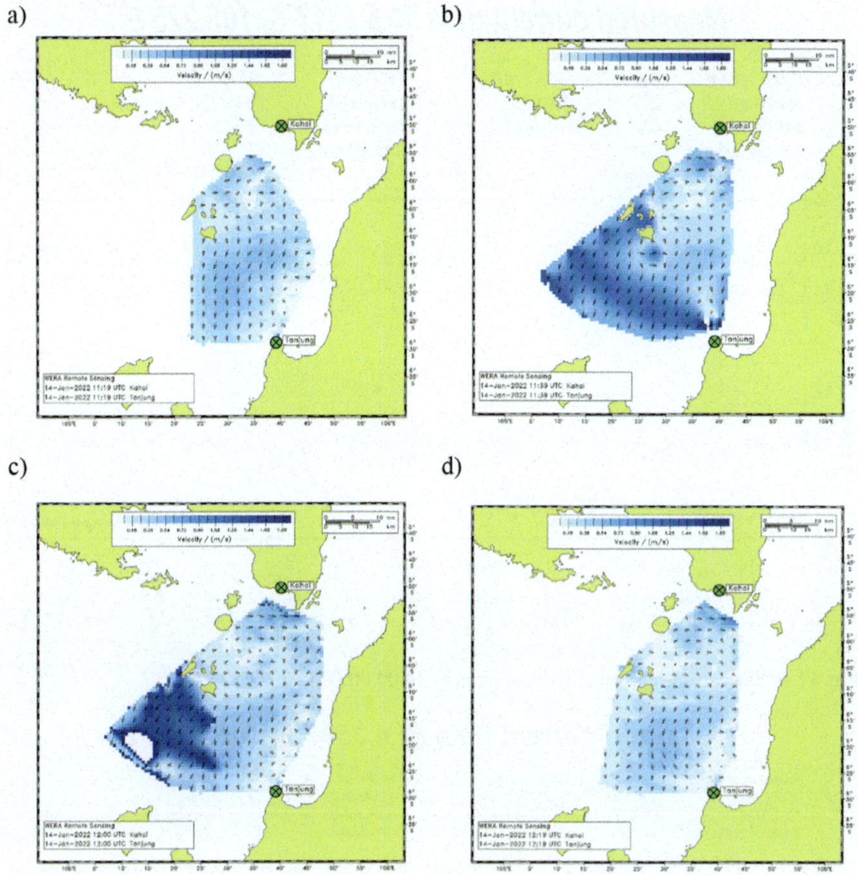

Fig. 11.3 Velocity maps **a** 11:19 UTC, **b** 11:39 UTC, **c** 12:00 UTC, **d** 12:19 UTC January 14, 2022

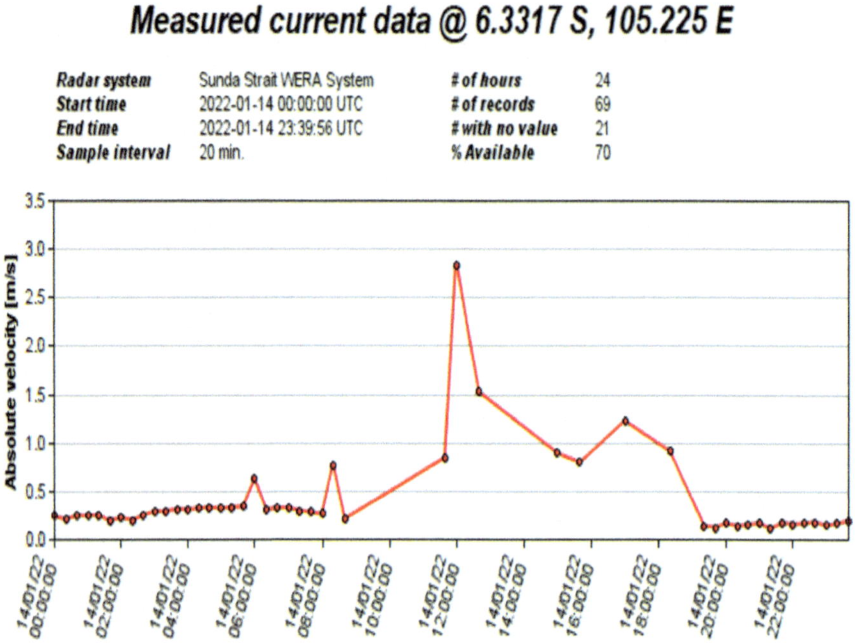

Fig. 11.4 Surface current velocity from January 14 00:00 UTC to 23:59 UTC

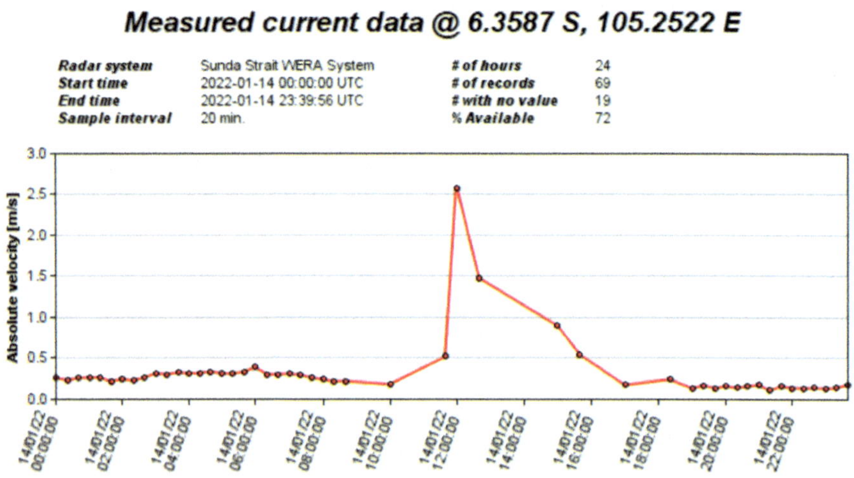

Fig. 11.5 Mount Anak Krakatau surface current velocity from January 14 00:00 UTC to 23:59 UTC

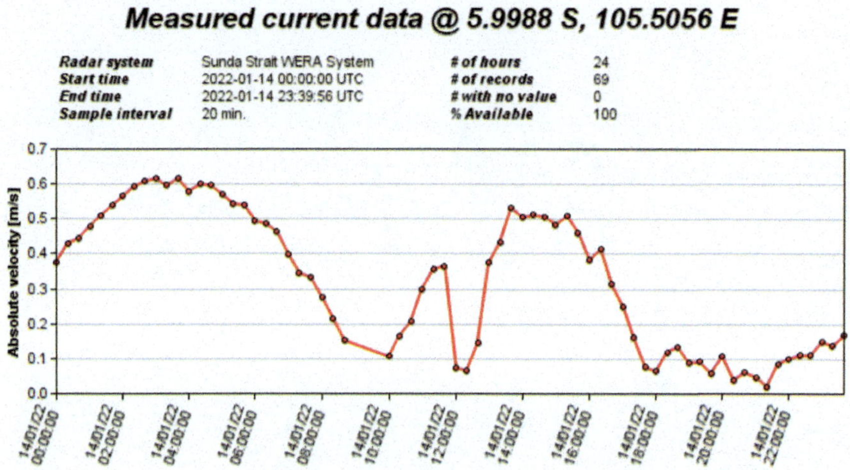

Fig. 11.6 Sebesi island surface current velocity from January 14 00:00 UTC to 23:59 UTC

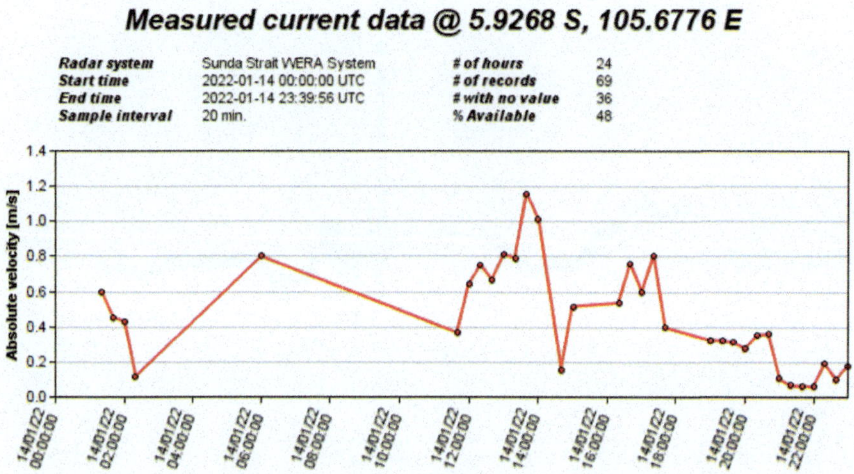

Fig. 11.7 Kahai water surface current velocity from January 14 00:00 UTC to 23:59 UTC Jan 14, 2022

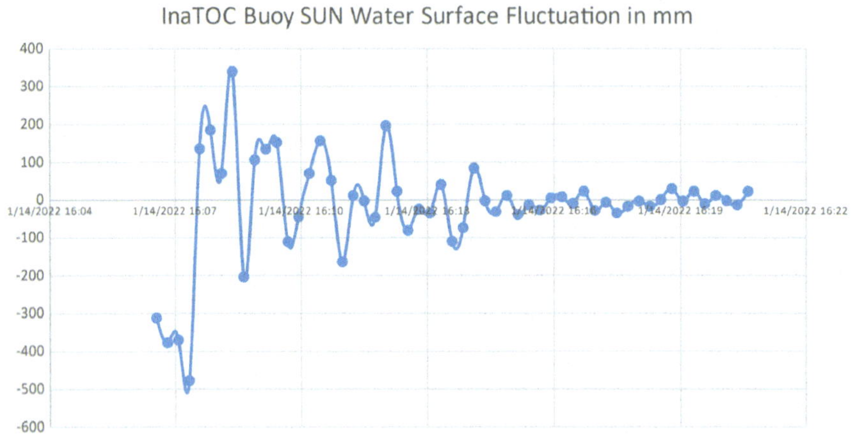

Fig. 11.8 InaTOC SUN buoy surface elevation from 15:50 WIB to 16:20 WIB Jan 14, 2022

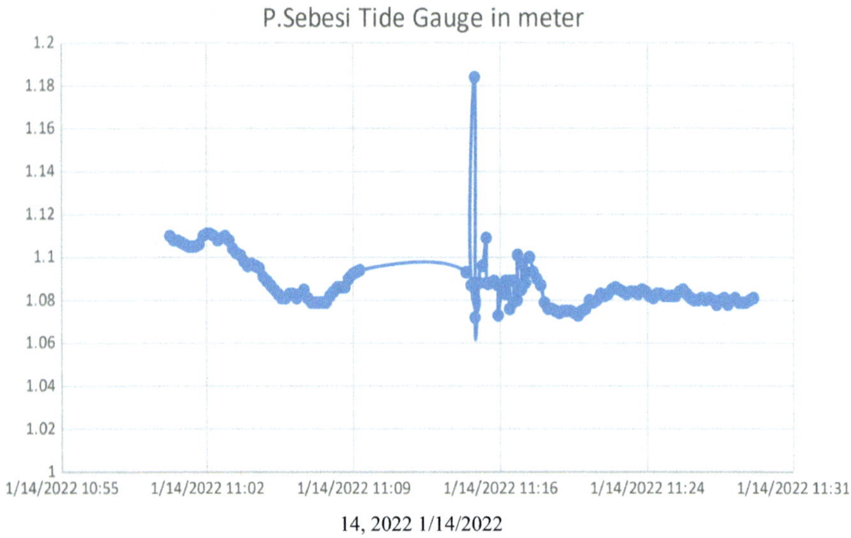

Fig. 11.9 Sebesi island IDSL tide gauge surface elevation from 10:55 UTC to 11:30 UTC Jan 14, 2022

11.4 Conclusion

HF ocean radar systems, such as the WERA radar system, were able to detect an unusual event on January 14, 2022. It was also able to generate tsunami detection maps within minutes.

Further analysis of data from nearby tide gauges and a hydro-oceanographic buoy supported the presence of tsunami waves. The WERA system was capable of detecting deviations in ocean currents caused by various mechanisms such as earthquakes and submarine landslide events.

The study highlighted the valuable role of HF phased-array radar technology in offshore tsunami monitoring and navigation safety. The WERA radar system proved effective in detecting and tracking tsunami events independently of their generating mechanisms.

Overall, this study emphasizes the significance of ocean radar systems like WERA in providing early warning and monitoring capabilities for potential tsunami events, even in complex environmental conditions.

Acknowledgements The authors appreciated PT Terrindo Bumi Raya and Helzel Messtechnik GmbH for providing free surface current and significant wave height observation data and graphs contributed to this article.

References

1. Riethmuller, R., et al.: COSYNA, an integrated coastal observation system for Northern and Arctic Seas. In: OCEANS 2009-EUROPE, pp. 1–7, Bremen (2009). https://doi.org/10.1109/OCEANSE.2009.5278153
2. Harlan, J., et al.: The integrated ocean observing system high frequency radar network: status and local, regional, and national applications. Mar. Technol. Soc. J. **44**(6), 122–132 (2010). https://doi.org/10.4031/MTSJ.44.6.6
3. Maresca, S., et al.: Maritime surveillance using multiple high-frequency surface-wave radars. IEEE Trans. Geosci. Remote Sens. **52**(8), 5056–5071 (2014)
4. Gurgel, K.W., et al.: Simulation and detection of tsunami signatures in ocean surface currents measured by HF radar. Ocean Dyn. **61**(10), 1495–1507 (2011). https://doi.org/10.1007/s10236-011-0420-9
5. Dzvonkovskaya, A.: Ocean surface current measurements using HF radar during the 2011 Japan tsunami hitting Chilean coast. In: Proceedings of IEEE IGARSS-2012, pp. 7605–7608, Munich (2012). https://doi.org/10.1109/IGARSS.2012.6351867
6. Dzvonkovskaya, A., Petersen, L., Insua, T.L.: Real-time capability of meteo tsunami detection by WERA ocean radar system. In: 18th International Radar Symposium (IRS), pp. 1–10, Prague (2017). https://doi.org/10.23919/IRS.2017.8008096
7. Dzvonkovskaya, A., et al.: High frequency ocean radar support for Tsunami Early Warning Systems. Geosci. Lett. **5**, 29 (2018). https://doi.org/10.1186/s40562-018-0128-5

Chapter 12
Comparative Analysis of Multi-storey Building with and Without Seismic Load

Ak Muhammad Saufi Ak Ali Putra and Ahmad Nurfaidhi Rizalman

Abstract Following to the 2015 Ranau earthquake, structural integrity of existing buildings in Sabah became a concern. This is because most structures in Sabah were constructed with British Standard which did not include any seismic provisions. Although Kota Kinabalu has low PGA with 0.04 g, it remains the state with the largest population in Sabah with around 550,000 people. Thus, this study compared the performance of multi-storey buildings of 5, 10, and 15 levels with and without seismic load. In this study, SAP2000 was employed to perform structural and pushover analyses on the multi-storey buildings. Then, the take-off approach was adopted in Microsoft Excel to compare the cost of buildings with and without seismic load. The results showed there were no changes in the size of the beam and column, and the volume of concrete. However, there was an increase in the weight of steel reinforcement bars of the buildings with seismic load compared to buildings without seismic load. Hence, the total cost of building materials with seismic load was higher than building without seismic load.

Keywords Seismic design · Non-seismic design · Pushover analysis · SAP2000 · Multi-storey buildings

12.1 Introduction

An earthquake is one of the nature's most unpredictable and destructive forces that cause thousands of deaths and property damages. The largest earthquake to ever be recorded had a magnitude of 9.5 and occurred in Chile on 22 May 1960. It killed approximately 1655 people, injured at least 3000 people, and left two million people homeless [1]. Another deadly earthquake was recorded in 2004 in Sumatra,

A. M. S. A. A. Putra · A. N. Rizalman (✉)
Faculty of Engineering, Universiti Malaysia Sabah, 88999 Kota Kinabalu, Sabah, Malaysia
e-mail: ahmadnurfaidhi@ums.edu.my

A. M. S. A. A. Putra
e-mail: bk17110298@student.ums.edu.my

A. Ismail et al. (eds.), *Exploring Diversity in Engineering and Technology for Knowledge and Innovation*, Advanced Structured Materials 215,
https://doi.org/10.1007/978-3-031-64330-9_12

Indonesia, with a magnitude of 9.1. The earthquakes claimed 227,898 lives and wounded more than 1.1 million people [2]. It also triggered tsunamis in fourteen countries across South Asia including Malaysia, Thailand, Myanmar, and Sri Lanka, among others.

According to Malaysian National Seismic Network (MNSN), Malaysia had experienced 59 earthquakes between 1922 and 2020 with magnitudes ranging from 2.5 to 3.5 [3]. Most local earthquakes in the Peninsular Malaysia were related to intraplate earthquakes. Thus, the majority of the earthquakes were lower about 3–4 magnitudes. Sabah, on the other hand, has experienced interplate earthquakes, which had higher magnitudes than the Peninsular Malaysia and Sarawak. It also caused the earthquakes in Sabah to be more frequent [4].

Following to the 2015 Ranau earthquake, structural integrity of existing buildings in Sabah became a concern. According to Jabatan Mineral and Geosains Malaysia [5], the peak ground acceleration (PGA) values in Kota Kinabalu and Ranau are 0.04 g and 0.12 g, respectively. Although Kota Kinabalu has low PGA with 0.04 g, it remains the state with the largest population in Sabah with around 550,000 people. This motivates to this research to compare the performance of multi-storey buildings in Kota Kinabalu under seismic and non-seismic design.

In this research, three buildings from 5 storey, 10 storey, and 15 storey were modelled using SAP2000. For seismic consideration, Kota Kinabalu's PGA value of 0.04 g was employed. First, the size of structural elements and displacement of the buildings under seismic and non-seismic design were compared. Then, the comparison of cost of materials (steel and concrete) between the buildings was investigated.

12.2 Methodology

12.2.1 Building Modelling

The buildings were modelled in the SAP2000 software based on two different design codes, Eurocode 2 (EC2) and Eurocode 8 (EC8). EC2 is for buildings under non-seismic design; meanwhile, EC8 is for buildings under seismic design. For each design code, three buildings were modelled which ranged from 5 storey, 10 storey, and 15 storey. Each floor was kept constant at 3 m. Table 12.1 shows the summary of the properties of the building, concrete, and steel. Meanwhile, Fig. 12.1 illustrates the modelled building in SAP2000.

Table 12.1 Properties of building, concrete, and steel

No.	Properties	Information
A. Building properties		
1	Number of stories	5, 10, 15
2	Story height (m)	3
3	Number of bays, X	3
4	Number of bays, Y	3
5	Bay width, X (m)	5
6	Bay width, Y (m)	5
B. Properties of concrete		
1	Grade	30/37
2	Modulus of elasticity, E (N/mm^2)	31,000
3	Density, ρ (kN/m^3)	25
4	Poisson's ratio, ν	0.2
5	Cylinder crushing strength, f_{ck} (N/mm^2)	30
6	Cube strength (N/mm^2)	37
C. Properties of steel reinforcement bars		
1	Grade	500
2	Modulus of elasticity, E (N/mm^2)	200,000
3	Specific weight, γ (kN/m^3)	77
4	Characteristic strength, f_{yk} (N/mm^2)	500

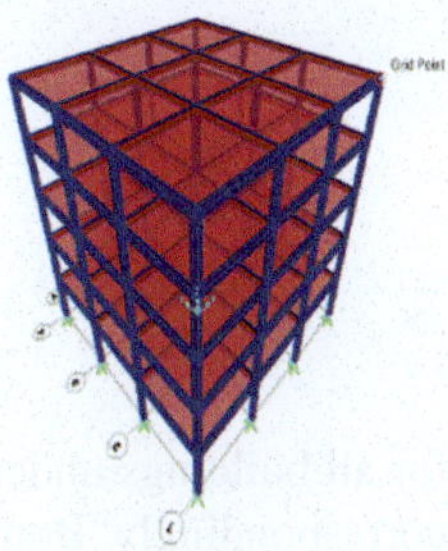
(a) 5-Storey

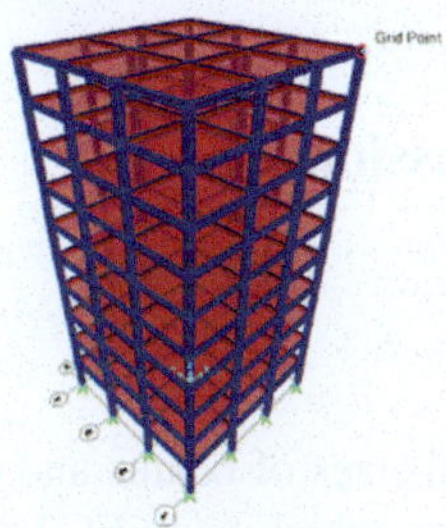
(b) 10-Storey

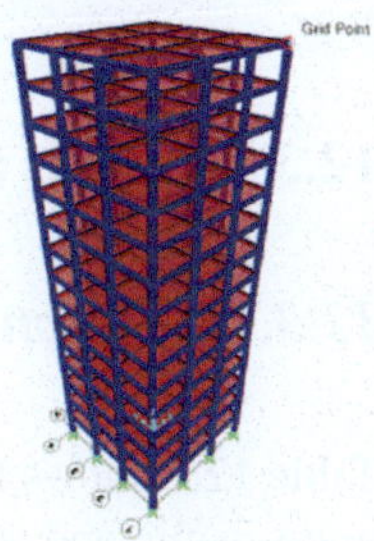
(c) 15-Storey

Fig. 12.1 Building models in SAP2000

12.2.2 Structural and Pushover Analysis

After the buildings were modelled, structural and pushover analyses were performed to compare the design and displacements (x and y-axis) of the buildings under seismic and non-seismic load. The outputs from the analysis include the size of structural elements, building displacements, volume of concrete, and area of steel reinforcement bars.

12.2.3 *Cost Evaluation*

The cost evaluation of the buildings was conducted using the take-off approach. Two components were considered in the price calculation including the volume of concrete and weight of steel reinforcement bars. The price of concrete and steel was taken from the 2019 Construction Industry Development Board (CIDB) Building Material Prices (Kota Kinabalu–Sabah). The equations used to calculate the cost of concrete and steel reinforcement bars of buildings with and without seismic load are shown below:

$$\text{Area } (A) = \text{Breadth } (B) \times \text{Height } (H) \tag{12.1}$$

$$\text{Volume } (V) = \text{Area } (A) \times \text{Length } (L) \tag{12.2}$$

$$\text{Cost } (c) = \text{Price } (P) \times \text{Volume } (V) \tag{12.3}$$

$$\text{Total Cost } (C) = \text{Cost of Beam } (C_b) + \text{Cost of Column } (C_c) \tag{12.4}$$

12.3 Results and Discussion

12.3.1 *Structural Size*

Table 12.2 shows the structural sizes of beams and columns for all buildings under non-seismic and seismic design, which were EC2 and EC8, correspondingly. Both 5-storey and 10-storey buildings used 250 mm × 350 mm beam size for both EC2 and EC8 designs, while the column used was 300 mm × 300 mm. As for the 15-storey building, both non-seismic and seismic designs used beam and column sizes of 300 mm × 350 mm and 450 mm × 450 mm, respectively.

The results show that there were no significant changes in the size of beam and column of buildings with and without seismic load. This indicates that the size of structural elements was sufficient to withstand the seismic load with PGA values of 0.04 g. The finding was in line with two previous researchers. The first one was Faisal et al. [6] who discovered that the size of beam and column was constant for 5-storey, 10-storey, and 20-storey buildings without seismic and with seismic load of

Table 12.2 Structural size comparison of buildings under EC2 and EC8 design

Storey	Structural size (beam)			Structural size (column)		
	EC2	EC8	Remarks	EC2	EC8	Remarks
5	250 × 350	250 × 350	Unchanged	300 × 300	300 × 300	Unchanged
10	250 × 350	250 × 350	Unchanged	300 × 300	300 × 300	Unchanged
15	300 × 350	300 × 350	Unchanged	450 × 450	450 × 450	Unchanged

0.165 g. The second one was Awaludin and Adnan [7] who found that there were no changes in the size of beam and column of buildings with different ductility classes.

12.3.2 Displacement

Table 12.3 shows the displacement of 5 storey, 10 storey, and 15 storey along the *x*- and *y*-axis. From the table, the displacement in the *x*- and *y*-axis for all buildings was similar due to the same number of bays in both directions. However, as the storey of the building increases, lower force is required to cause displacement on the building. For instance, 3958.51 kN was needed to cause a displacement of 0.226 m on a 5-storey building, while 1177.66 kN was needed to cause displacement of 0.2375 m on a 15-storey building.

The performance level of the buildings under seismic load was under immediate occupancy (IO), as indicated in the green colour in Fig. 12.2. This indicates that the building can still function and only incurs minor damages in the immediate occupancy state.

Table 12.3 Displacement along the *x*-axis and *y*-axis

Step	5 storey		10 storey		15 storey	
	U_x and U_y	V (kN)	U_x and U_y	V (kN)	U_x and U_y	V (kN)
0	0.0000	0.00	0.0000	0.00	– 5E–18	0.00
1	0.0300	838.25	0.0300	0.00051	0.1125	731.07
2	0.0367	1026.21	0.0367	31.70	0.2375	1177.66
3	0.0671	1723.88	0.0671	735.20		
4	0.1066	2433.94	0.1066	710.75		
5	0.1384	2975.31				
6	0.1724	3505.29				
7	0.1926	3791.09				
8	0.2240	3953.19				
9	0.2260	3958.51				

Indicator: U_x = displacement along the *x*-axis, U_y = displacement along the *y*-axis, V = base force

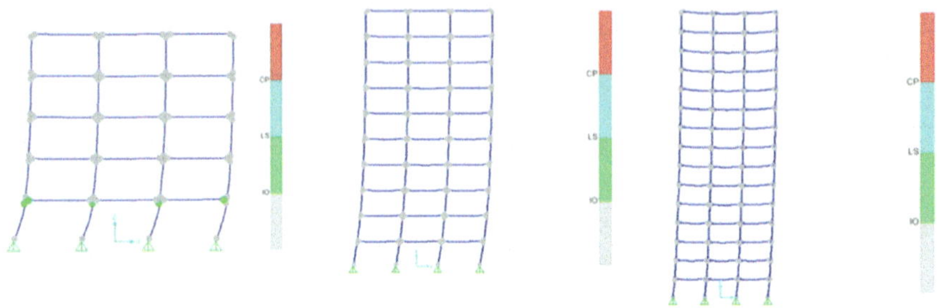

a) 5 storey building b) 10 storey building c) 15 storey building

Fig. 12.2 Deformed shape of buildings after pushover analysis

12.3.3 Volume of Concrete

Figure 12.3 shows the volume of concrete for multi-storey buildings under non-seismic and seismic design. The total volume of concrete for 5-storey, 10-storey, and 15-storey buildings were 74.1 m^3, 148.2 m^3, and 334.2 m^3, respectively. It remains unchanged for both buildings with and without seismic and non-seismic load. This is because the size of structural elements, as pointed in the previous section, for all buildings remains unchanged. These results are aligned with findings by Faisal et al. [6] and Awaludin and Adnan [7] who discovered similar result patterns as presented in Fig. 12.3.

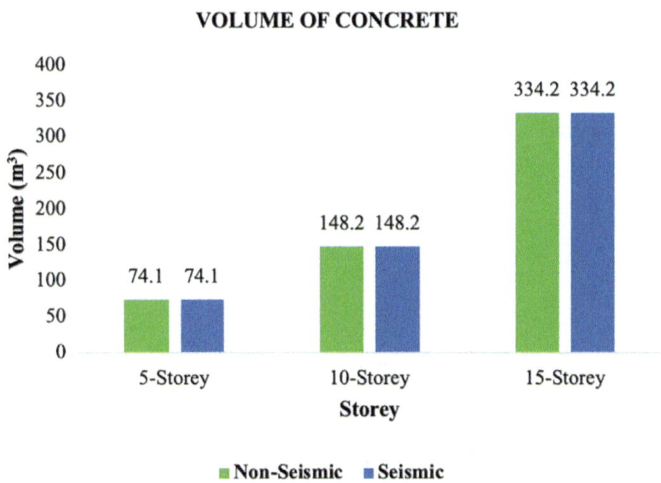

Fig. 12.3 Comparison of the total volume of concrete under EC2 and EC8 design

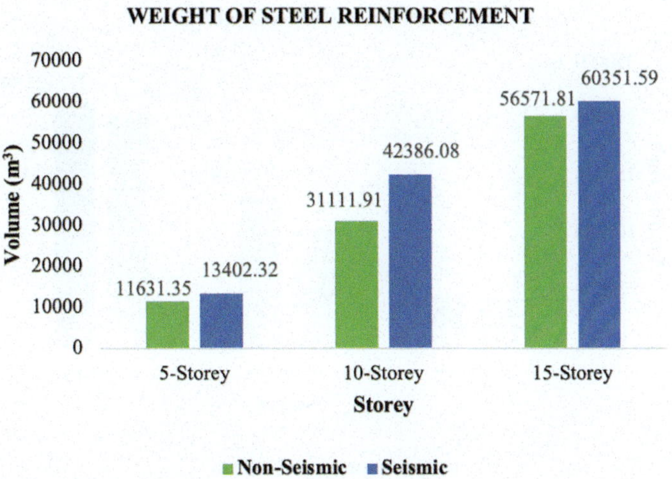

Fig. 12.4 Comparison of the total volume of steel under EC2 and EC8 design

12.3.4 Weight of Steel Reinforcement

Figure 12.4 shows the weight of steel reinforcement bars of the buildings with and without seismic load. It shows that the steel weight for buildings under EC8 design is higher than the buildings under non-seismic design. The steel weight for 5-, 10-, and 15-storey buildings under EC8 design increased by 15.23%, 36.24%, and 6.68%, respectively. The percentage increase was found higher in the columns than for the beams. This finding agrees with the concept of strong column-weak beam in EC8 which stated that the steel reinforcement used in columns was greater than the steel reinforcement used in beams [8]. This is because columns are the primary load-resisting members in buildings and thus play a key role to ensure the structural stability when earthquakes strike [9].

12.3.5 Cost of Materials

Figure 12.5 shows the comparison of the material costs between buildings with and without seismic load. It shows that the cost of building materials of 5, 10, and 15 storeys with seismic load increased by 10.36%, 26.81%, and 4.53%, respectively. This was due to the increase in the weight of steel reinforcement bars of buildings with seismic load. According to Adiyanto et al., buildings that were designed to sustain seismic load required a high number of shear reinforcement bars in beams and columns [8]. Since the weight of steel reinforcement bars was increased, the cost of building materials also increased. The results were aligned with the results found by Faisal et al. [6].

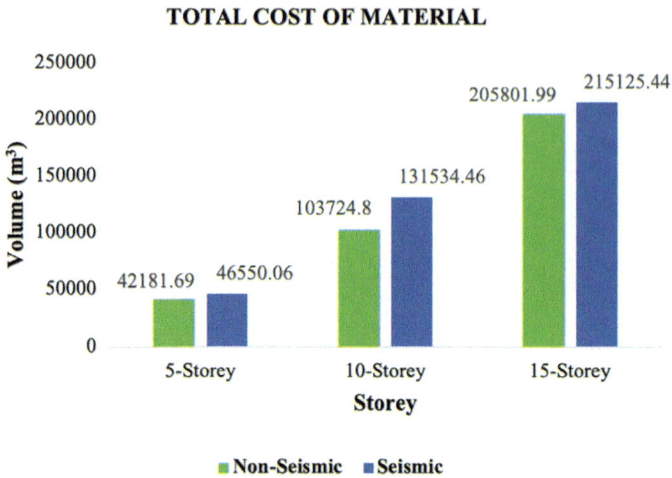

Fig. 12.5 Comparison of the total of material cost of building with and without seismic load

12.4 Conclusions

The following conclusions have been drawn from this research:

(a) There were no changes in the size of the beam and column, and the volume of concrete of the multi-storey buildings with and without seismic load.
(b) However, there was an increase in the weight of steel reinforcement bars of buildings with seismic load compared to building without seismic load.
(c) The total cost of material in buildings under seismic design was higher than buildings under non-seismic design. This is because there was an increase in the cost of steel reinforcement when buildings were designed under seismic load.

Acknowledgements This study was supported by the Dana Kluster Penyelidikan (DKP0087) Grant, Universiti Malaysia Sabah, under "Dispersion and Mixing Method of Synthetic Fibre Reinforced Cementitious Mortar".

References

1. Melgar, D., Riguelme, S., Xu, X., Baez, J.C., Geng, J., Moreno, M.: The first since 1960: a large event in the Valdivia segment of the Chilean subduction zone, the 2016 M7.6 Melinka earthquake. Earth Planet. Sci. Lett. **474**, 68–75 (2017). https://doi.org/10.1016/j.epsl.2017.06.026
2. Duputel, Z., Rivera, L., Kanamori, H., Hayes, G.P.: W phase source inversion for moderate to large earthquakes (1990–2010). Geophys. J. Int. **189**(2), 1125–1147 (2012). https://doi.org/10.1111/j.1365-246X.2012.05419.x

3. Nazaruddin, D.A., Duerrast, H.: Intraplate earthquake occurrence and distribution in Peninsular Malaysia over the past 100 years. SN Appl. Sci. **3**(693) (2021)
4. Tongkul, F.: An overview of earthquake science in Malaysia. ASM Sci. J. **14** (2021). https://doi.org/10.32802/asmscj.2020.440
5. Jabatan Mineral dan Geosains Malaysia (JMG): Seismic Hazard Map of Malaysia (1st edn.). Putrajaya (2018)
6. Faisal, F., Zaini, S., Selokumar, T.: Evaluation of cost analyses for earthquake resistant reinforced concrete buildings based on Malaysian National Annex to Eurocode 8. In: Mohamed Nazri, F. (ed.) Proceedings of AICCE' 19. Lecture Notes in Civil Engineering, vol. 53, pp. 1485–1492. Springer, Heidelberg (1996). https://doi.org/10.1007/978-3-030-32816-0_113
7. Awaludin, S.Z., Adnan, A.: Cost implementation on earthquake resistance design for building in Malaysia. Engineering (2016)
8. Adiyanto, M.I., Ramadhansyah, P.J., Ramli, N.I., Harith, N.S.H., Razak, S.M.S.A., Mustapa, S.A.H.S.: Increment of material usage in construction of four storey reinforced concrete building due to seismic design. IOP Conf. Ser.: Earth Environ. Sci. **682** (012051) (2021). https://doi.org/10.1088/1755-1315/682/1/012051
9. Noor, S.M., Halim, N.S.M., Ibrahim, A., Majid, T.A., Hassan, S.H.: Fundamental period of vibrations influencing characteristics of torsional irregularity in reinforced concrete buildings. IOP Conf. Ser.: Earth and Environ. Sci. **244** (012021) (2019). https://doi.org/10.1088/1755-1315/244/1/012021

Chapter 13
Seismic Analysis of Tun Mustapha Residential College of UMS and Staff Quarters of Desa Wawasan Tambunan Secondary School

Muhammad Syamim Fattah Mohamad Juhari and Ahmad Nurfaidhi Rizalman

Abstract An earthquake is defined as a rapid movement of the earth's surface, which caused the plate tectonic boundaries to shake violently. Earthquakes had caused various damages, which had led major impact on humans in many ways: finances, safety, and health. Most existing buildings in Sabah were designed under British Standards in which the earthquake element was not taken into consideration. Hence, this research aimed to investigate the performance of two existing buildings in Sabah when subjected to seismic loads. The buildings were Tun Mustapha Residential College of the Universiti Malaysia Sabah and Staff Quarters of Desa Wawasan Tambunan Secondary School. In the simulation, three peak ground acceleration (PGA) values were employed including 0.145 g, 0.097 g, and 0.048 g. First, the structural deformation of the buildings was analysed using the PROTA software. Then, the safety status of the building was examined using standard FEMA 356. It was found that the displacement of the buildings in the *x*- and *y*-axis increased as the storey height and PGA values increased. Based on the results, it can be concluded that both buildings were classified as immediate occupancy (IO) which means the buildings are safe to be used.

Keywords Seismic analysis · Performance level · Interstorey drift ratio · PROTA

M. S. F. M. Juhari · A. N. Rizalman (✉)
Faculty of Engineering, Universiti Malaysia Sabah, Kota Kinabalu, Sabah, Malaysia
e-mail: ahmadnurfaidhi@ums.edu.my

M. S. F. M. Juhari
e-mail: bk18110258@student.ums.edu.my

A. Ismail et al. (eds.), *Exploring Diversity in Engineering and Technology for Knowledge and Innovation*, Advanced Structured Materials 215,
https://doi.org/10.1007/978-3-031-64330-9_13

13.1 Introduction

An earthquake is defined as a rapid movement of the earth's surface, which caused the plate tectonic boundaries to shake violently. Earthquake poses catastrophic threats to the industrial society and has caused significant loss of life, damage to infrastructure and economic instability [1].

Malaysia consists of two regions separated by the South China Sea namely West and East Malaysia. Therefore, the earthquake classification in Malaysia is based on these regions, where the West Malaysia (also known as Peninsular Malaysia) is classified as intraplate earthquake, and the East Malaysia (also known as Sabah and Sarawak) is classified as interplate and intraplate earthquake [2]. There are three seismic zones in Malaysia, namely zone 1 for Central, Northern, and Eastern Sabah, zone 2 for Southwestern Sabah and Sarawak, and zone 3 for the Peninsular Malaysia. Each zone is characterized by the peak ground acceleration (PGA) which is the intensity of ground shaking. The peak ground acceleration (PGA) value increases from zone 1 to 3, hence the earthquake intensity.

On June 5, 2015, the Sabah earthquake struck Ranau, Sabah with a magnitude of 6.0. 18 people were confirmed death, while the other 11 people were wounded. The earthquake has caused $2.84 billion of damage to the building and infrastructure, as well as changes to the geological structures [3]. Universiti Malaysia Sabah (UMS) geologist, Prof. Dr. Felix Tongkul predicted that more minor earthquakes would occur in Sabah in future [4]. In fact, two minor earthquakes occurred in Tawau and Ranau in early September 2017 [5]. Therefore, the aftereffect and precautions should be investigated especially when most buildings in Sabah were designed using British Standards where the earthquake component in design was not taken into consideration.

In this research, two existing buildings, namely Tun Mustapha Residential College of Universiti Malaysia Sabah (UMS) and Staff Quarters of Desa Wawasan Tambunan Secondary School, were modelled using the PROTA software. Three PGA values were employed in the study which include 0.145, 0.097, and 0.048 g. This research consisted of two objectives. The first objective was to examine the structural deformation of the building by measuring the lateral displacement and the interstorey drift ratio (IDR) of the buildings. Then, the performance level of the buildings was examined using FEMA 356.

13.2 Methodology

13.2.1 Building Description

The two buildings investigated in this study include the 3-storey building of Tun Mustapha Residential College of UMS and the 5-storey building of Staff Quarters of Desa Wawasan Tambunan Secondary School. Table 13.1 shows the properties of the buildings. The map locations of Tun Mustapha Residential College of UMS and Staff Quarters of Desa Wawasan Tambunan Secondary School are shown in Figs. 13.1 and 13.2, respectively. Meanwhile, the 3D models of the buildings are shown in Fig. 13.3.

Table 13.1 Building properties

Building description	Tun Mustapha Residential College of UMS	Staff Quarters of Desa Wawasan Tambunan Secondary School
No. of storey	3	5
Type of construction	RC frame structure	RC frame structure
Building function	Residential	Residential
Concrete strength	30 N/mm^2	30 N/mm^2
Steel strength	275 N/mm^2	275 N/mm^2
Building area	290.4 mm^2	268.0 m^2
Height of building	9.6 m	16.0 m
Beam size	Range from (150–230) mm × (300–600) mm	Range from (150–230) mm × (450–600) mm
Column size	Range from 230 mm × (125–300) mm	Range from (225–1300) mm × (225–825) mm
Slab thickness	Range from 100 to 150 mm	125 mm

Fig. 13.1 Map location of Tun Mustapha Residential College of UMS

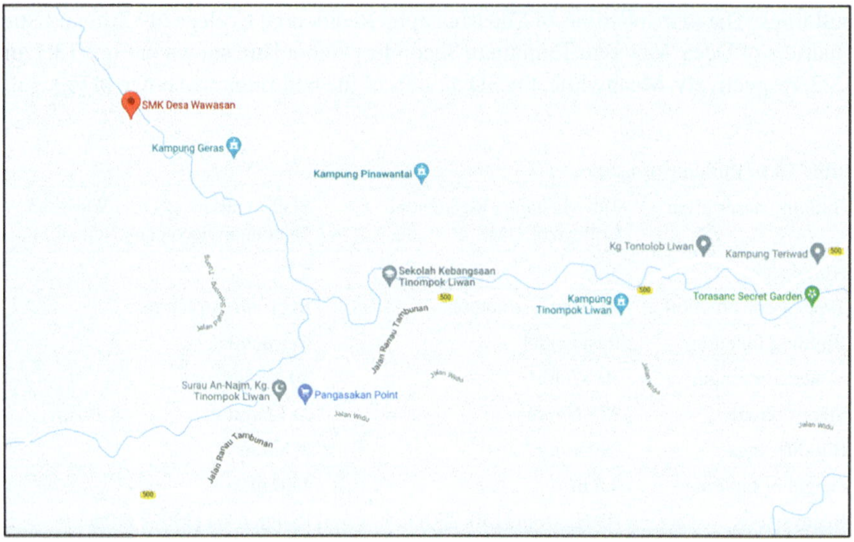

Fig. 13.2 Map location of Staff Quarters of Desa Wawasan Tambunan Secondary School

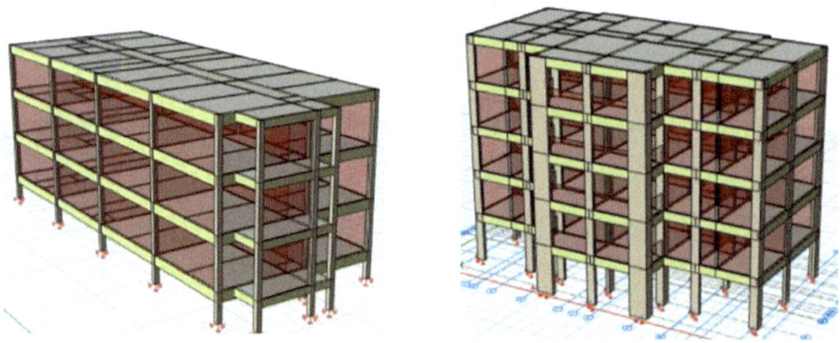

(a) Tun Mustapha Residential College of UMS

(b) Staff Quarters Desa Wawasan Tambunan Secondary School

Fig. 13.3 3D modelling from PROTA

13.2.2 Development of Buildings

After the buildings were modelled, structural analysis was performed to compare the design and displacement (x and y-axis) of the buildings under seismic load. Three PGA values were used including 0.145, 0.097, and 0.048 g. The performance level of the buildings was examined through the lateral displacement in the x- and y-axis directions.

Table 13.2 Lateral displacement of Tun Mustapha Residential College of UMS

PGA	0.048 g		0.097 g		0.145 g	
Storey	*x*-axis	*y*-axis	*x*-axis	*y*-axis	*x*-axis	*y*-axis
0	0.000	0.000	0.000	0.000	0.000	0.000
1	0.284	5.780	0.574	11.680	0.858	17.460
2	0.558	11.243	1.127	22.720	1.685	33.964
3	0.685	13.583	1.384	27.448	2.068	41.031

13.2.3 Examination of the Performance Level of Building

Examination of the performance level of a building after an earthquake is crucial to determine whether the building is still safe to be used, requires repair work, or must be demolished. Therefore, in the second part of this study, FEMA 356 (2000) was used to examine the performance level of the buildings. There are three categories of performance level according to FEMA 356 (2000) which are immediate occupancy (IO), life safety (LS), and collapse prevention (CP). Every performance level was measured based on the interstorey drift ratio (IDR) generated from the response spectrum analysis in PROTA.

13.3 Results and Discussion

13.3.1 Lateral Displacement of Tun Mustapha Residential College of UMS

The lateral displacement of Tun Mustapha Residential College of UMS at the *x*- and *y*-axis for different PGA values is summarized in Table 13.2. The table shows that the lateral displacement increased from the bottom to the top of the storey for both axes. It was also discovered that the displacement of the building increased as the PGA value increased. From the table, the highest lateral displacement occurred at a PGA value of 0.145 g where the 3rd storey was affected the most with 2.068 mm and 41.031 at the *x*- and *y*-axis, respectively. The deformation of the building under seismic load in the *x*- and *y*-axis is presented in Fig. 13.4.

13.3.2 Lateral Displacement of Staff Quarters of Desa Wawasan Tambunan Secondary School

Table 13.3 shows the lateral displacement of Staff Quarters of Desa Wawasan Tambunan Secondary School for different PGA values. A similar with the previous

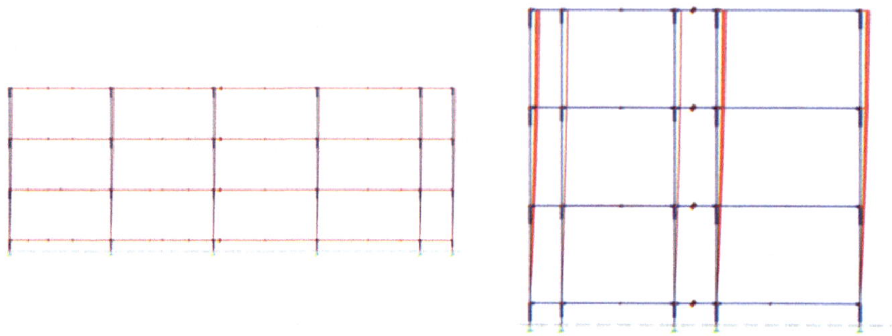

Fig. 13.4 Analytical model of Tun Mustapha Residential College of UMS

Table 13.3 Lateral displacement of Staff Quarters of Desa Wawasan Tambunan Secondary School

PGA	0.048 g		0.097 g		0.145 g	
Storey	*x*-axis	*y*-axis	*x*-axis	*y*-axis	*x*-axis	y-axis
1	0.151	2.898	0.403	7.835	1.153	18.732
2	0.288	7.894	1.177	11.889	1.865	34.783
3	0.300	11.664	1.202	18.487	2.502	45.151
4	0.594	19.709	2.190	26.619	3.284	49.328
5	1.063	20.920	3.127	35.067	4.190	52.987

building, the lateral displacement increased from the bottom to the top of the storey for both axes. Similar pattern was also observed with the displacement as the PGA value increased from 0.048 to 0.145 g. The highest lateral displacement occurred at a PGA value of 0.145 g where the 5th storey was affected the most with 4.19 mm and 52.987 mm at the *x*- and *y*-axis, respectively. The deformation of the building under seismic load in the *x*- and *y*-axis is presented in Fig. 13.5.

13.3.3 Result Summary of Lateral Displacement of the Buildings

The lateral displacement of both buildings in the *x*- and *y*-axis is illustrated in Figs. 13.6 and 13.7, respectively. It shows that the displacement in the Staff Quarters of Desa Wawasan Tambunan Secondary School was slightly higher than the Tun Mustapha Residential College of UMS. This is because the Staff Quarters of Desa Wawasan Tambunan Secondary School are taller; hence, they tend to deform more when subjected to seismic load. This finding is similar with the studies carried out

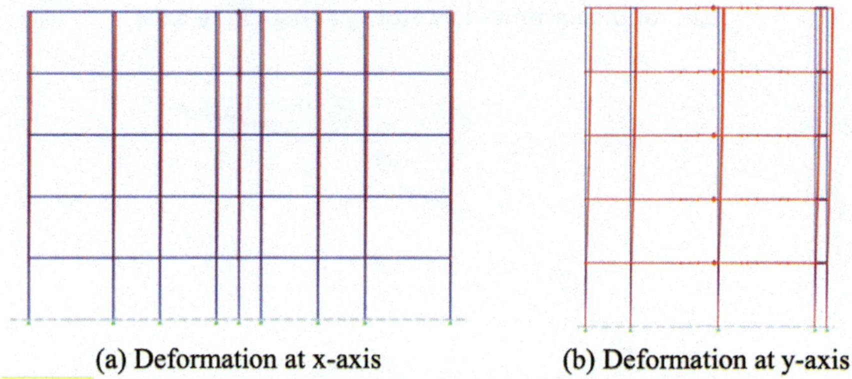

Fig. 13.5 Analytical model of Staff Quarters of Desa Wawasan Tambunan Secondary School

by Gaikwad and Pujari [6] and Khan and Raza [7] where they said that the storey drifts increased as height of the structure increased.

On the other hand, the displacement in the *y*-axis is larger than in the *x*-axis. As shown in Figs. 13.6 and 13.7, there were more bays in the *x*-axis than the *y*-axis. In addition, according to Khan and Raza [7], the displacement and interstorey drift ratio decreased with the increased number of bays. This is related to the present findings where there was a larger displacement occurred in the *y*-axis than the *x*-axis.

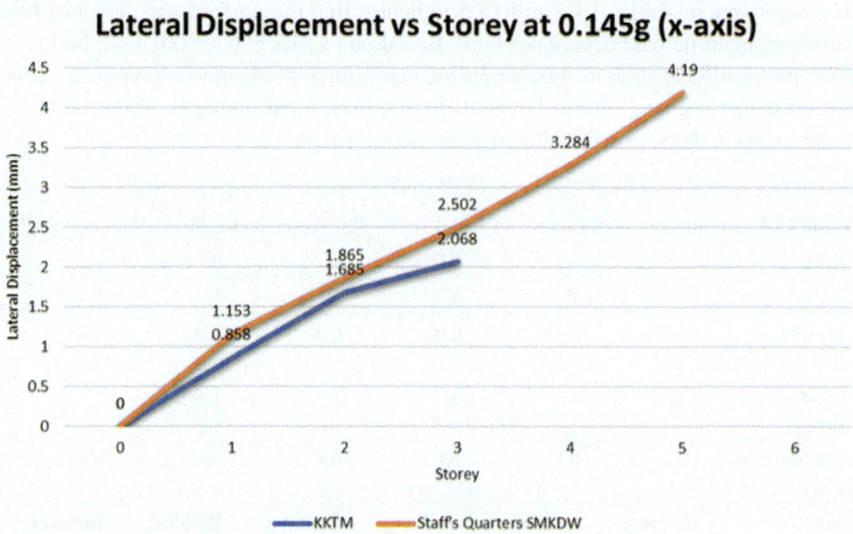

Fig. 13.6 Lateral displacement versus storeys at 0.145 g (*x*-axis)

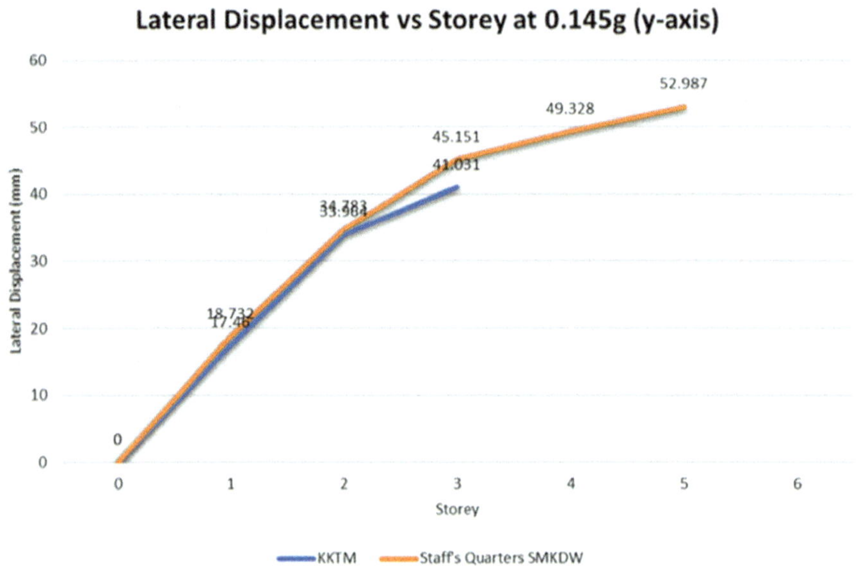

Fig. 13.7 Lateral displacement versus storeys at 0.145 g (*y*-axis)

13.3.4 Interstorey Drift Ratio (IDR)

The summary in Tables 13.4 and 13.5 indicates that the performance level of both buildings is immediate occupancy (IO). Based on FEMA 356 (2000), both buildings have minor effects such as minor hairline cracking at the beam-column area. Thus, the buildings are still safe to be used. In addition, most damages in the buildings were found at the area where the highest value of IDR was located.

Table 13.4 Summary of performance level of 3-storey Tun Mustapha Residential College of UMS

PGA	0.048 g		0.097 g		0.145 g	
Axis	*X*	*Y*	*X*	*Y*	*X*	*Y*
Maximum IDR value (%)	0.009	0.18	0.04	0.37	0.03	1.00
Most affected storey	1 & 2	1	3	1	1 & 2	1
Performance level	IO	IO	IO	IO	IO	IO
Note	Building safe to be used	Building safe to be used	Building safe to be used	Building safe to be used	Building safe to be used	Building safe to be used

IO Immediate occupancy

Table 13.5 Summary of performance level of 5-storey Staff Quarters of Desa Wawasan Tambunan Secondary School

PGA	0.048 g		0.097 g		0.145 g	
Axis	*X*	*Y*	*X*	*Y*	*X*	*Y*
Maximum IDR value (%)	0.01	0.25	0.03	0.26	0.04	0.59
Most affected storey	1, 4 and 5	4	3, 4, and 5	5	1	1
Performance level	IO	IO	IO	IO	IO	IO
Note	Building safe to be used	Building safe to be used	Building safe to be used	Building safe to be used	Building safe to be used	Building safe to be used

IO Immediate occupancy

Since the number of bays was less at the *y*-axis than the *x*-axis for both buildings, minor damage was easier to identify at the *y*-axis. It was also clear that the range of IDR values at the *y*-axis is larger than the IDR value at the *x*-axis for both buildings.

13.4 Conclusions

From the analysis of the buildings, it was found that the lateral displacement was directly proportional to the height of the building. It was also discovered that bay plays a significant role in increasing the strength of building where the higher the number of bays, the higher the seismic resistance of the building. Lastly, the performance level for both buildings was immediate occupancy (IO) where the maximum IDR value of the buildings was below 2.0%.

Acknowledgements This study was supported by the Dana Kluster Penyelidikan (DKP0087) Grant, Universiti Malaysia Sabah, under "Dispersion and Mixing Method of Synthetic Fibre Reinforced Cementitious Mortar".

References

1. Razak, S.K.S.A., Adnan, A., Abas, M.R.C., Wong, S.L., Zainol, N.Z., Yahya, N., Rizalman, A.N., Mohamad, M.E.: A historical review of significant earthquakes in region surrounding Malaysia. In: International Conference on Durability of Building and Infrastructures (2018)
2. Loi, D.W., Raghunandan, M.E., Swamy, V.: Revisting seismic hazard assessment for Peninsular Malaysia using deterministic and probabilistic approaches. Nat. Hazards Earth Syst. Sci. **18**, 2387–2408 (2018). https://doi.org/10.5194/nhess-18-2387-2018

3. Adiyanto, M.I., Majid, T.A., Nazri, F.M.: Nonstructural damages of reinforced concrete buildings due to 2015 Ranau earthquake. AIP Conf Proceed (2017). https://doi.org/10.1063/1.4993403
4. Tongkul, F.: An overview of earthquake science in Malaysia. ASM. Sci. J. **14**, 12 (2021). https://doi.org/10.32802/asmscj.2020.440
5. Inus, K.: More earthquakes expected for Sabah in near future: UMS. New Starits Time (2017)
6. Gaikwad, A.A., Pujari, A.A.B.: Seismic analysis of low rise, mid rise and high rise RCC structure on sloping ground. Int. J. Eng. Technol. **06**(07), 1357–1365 (2019)
7. Khan, K., Raza, S.: Seismic performance assessment of masonry infilled reinforced (RC) frame structures. Int. J. Comput. Civ. Struct. Eng. **6**(1), 40–57 (2015). https://doi.org/10.6088/ijcser.6005

Chapter 14
Review of the Use of Wind and Solar Energy for Powering Small- and Medium-Sized Vessels, the Case of Indonesia

Ahmad Nasirudin, I Ketut Aria Pria Utama, and Yuda Apri Hermawan

Abstract Decarbonization has become a major issue in the last three decades following the increase of atmospheric temperature worldwide. Among others, transportation sectors have contributed to the disaster quite significantly; land and air transportation have dominated the cause of about 22% contribution of air pollution and emissions. However, sea transportation has shown an increase trend and it is reported that it is about 3–4% in 2014 and this will increase sharply if proper actions are not taken seriously which are predicted 2–3 times in 2050. Several initiatives have been suggested to decrease emissions by adopting non-fossil fuel sources such as wind and solar energy. This includes the installation of sails, solar panels, wind turbines, and similar technologies. This paper reviews the potential of wind and solar energy for small- and medium-sized vessel and describes the current condition and its development of the use of wind and solar energy for small- and medium-sized vessels in the case of Indonesia.

Keywords Decarbonization · Wind and solar energy · Small- and medium-sized vessel

---

A. Nasirudin (✉) · I. K. A. P. Utama · Y. A. Hermawan
Faculty of Marine Technology, Department of Naval Architecture, Institut Teknologi Sepuluh Nopember, Surabaya 60111, Indonesia
e-mail: anasirudin@na.its.ac.id

I. K. A. P. Utama
e-mail: kutama@its.ac.id

Y. A. Hermawan
e-mail: yuda.apri@its.ac.id

A. Ismail et al. (eds.), *Exploring Diversity in Engineering and Technology for Knowledge and Innovation*, Advanced Structured Materials 215,
https://doi.org/10.1007/978-3-031-64330-9_14

14.1 Introduction

Climate change has become a very important issue in the last few decades. Starting in 1988, the United Nations (UN) established the Intergovernmental Panel on Climate Change (IPCC) with the task of conducting academic studies related to this issue. In 1992, the United Nations Framework Convention on Climate Change (UNFCCC) was formed which is a forum for drafting agreements related to climate change in the form of an annual conference. In 1995, the first conference was held in Berlin, Germany, called COP-1 (Conference of the Parties). Important agreements were made during COP-3 which was held in 1997 in Kyoto, Japan, that produced the Kyoto Protocol. This Protocol contains the commitments of member countries to reduce greenhouse gas (GHG) emissions. Further important agreements were made at COP-21 2015 in Paris, France, that produced the Paris Agreement. This agreement contains a joint commitment so that by 2040 the earth would not experience a temperature increase of 2 °C or even try not to reach a temperature increase of 1.5 °C compared to the pre-industrial period (1850–1900) [1].

The Paris Agreement requires each member to establish and implement a National Determined Contribution (NDC) which includes three things, namely mitigation, adaptation, and funding. The latest conference, COP-26, was held in 2021 in Glasgow, England, with the result of reaffirming the commitment to the implementation of the Paris Agreement related to the NDC that has been set by each member. Indonesia has set the NDC with a commitment that by 2030 it will reduce GHG by 29% which is equivalent to 834 million tons of CO_2 independently or 41% which is equivalent to 1185 million tons of CO_2 if there is international assistance. This is also stated in Presidential Regulation number 98 of 2021 concerning the Implementation of Carbon Economic Values for Achieving Nationally Determined Contribution Targets and Control of Greenhouse Gas Emissions in National Development [2]. In addition, Indonesia's commitment to reducing GHG emissions will also be supported by a funding scheme in the G20 program, which in this year (2022) Indonesia as presidency.

The sea transportation sector is one of the sectors projected in the GHG emission reduction program in Indonesia. Some of the Indonesian government's initiatives to reduce GHG emissions stipulated in the NDC, particularly in the sector of sea transportation, include providing electrical facility for ships berthing at ports, implementing biofuels, and utilizing renewable energy sources. This paper reviews the potential of wind and solar energy and the potential of the use of wind and solar energy for small- and medium-sized vessels and describes the current condition and its development of the use of wind and solar energy for small- and medium-sized vessels in Indonesia.

14.2 Methodology

The method used in this paper is to conduct a literature review related to energy potential, especially solar and wind energy in Indonesia. Furthermore, a review is also carried out on the potential utilization of solar energy and wind energy on small- and medium-sized vessels operating in Indonesia. Lastly, a review is conducted of the current condition of small- and medium-sized vessels that have been operating in Indonesia as well as studies that are being developed. The literatures that are reviewed includes the journal papers, reports, articles, and news. Some conclusions based on the review are then drawn.

14.3 Results and Discussion

Indonesia is located at the equator and has a great advantage in terms of wind and solar energy potential. Based on [3, 4] as given in Table 14.1, it is known that from 153 wind measurement sites in Indonesia, around 19 sites (covering Banten, Jakarta, Central Java, Yogyakarta, East Nusa Tenggara, West Nusa Tenggara, Southeast Sulawesi, North Sulawesi, and South Sulawesi) have very good potential with wind speed above 5 m/s or equivalent to a wind power density of more than 150 W/m^2. Around 34 sites (covering Bali, Bengkulu, East Nusa Tenggara, South Sulawesi, and Central Sulawesi) have wind speed between 4–5 m/s which is equivalent to a wind power density of 75–150 W/m^2. Regarding to the potential of solar energy in Indonesia, based on the study [5] as given in Table 14.2, it is known that the average daily solar irradiation in Indonesia is about 5.12 kWh/m^2, exceeding the world average of 5 kWh/m^2. Even the maximum solar irradiation can reach 7.54 kWh/m^2 for Kupang city which is on the equator.

As an archipelagic country, Indonesia requires domestic vessels to transport goods and people that connect the islands for supporting economic growth. Small- and medium-sized vessels are dominated where around 42% of domestic shipping departures are represented by small passenger and vehicle carriers, fishing boats, tugs, and offshore service vessels which are equal to around 28% of total energy demand of vessel departure [6]. Besides that, Indonesia is the second largest fishing nation in the world after China [7].

The current condition, in Indonesia, vessels that use wind and solar energy as energy sources are not many. The small- and medium-sized vessels that previously used wind energy as an energy source have now switched to combustion engines. Some vessels that are still using wind energy as power are limited to sports or competition purposes only. For vessels that utilize solar energy, currently, it has been widely applied for fishing vessels which are used as an energy source of electrical equipment such as lamps. The example of those vessels is fishing vessels at Natuna [8] and Bangka Island (Fig. 14.1). In addition, currently vessels start to apply solar energy for propulsion energy sources. Small fishing vessels at Pekalongan Central

Table 14.1 Wind speed and power density [3, 4]

Potential	Wind speed at 50 m (m/s)	Wind power density at 50 m (W/m^2)	Total sites	Province
Poor	< 3	< 45	66	Maluku, Papua, Sumba, Mentawi, Bengkulu, Jambi, East Nusa Tenggara, West Nusa Tenggara, Southeast Sulawesi, North Sulawesi, and North Sumatra
Low	3–4	< 75	34	Central Java, Maluku, Yogyakarta, Lampung, South Kalimantan, East Nusa Tenggara, West Nusa Tenggara, South Sulawesi, North Sulawesi, Central Sulawesi, and North Sumatra
Fair	4.1–5	75–150	34	Central Java, East Java, Yogyakarta, Bali, Bengkulu, East Nusa Tenggara, West Nusa Tenggara, South Sulawesi, and Central Sulawesi
Good	> 5	> 150	19	Banten, Jakarta, Central Java, Yogyakarta, East Nusa Tenggara, West Nusa Tenggara, Southeast Sulawesi, North Sulawesi, and South Sulawesi

Java and Bali are those examples. At Pekalongan, a fishing vessel has a length of 7.25 m, a width of 1.8 m, and a tonnage of 3 GT. The vessel was installed with one 160 W and eight 120 W solar panels. The vessel is equipped with a 3 kW 48 V DC motor and a lithium battery configuration consisting of one 12 V/80 Ah battery and four 48 V/50 Ah batteries. This system configuration is believed to be able to operate for 8 h under normal operating conditions [9]. Another small fishing vessel that uses solar panels as a source of energy for propulsion is the "Surya Namaskar." It is Bali's traditional fishing boat which was developed as a solar-powered fishing boat by the Azura Marine company. It is reported that the boat has been sailed over 2000 km around Indonesia without using a single drop of fuel [10].

Research related to the application of wind and solar energy, especially for small and medium ships, is still being developed such as studies regarding vessel that fully utilizes wind energy as a propulsion energy source for fishing vessels [11, 12], solar energy as a propulsion system for a small passenger and tourism vessel [13, 14], and solar energy as energy source for electricity of small fishing vessels [15–20].

Table 14.2 Solar irradiation (kWh/m^2/day) [5]

City	Jan	Feb	Mar	Apr	May	Jun	Jul	Aug	Sept	Oct	Nov	Dec
Aceh	4.76	4.91	4.94	4.88	4.84	4.68	4.58	4.62	4.56	4.32	4.19	4.74
Medan	4.65	4.77	4.83	4.6	4.42	4.34	4.22	4.38	4.38	4.23	4.09	4.37
Padang	4.89	4.82	4.82	4.93	4.94	4.87	4.94	4.78	4.69	4.57	4.54	4.34
Riau	5.41	5.85	6.06	5.54	4.87	5.02	5.21	5.15	4.75	4.39	3.99	4.63
Jambi	4.85	4.86	4.88	4.69	4.69	4.61	4.71	4.88	4.85	4.59	4.18	4.64
Palembang	4.57	4.57	4.78	4.66	4.73	4.49	4.79	4.8	4.6	4.46	4.39	4.47
Bengkulu	4.54	4.69	4.69	4.71	4.7	4.73	4.83	5.24	5.13	4.8	4.47	4.52
Lampung	4.64	4.77	4.94	4.85	4.92	4.87	5.06	5.12	5	4.67	4.48	4.43
Belitung	4.74	4.79	4.69	4.42	4.37	4.75	5.18	5.14	4.84	4.54	4.46	4.44
Jakarta	4.57	4.65	4.85	4.95	4.96	5	5.07	5.21	5.42	5.4	4.84	4.74
Bandung	4.57	4.75	4.87	4.95	5.02	4.97	5.17	5.35	5.11	4.77	4.7	4.96
Semarang	4.85	5.04	5.14	5.15	5.21	5.59	6.1	6.64	6.21	5.05	4.9	5.15
Yogyakarta	4.37	4.72	4.8	4.65	4.52	4.56	4.93	5.4	5.61	5.13	4.98	4.52
Surabaya	4.64	4.84	4.9	4.81	4.64	4.71	5.24	5.81	5.83	5.03	4.85	4.79
Banten	4.74	4.96	4.94	4.77	4.88	4.98	5.43	5.77	5.52	4.88	4.8	4.95
Bali	5.21	5.5	5.64	5.14	5	5.29	5.84	6.11	6.1	5.55	5.29	4.9
Lombok	4.99	5.29	5.46	5.04	5.05	5.33	5.82	6.16	6.19	5.66	5.4	4.67
Kupang	5.56	5.96	6.37	5.78	5.96	5.88	6.7	7.16	7.54	7.41	6.68	4.6
Pontianak	5.17	5.17	5.11	5.08	5.03	4.98	5.31	5.3	5.2	4.99	4.85	5.27
Palangkaraya	4.97	4.92	4.86	4.81	4.8	4.77	5.01	4.96	4.95	4.7	4.64	5.01
Banjarmasin	5.04	5.05	5.03	4.92	4.84	4.88	5.29	5.51	5.27	4.66	4.75	4.77
Samarindaa	4.66	4.88	4.99	4.98	4.89	4.76	4.76	4.87	4.92	5.04	4.8	4.42
Manado	5.61	5.77	6.04	6.24	6	5.65	5.87	6.53	6.61	6.19	5.69	5.59
Palu	5.24	5.34	5.43	5.28	5.46	5.2	5.7	5.84	5.6	5.22	4.98	5.67
Makassar	5.3	5.47	5.74	5.99	5.96	5.92	6.41	6.74	6.65	5.51	4.92	5.36
Kendari	4.64	4.8	4.66	4.63	4.44	4.27	4.69	6.05	6.3	5.46	4.8	5.61
Gorontalo	5.26	5.38	5.43	5.31	5.1	5.15	5.48	5.6	5.42	5.13	5	4.7
Ambon	5.52	5.57	5.49	5.37	5.17	5.16	5.3	6	6.02	6.25	6.2	6.04
Ternate	5.73	6	6.08	5.73	5.36	5.4	6.04	6.32	6.23	6	5.75	5.14
Jayapura	4.95	5	4.97	4.9	4.8	4.76	4.89	4.99	5	4.93	4.87	4.57

Fig. 14.1 Fishing vessel with solar panel at Bangka Island

14.4 Conclusion

The potency of solar and wind energy in Indonesia is very huge. With around 42% of domestic shipping departures are dominated by small and medium size vessels which are equal to around 28% of total energy demand of vessel departure, it means that the utilization of an alternative energy such as solar and wind energy is one way out in reducing GHG. Currently, there are still not many, especially for small- and medium-sized vessels that have implemented the use of wind and solar energy, but with initial implementation on some fishing vessels and some research that continues to be developed, it is a good first step for further development.

References

1. IPCC: Global warming of 1.5 °C. https://www.ipcc.ch/sr15/ (2018). Accessed 10 Aug 2022
2. Ministry of the State Secretariat Republic of Indonesia: Perpres no. 98 tahun 2021, Penyelenggaraan Nilai Ekonomi Karbon untuk Pencapaian Target Kontribusi yang Ditetapkan Secara Nasional dan Pengendalian Emisi Gas rumah Kaca dalam Pembangunan Nasional. https://jdih.setneg.go.id/Produk (2021). Accessed 10 Aug 2022
3. Martosaputro, S., Murti, N.: Blowing the wind energy in Indonesia. Energy Procedia **47**, 273–282 (2014)
4. Pristiandaru, D.L., Pambudi, N.A.: Wind energy in Indonesia: current status, potential, challenge, opportunities, and future policy. Indones. J. Energy. **2**(2), 65–73 (2019)
5. Rumbayan, M., Abudureyimu, A., Nagasaka, K.: Mapping of solar energy potential in Indonesia using artificial neural network and geographical information system. Renew. Sustain. Energy Rev. **16**, 1437–1449 (2012)

6. Global Maritime Forum: Indonesia: fuelling the future of shipping. https://www.globalmaritimeforum.org/publications/indonesia-fuelling-the-future-of-shipping (2021). Accessed 10 Aug 2022
7. Global Marine Commodities: Indonesia. https://globalmarinecommodities.org/en/indonesia-2/ (2022). Accessed 10 August 2022.
8. Sahputra, Y. E.: Panen Matahari di Atas Kapal Nelayan Natuna. https://www.mongabay.co.id/2021/12/01/panen-matahari-di-atas-kapal-nelayan-natuna/ (2021). Accessed 10 Aug 2022
9. Bernardi, R.: Melihat Kapal Nelayan Bertenaga Surya di Kota Pekalongan. https://news.detik.com/berita-jawa-tengah/d-5077245/melihat-kapal-nelayan-bertenaga-surya-di-kota-pekalongan (2020). Accessed 10 Aug 2022
10. Mélot, J.: This Balinese traditional fishing boat runs on solar power. https://makeachange.world/read/balifishingboatrunsonsolar (2019). Accessed 10 Aug 2022
11. Nugroho, A.A., Trimulyono, A., Rindo, G.: Studi Penggunaan Layar Sebagai Sumber Utama Penggerak Kapal. J. Teknik Perkapalan. **4**(1), 172–181 (2016)
12. Manik, P., Hadi, E.S.: Desain Kapal Ikan Dengan Bentuk Lambung Catamaran Yang Menggunakan Sistem Penggerak Layar Dan Mesin Untuk Muatan Ikan Hidup. Teknik **31**(1), 75–83 (2010). https://doi.org/10.14710/teknik.v31i1.1758
13. Nasirudin, Hamdan, A.: Design concept of catamaran passenger solar power boat for Gili Ketapang Island, Probolinggo—Indonesia. In: Proceeding of the 3rd international conference on marine technology, Institut Teknologi Sepuluh Nopember, Surabaya, 5–6 Dec 2018
14. Nasirudin, A., Hasanudin, H., Utama, D., Tahwoto, L.P.: Determination of PV power and battery capacity size for a leisure solar powered boat at Kalimas River, Surabaya, Indonesia, Kapal. J. Mar. Sci. Tech. **17**(3), 123–129 (2020). https://doi.org/10.14710/kapal.v17i3.31965
15. Manullang, S., Eneste, M.A., Buwono, A.: Analisis Potensial Pemanfaatan Energi Surya Pada Kapal Ikan di Pantai Selatan Pulau Jawa Sebagai Sumber Energi. J. Riset Kapal Perikanan **11**(1), 49–47 (2021). https://doi.org/10.29244/jrisetkapal.1.1.49-57
16. Asri, P., Widodo, H.A., Nugraha, A.T., Rachman, I.: Desain hybrid panel Surya Dan generator set Pada Kapal Ikan Pesisir Selatan Jawa. J. Inovtek Polbeng **12**(1), 46–53 (2022). https://doi.org/10.35314/ip.v12i1.2387
17. Sudjasta, B., Montreano, D., Prayito, S.: Pemanfaatan Energi Surya pada Kapal Penangkap Ikan 10 GT sebagai Sarana Perlistrikan Alternatif di PPI Cituis Kabupaten Tangerang. J. Ilmiah Giga. **22**(2), 60–67 (2019). https://doi.org/10.47313/jig.v22i2.768
18. Pulungan, A.B., Sardi, J., Hastuti, H., Islami, S., Hamdani, H.: Pemasangan Solar Cell Untuk Kapal Nelayan, Itecoms. J. Inf. Tech. Comp. Sci. **2**(2), 53–58 (2019). https://doi.org/10.31539/intecoms.v2i2.811
19. Saputra, I.P.I., Kumara, I.N.S., Partha, C.G.I.: Perancangan PLTS Untuk Perahu Nelayan Tradisional Sebagai Pengganti Genset. J. Spektrum. **6**(4), 102–109 (2019). https://doi.org/10.24843/SPEKTRUM.2019.v06.i04.p15
20. Daging, I.K., Alirejo, M.S., Antara, I.P.W., Dwiyatmo, E.F., Wahyu, T.: Design of solar power plants as a source of electricity for small scale fishery, in Pangkep District, South Sulawesi. J. Kelautan dan Perikanan Terapan. **2**(1), 33–40 (2019)

Chapter 15
A Study of Ramp Shapes Configuration for Wave Energy Converter

Muhammad Faris Roslan, Mohd Azlan Musa, Mohamad Fadhli Ahmad, Mohd Asamuddin A Rahman, Yasser Waleed Sayed Ahmed Eissa, and Mohd Huzmin Mohamed Salleh

Abstract Ocean wave energy has a positive impact on the development of wave energy converters (WECs). Overtopping breakwater for energy conversion (OBREC) is a current study device that integrates the wave energy converter with the existing breakwater. The efficiency of the device is determined by the overtopping wave collected in the reservoir which drives the low-head turbines. Previous researchers thoroughly investigated the effectiveness of devices with linear ramp shape configuration. However, the ramp has provided a low energy performance, particularly in mild wave conditions such as Malaysia, and this provides opportunities to further research into a variety of topics, including ramp shapes. Thus, the purpose of this study is to extend existing knowledge by experimentally investigating the performance of various geometrical ramp configurations in Malaysia environmental characteristics using polynomial equations. Various wave parameters and the effect of relative depth are used to obtain data on the device's hydraulic performance. The results indicate that the cubic ramp shape is more effective in capturing overtopping wave and is 21.5% more efficient in reducing reflected wave compared to the linear

M. F. Roslan · M. A. Musa (✉) · M. F. Ahmad · M. A. A. Rahman · Y. W. S. A. Eissa · M. H. M. Salleh
Faculty of Ocean Engineering Technology, University Malaysia Terengganu, 21300 Kuala Terengganu, Terengganu, Malaysia
e-mail: mohdazlan@umt.edu.my

M. F. Roslan
e-mail: m.farisroslan@gmail.com

M. F. Ahmad
e-mail: fadhli@umt.edu.my

M. A. A. Rahman
e-mail: mohdasamuddin@umt.edu.my

Y. W. S. A. Eissa
e-mail: yasserws998@gmail.com

M. H. M. Salleh
e-mail: p3154@pps.umt.edu.my

A. Ismail et al. (eds.), *Exploring Diversity in Engineering and Technology for Knowledge and Innovation*, Advanced Structured Materials 215,
https://doi.org/10.1007/978-3-031-64330-9_15

shape. In addition, to compare the results with previous studies, this research used a non-dimensionless parameter; the breaker parameter and described the behavior of the structure against wave steepness. As a result, this research provides a good potential concept with a new finding on ramp parameter designs to empower wave energy and improve coastal defense structure, especially in Malaysian conditions.

Keywords Wave energy converter · Ramp shape · Breakwater · Overtopping wave

15.1 Introduction

The recent challenge to reduce the emission of carbon and dependence on fossil fuels has triggered the great interest of researchers for renewable energy [1]. The energy consumption management depicts society's culture, and most of the country today, by studying their probability, is aimed at extracting all potential renewable energy [2]. Renewable energy is the energy that can be obtained from biomass, hydropower, solar, geothermal, wind and oceans that is constantly replenished from natural sources [3]. Among them, ocean energy has enormous potential to be extracted in the form of currents, waves, tides and heat to satisfy the total global power demand [4].

The wave energy converter (WECs) is a power extraction device for ocean waves. Several concepts, including overtopping devices (see Table 15.1), oscillating water systems and wave-activated bodies, have been developed and explored for the maximization of wave energy converters [5]. It is currently under development and remains an immature phase in which over 1000 WECs are worldwide patterned [6]. The main obstacle in wave power harvest is the structure's reliability to withstand harsh weather at sea and reduce costs [4].

A recent innovative idea has been proposed to reduce costs while at the same time enhancing the reliability of WECs through integration into coastal defense structures, such as rubble-mound breakwater and vertical structures. The hybrid system called as overtopping breakwater for energy conversion (OBREC), has enhanced

Table 15.1 Current (OTD) concept of devices

Types	Location
Wave plane (WP)	Floating—offshore
Tapper channel energy plant (TAPCHAN)	Fix—coastal
Spiral reef overtopping device (OWEC)	Floating—offshore
Wave Dragon (WD)	Floating—offshore
Wave catamaran (WAVECAT)	Floating—offshore
Composite sea wall (CSWEC)	Fix—coastal
Sea slot cone generator (SSG)	Fix—coastal and floating—offshore
Overtopping breakwater for energy conversion (OBREC)	Fix—coastal

Fig. 15.1 OBREC concept [14]

the coastal defense principle by sharing advantages, utilizing wave energy in order to generate income and strengthening WECs' reliability by means of the strong shoreline protection structure [7].

OBREC captured wave power by using concrete ramps to enhance the overtopping wave and stored it in the reservoir, turning the low-head hydraulic turbines into electricity generation [8]. The performance is driven by a strongly linked overtopping activity; high overtopping will reduce reflection of the wave because more dissipation occurs [9]. Since OBREC utilized the overtopping phenomenon to accumulate wave power, it is extremely useful to enhance the device performance. An improvising approach is to optimize OBREC's geometric shape or configuration [10, 11]. In addition, overtopping is significantly related to the wave run-up activity over the front structure [12]; roughness, angle of slope, length, permeability and shapes [13]. Heterogeneous ramp designs lead to dissimilar energy outputs from surges and influence the capture of the incoming wave [12] (Figs. 15.1, 15.2 and 15.3).

Hence, this study aims to analyze the hydraulic performance of different geometrical ramp shapes of OBREC that could enhance the performance of the structure. Hydraulic performance is focused on the overtopping discharge and the reflection coefficient caused by the structure. The overtopping discharge can be calculated by measuring the volume of incoming wave captured in the reservoir while for the reflection coefficient, the resolution technique was used. This study employed the CFD technique and experimental work to achieve its goals by referring previous researchers (Table 15.2).

15.2 Methodology

In this study, the improvement of the OBREC's designed is through the modification of the geometrical ramp shape. Based on previous research, the ramp designs are linear, convex, concave, [15] and the study innovatively proposes the use of cubic ramps, foreseeing enhanced performance outcomes (Table 15.3).

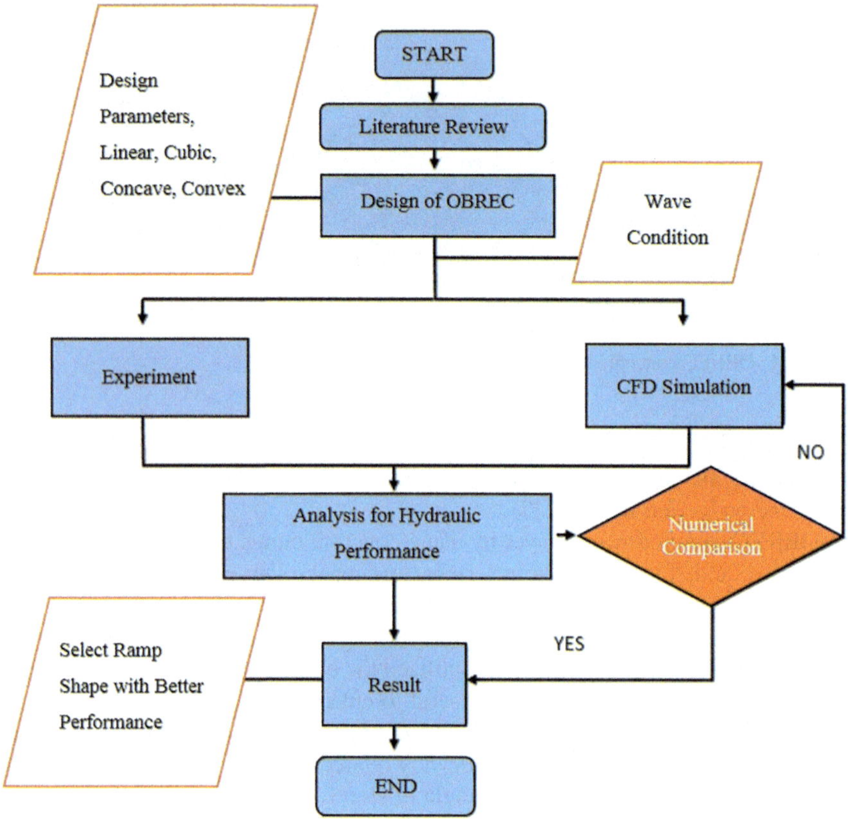

Fig. 15.2 Flow chart of the study

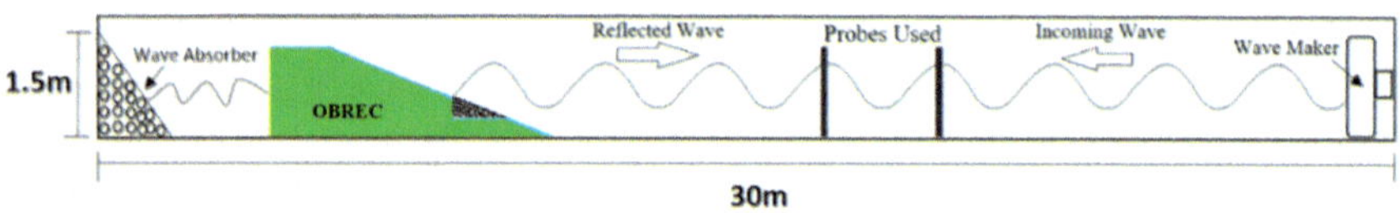

Fig. 15.3 Illustration for experimental setup

Table 15.2 Wave parameter

Type of wave	Wave period Tp (s)	Significant wave height Hs (m)	Depth (D) min–max (m)
R1	1.43	0.0527	0.35–0.45
R2	1.91	0.083	
R3	2.04	0.102	
R4	2.22	0.1173	

Table 15.3 OBREC parameters and ramp shapes

Type of shape	OBREC parameters (1:15)		2D illustration	Polynomial degree
	Crest level (R_r) m	Reservoir width (B_r) m		
Linear	0.12	0.415	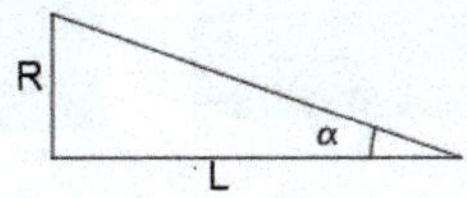	$f(x) = -a_1x + a_0$
Convex			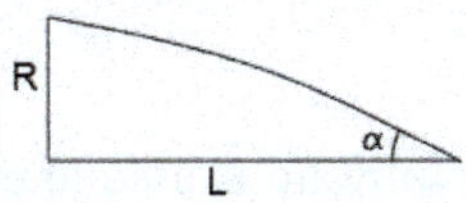	$f(x) = -a_2x^2 - a_1x + a_0$
Concave			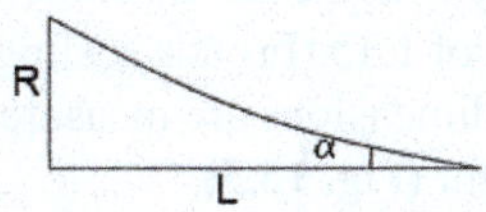	$f(x) = a_2x^2 - a_1x + a_0$
Cubic			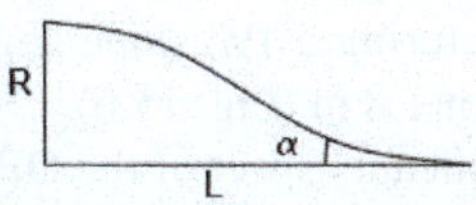	$f(x) = a_3x^3 - a_2x^2 + a_1x + a_0$

15.2.1 *Experimental Setup*

The experimental procedure was undertaken at the Port and Harbor 3D wave basin of the National Hydraulic Research Institute Malaysia (NAHRIM). This basin, characterized by dimensions of 30 m × 30 m × 1.5 m, is outfitted with a series of 30 flat-type paddle wave generators, strategically employed for the generation of linear waves. Additionally, a wave absorber has been affixed behind the basin's infrastructure with the primary objective of mitigating the wave reflection. The acquisition of wave elevation data and the measurement of the coefficient for reflective waves are accomplished through the utilization of a pair of historical probes. Concurrently, water level sensors have been deployed within the reservoir to ensure the ongoing monitoring of the volume of water accrued as a result of overtopping actions. In the endeavor to enhance the performance of the Overtopping breakwater for energy converter (OBREC), this study encompasses an examination of various wave factors and water depth variations (Fig. 15.4).

An Arduino ultrasonic sensor (HC-SR04) and data logger were used to record the changes of water level (h) in the reservoir. The data logger was configured to gather the changes every 0.01 s and could accurately identify changes up to 1 mm in height (Table 15.4).

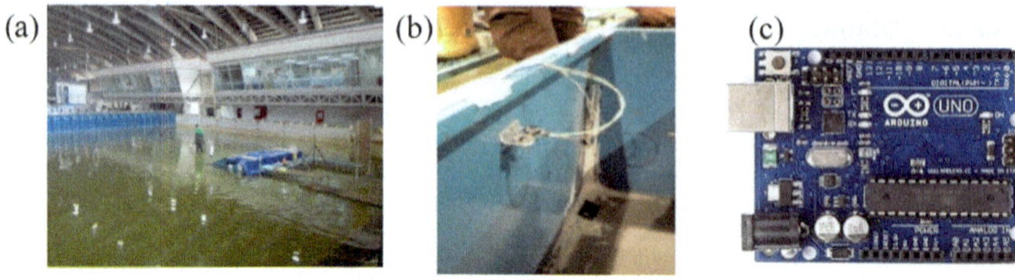

Fig. 15.4 **a** Port and harbor 3D wave basin. **b** Ultrasonic sensor HC-SR04. **c** Arduino Uno data logger

The AutoCAD software is used to create the OBREC model and ramp shapes before being built with a CNC machine. Based on an earlier model created by Vicinanza [6], a scale of 1:15 (Froude scaling) was employed. The experimental model's major structural dimensions are measured as follows: 4 m in width, 3.4 m in length, and 0.7 m in height (Fig. 15.5).

The configurations of the ramps are designed through the utilization of a polynomial equation, featuring a 19° slope angle to enhance performance [5] and a crest freeboard height of 1.8 m (Fig. 15.6).

The most significant aspect of the OBREC research is to comprehend the potential for harnessing power and overtopping wave. The general models accepted for OBREC are shown in Eqs. (15.1) and (15.2) where $q^*_{reservoir}$ is the non-dimensional average overtopping discharge in the reservoir and R_r^* is the relative crest freeboard.

$$q^*_{\text{reservoir}} = \frac{q_{\text{reservoir}}}{\sqrt{gH_s^3}} \tag{15.1}$$

$$R_r^* = \frac{R_r}{H_s} \tag{15.2}$$

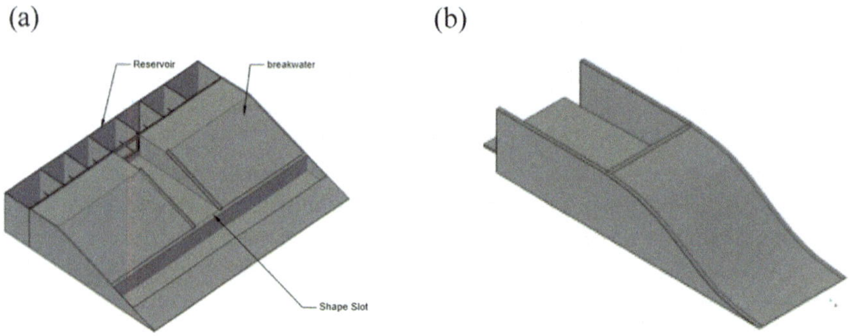

Fig. 15.5 AutoCAD design of: **a** OBREC model and **b** ramp shape

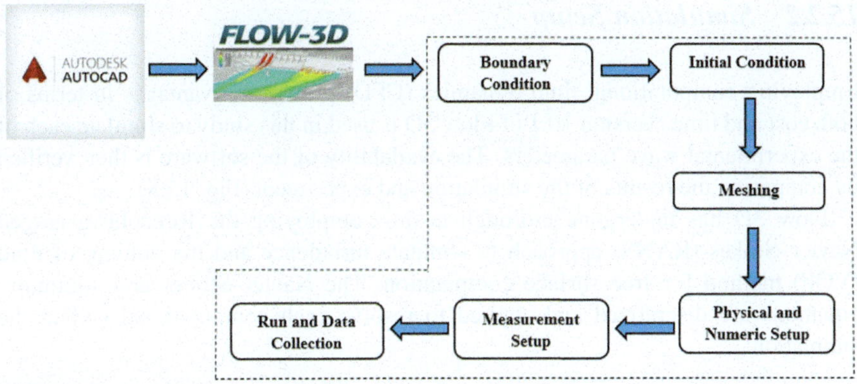

Fig. 15.6 Flow chart for simulation setup

Multiple studies have strived to differentiate incoming and reflected wave elements, offering valuable insights into the effectiveness of various coastal defense structures [16]. This research employed the suggested method by Goda and Suzuki [17] in which pair of historical probes were installed at least one λ from the breakwater's toes to serve as the wave gauge [18]. In order to determine the energies of the resolved incident and reflected wave (E_1 and E_2), wave spectra are formed through the use of the fast Fourier transform (FFT) for the examination and estimation of wave amplitudes (Fig. 15.7).

$$K_R = \sqrt{E_1/E_2}. \tag{15.3}$$

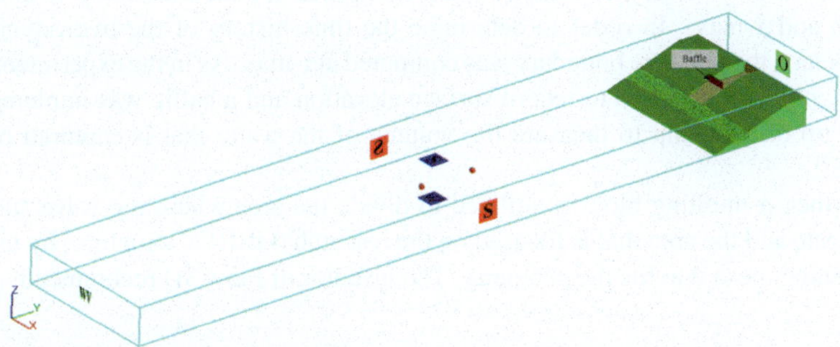

Fig. 15.7 Boundary condition setup and baffle

15.2.2 Simulation Setup

Employing computational fluid dynamics (CFD) provides advantages in terms of both cost and time. Version 10.1 of Flow 3D is used in this study to simulate each of the experimental wave parameters. The availability of the software is then verified by comparing the results of the simulation and experiment (Fig. 15.8).

Flow 3D has undergone thorough testing, employing the Reynolds-averaged Navier–Stokes (RANS) approach to simulate turbulence and the volume of fluid (VOF) method for free surface computation. The Navier–Stokes and continuity equations are discretized and applied to resolve each computational cell in the simulation.

$$\frac{V_f}{\rho}\frac{\partial \rho}{\partial t} + \frac{1}{\rho}\nabla \cdot \left(\rho\, \vec{u}\, A_f\right) = -\frac{\partial V_f}{\partial t} \tag{15.4}$$

$$\frac{\partial \vec{u}}{\partial t} + \frac{1}{V_f}\left(\vec{u}\, Af \cdot \nabla \vec{u}\right) = -\frac{1}{\rho}\left[\nabla \rho + \nabla \cdot \left(\tau A_f\right)\right] + \vec{G} \tag{15.5}$$

$$\frac{\partial F}{\partial t} + \frac{1}{V_f}\nabla \cdot \left(F \vec{u}\, A_f\right) = -\frac{F}{V_f}\frac{\partial V_f}{\partial t}. \tag{15.6}$$

The basin served as the simulation's main component, with the breakwater and ramp serving as its two subcomponents. The basin's boundaries were initially built up without a wave generator and measured (30 m length × 4 m width × 1 m height). It was set up to simulate the real condition of the experiment. Wave boundaries were established to initiate with regular waves at the *y*-min (starting point), extending to an outflow boundary at the *y*-max (endpoint). The boundary wall was designated at *Z*-min, while symmetry conditions were configured for the remaining boundaries (*x*-min and *x*-max). In order to determine the time history of the hydrodynamics parameters, the pressure boundary was computed at *z*-max. As in the experiment, two history probes were used to record surface elevation and a baffle was implemented at the tip of the ramp to measure the volume of the water that is captured by the OBREC.

Further, a meshing block is utilized to divide the area where the wave starts to propagate and the area that is focused on the research (OBREC structure) by using a confirming mesh that fits the geometry. The inventor of Flow 3D recommends using

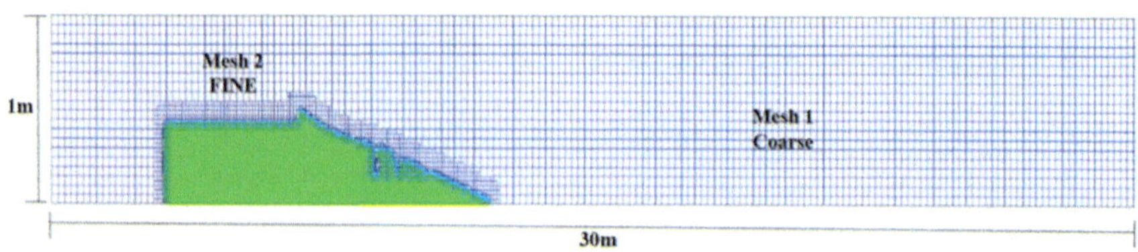

Fig. 15.8 Mesh block setup

Table 15.4 Boundary setup details

Boundary	Type of boundary condition
X_{min}	Symmetry
X_{max}	Symmetry
Y_{min}	Wave
Y_{max}	Outflow
Z_{min}	Wall
Z_{max}	Specified pressure

a ratio of 1:2 between fine and course mesh to get the best results from the simulation and save computing load [19].

The simulation ran for 60 s with a time step of 0.01 s and a RNG turbulence closure was assigned. The fluid initial condition was defined using the fluid region, and the water levels employed were fixed at 0.4, 0.35, and 0.45 m.

15.3 Results and Discussion

In the experiment, the measurement of overtopping started in the tenth number of waves and the time was extended until a stable reading in the reservoir was obtained. Figure 15.9 demonstrates the overtopping wave discharge time series for various ramp designs under wave condition R2. At $t = 0$, the first overtopping wave entered the reservoir, and the water level then progressively elevated. The result shows that, when compared to other designs, the cubic ramp provides maximum overtopping volume discharge followed by convex, linear, and concave. For a 60 s running duration, the total overtopping discharge volume for a cubic ramp shape reaches up to 0.05 m^3. The outcome suggests that geometrical polynomial orders of ramp shapes and overtopping volume discharge have reasonable connections. The concave design captured less overtopping volume compared to the others, probably due to the high resistance or reflection for wave run-up.

Figure 15.10a and b demonstrate the relationship between the overtopping discharge and the wave condition in a non-dimensionless unit. The finding indicated that, even though wave parameter R4 has the highest wave height, overtopping will not increase gradually as wave height increases (R1–R3). This is due to the fact that wave parameter R4 has the highest wave period in the wave parameters assigned which increase wave length and decrease the number of waves created within the first 60 s of running time. Both factors indirectly decrease overtopping waves. This result strongly suggests a correlation between wave height and wave period for the overtopping phenomenon to occur, which is consistent with the majority of the general hypotheses made forth by past researches.

The reflection result starts by contrasting experimental findings for various ramp shapes to the various wave parameter tests, as shown in Fig. 15.11. According to

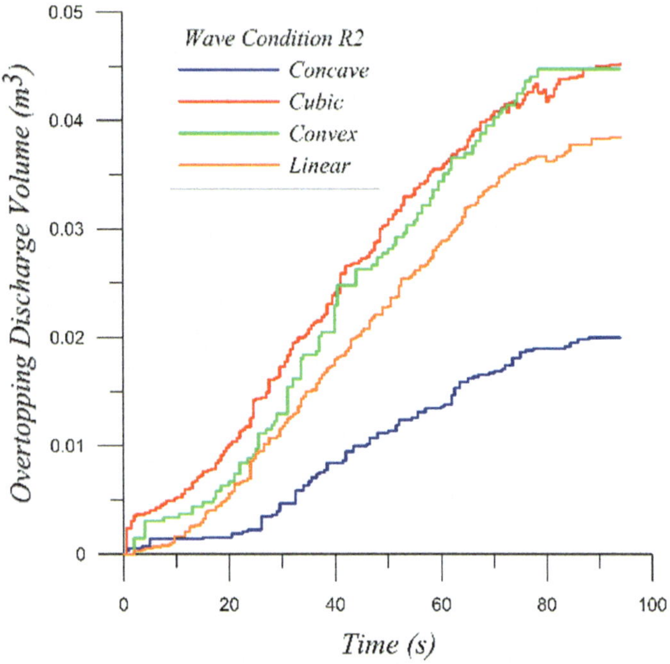

Fig. 15.9 Total cumulative of overtopping discharge volume in time series

prior studies, the use of a linear ramp shape by OBREC allows it to reduce refraction against conventional breakwaters by up to 22% [6]. The study's most surprising finding is that cubic ramp shapes perform better in all tests than linear, concave, and convex shapes. Figure 15.11 illustrates the considerable gap between the cubic ramp shape and the linear ramp shape on the wave parameter R1, where the cubic ramp shape is ahead by 21.5%, the convex ramp shape is ahead by 10% better, while the linear ramp shape is ahead by 13% of the concave shape. Eurotop asserts that the curve shape has greater overtopping capabilities than the linear shape; however, this work has improved OBREC's capability by revealing a new cubic ramp to maximize overtopping and lower the wave reflection [10]. The OBREC's overtopping performance and reflection are closely associated [20], with the run-up and overtopping being inversely correlated with the reflection [9, 21].

Figure 15.12 compares the reflection coefficient results from simulation and experimental studies, showing that both methods have the same trend across all experiments. For the wave parameters R1, R2, and R3, the difference is less than 10% and is not significant, but for R4, the difference is 22%. An explanation for this could be interpolation data errors within the meshing block or fluctuations in the splashing factor on the OBREC ramp, which causes a certain volume of water to be computed incorrectly and alter the outcome [22].

In Fig. 15.13, the relationship between the reflected wave and relative depth (*Hm0/h*) is depicted. The reduction in wave reflection is observed as the depth decreases

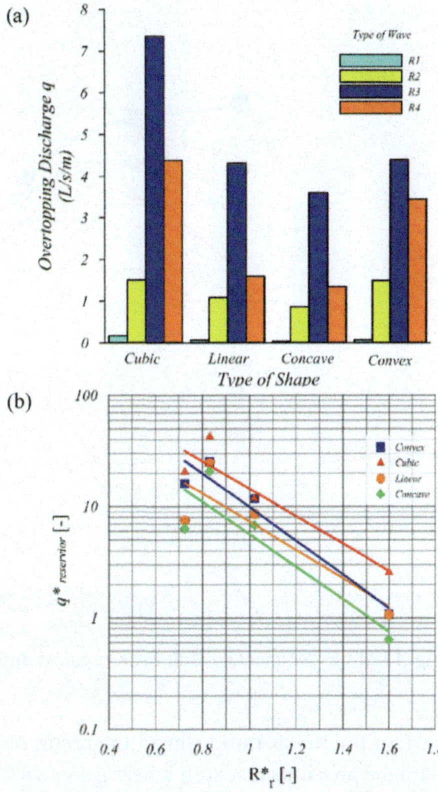

Fig. 15.10 **a** Overtopping discharge vs wave condition. **b** Non-dimensionless overtopping discharge

for cubic and convex shapes, whereas the opposite trend is observed for linear and concave shapes. One possible explanation for this phenomenon is the decrease in water level, leading to a loss of momentum in the wave flux due to a reduction in the weight of water in the swash zone. The cubic and convex shapes exhibit a notable impact on reflection reduction, particularly concerning the curvature angle. The reflection coefficient tends to increase with the slope [8]. At the maximum water level, the area for wave run-up activity diminishes, rendering the differences in the reflection coefficient among various shapes negligible. Despite the anticipation of increased overtopping in this scenario, the design of the ramp does not contribute to reflection reduction. Conversely, an increase in the submerged ramp length results in elevated reflection of waves [8].

The result from this study is compared with the pioneer researcher using a non-dimensionless parameter which is the breaker parameter (ξ). It is also known as surf similarity or Iribarren number to define relations between the coastal defense structure and wave steepness. Figure 15.14 illustrates the comparison of the OBREC model used in this study with previous researchers and the traditional breakwater. The trend for all shapes of this study is in line with [3] where the coefficient increases as the breaker parameter increases and this is supported by other researchers [12]. In the

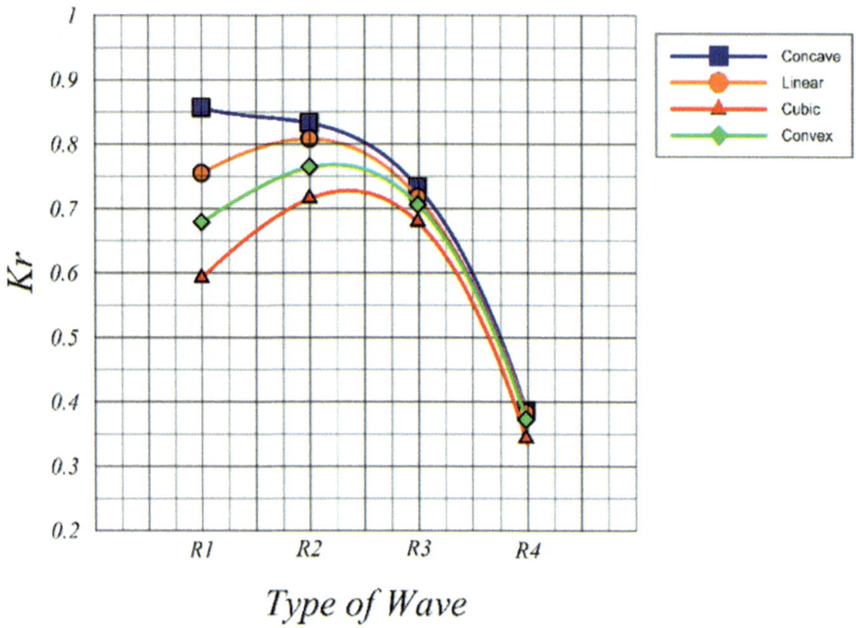

Fig. 15.11 Experimental result for Kr against different types of wave

case of the linear ramp shape, the result obtained from this study shows differences with the previous research where a significant difference is observed for the low ramp and slightly different for the high ramp. The difference is significantly influenced by the height of the front ramp (R_r) which is the vertical distance between the still water level and the tip of the front ramp. The length of ramp or crest height (R_r) used in this study is 0.12 m while the previous researcher used 0.075 m for the low ramp and 0.125 m for the high ramp. Crest height (R_r) is one of the factors that influence the reflection of wave [23]. However, the cubic ramp shape in this study leads the linear ramp with height crest which proves that the curve shape provides high impact on reducing wave reflection and improves the hydraulic efficiency of the OBREC model [24–27].

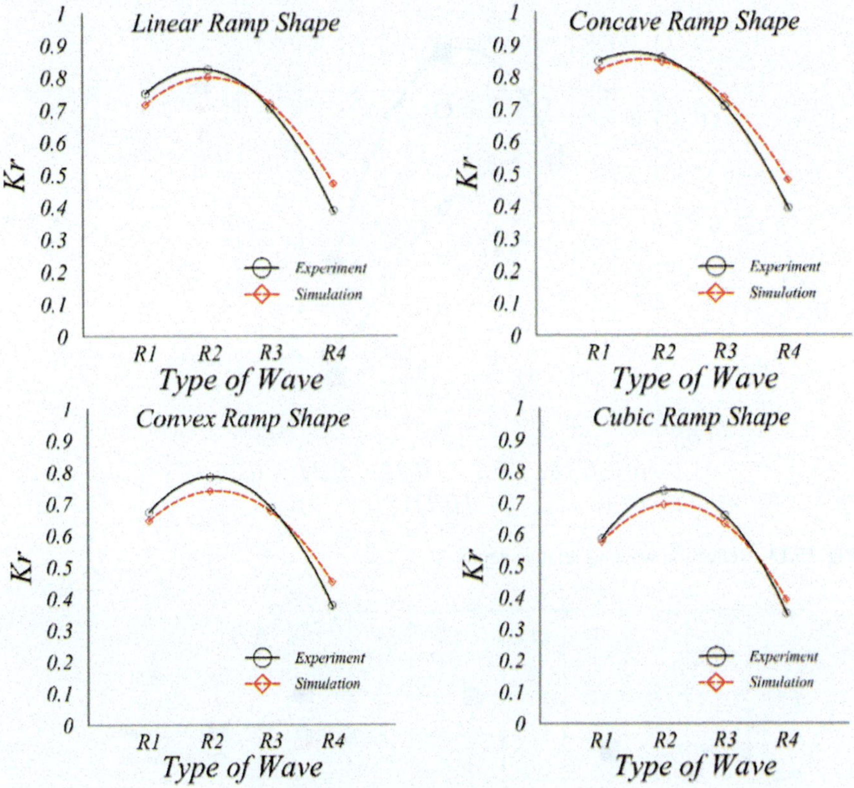

Fig. 15.12 Comparison of experimental and simulation

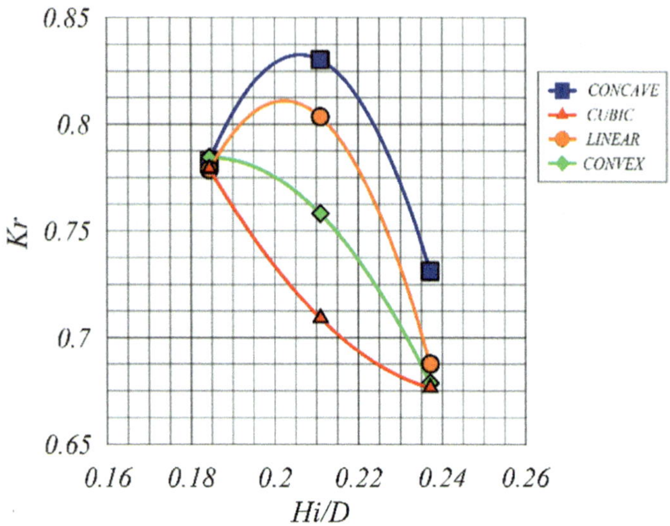

Fig. 15.13 Reflection wave vs relative depth

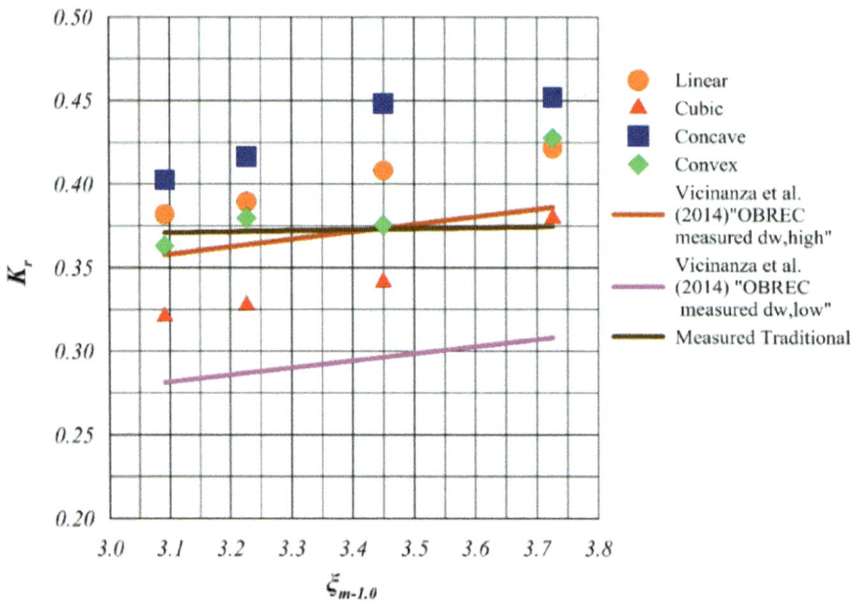

Fig. 15.14 Comparison between reflection coefficient OBREC model in this study with prior model by Vicinanza and traditional rubble-mound breakwater [3]

15.4 Conclusion

This paper presents the experimental work of different geometrical ramp shapes of the OBREC device tested under various circumstances of wave parameters and depth with the comparison of CFD approach. Flow 3D was used appropriately and compared with the experimental outcome to verify the results consent and confirm CFD's accessibility. The relationship of the wave height and wave period on overtopping discharge is acknowledged which then has an effect on the reflection phenomenon. Geometrical ramp shapes play a significant role for hydraulic performance of OBREC, and in this study, the cubic ramp shape has been revealed to have better performance compared to the linear, concave and convex. Although there is an expected rise in overtopping with the elongation of the immersed ramp length, the role of ramp shape in diminishing reflected waves is minimal, particularly at maximum depth. The comparison with previous study on the breaker parameter ($\boldsymbol{\xi}$) also shows the positive outcomes of the curve designed where further studies are required.

Acknowledgements The authors appreciate the financial help provided by the Fundamental Research Grants Scheme (FRGS)/1/2021/TK0/UMT/02/3 of the Ministry of Higher Education of Malaysia, Department of Higher Education. We also want to express our gratitude to the University of Malaysia Terengganu's Institute of Oceanography and Environment (INOS) for the data they have supplied, as well as to the entire personnel of the department's Marine Technology for their invaluable support.

References

1. Pelc, R., Fujita, R.M.: Renewable energy from the ocean. Mar. Policy **26**, 471–479 (2002). https://doi.org/10.1016/S0308-597X(02)00045-3
2. Borthwick, A.G.L.: Marine renewable energy seascape. Engineering **2**, 69–78 (2016). https://doi.org/10.1016/J.ENG.2016.01.011
3. Wan Nik, W., Sulaiman, O., Rosliza, R., Prawoto, Y., Muzathik, A.: Wave energy resource assessment and review of the technologies. Int. J. Energy Environ. **2**, 1101–1112 (2011). https://ijee.ieefoundation.org
4. Mofor, L., Goldsmith, J., Jones, F.: Ocean Energy: Technologies, Patents, Deployment Status and Outlook. International Renewable Energy Agency report. (2014). https://www.ourenergypolicy.org/wp-content/uploads/2015/06/IRENA_Ocean_Energy_report_2014
5. Vicinanza, D., Margheritini, L., Kofoed, J.P., Buccino, M.: The SSG wave energy converter: performance, status and recent developments. Energies **5**, 193–226 (2012). https://doi.org/10.3390/en5020193
6. Vicinanza, D., Contestabile, P., Quvang Harck Nørgaard, J., Lykke Andersen, T.: Innovative rubble mound breakwaters for overtopping wave energy conversion. Coastal Eng. **88**, 154–170 (2014). https://doi.org/10.1016/j.coastaleng.2014.02.004
7. Vicinanza, D., Stagonas, D., Müller, G., Nørgaard, J.H., Andersen, T.L.: Innovative Breakwaters Design for Wave Energy Conversion, pp. 1–12 (2012)
8. Contestabile, P., Vincenzo, F., Di Lauro, E., Vicinanza, D.: Full-scale prototype of an overtopping breakwater for wave energy conversion. Coastal Eng. Proc. **1**, 12 (2017). https://doi.org/10.9753/icce.v35.structures.12

9. Zanuttigh, B., Margheritini, L., Gambles, L., Martinelli, L.: Analysis of wave reflection from wave energy converters installed as breakwaters in harbour. In: Proceedings of the 8th European Wave and Tidal Energy Conference: EWTEC 2009, pp. 384–392 (2009)
10. Goggins, J., Finnegan, W.: Shape optimisation of floating wave energy converters for a specified wave energy spectrum. Renew. Energy **71**, 208–220 (2014). https://doi.org/10.1016/j.renene.2014.05.022
11. Margheritini, L.: R&D Towards Commercialization of Sea Wave Slot Cone Generator (SSG) Overtopping Wave Energy Converter: Selected Topics in the Field of Wave Energy. Department of Civil Engineering, Aalborg University (2009). https://vbn.aau.dk/ws/portalfiles/potal/549441785/R_D_Towards_Commercialization_of_Sea_Wave_Slot_Cone_Generator_SSG_Overtopping_Wave_Energy_Converter
12. Pullen, T., Allsop, N.W.H., Bruce, T., Kortenhaus, A., Schüttrumpf, H., van der Meer, J.W.: Wave Overtopping of Sea Defences and Related Structures: Assessment Manual (2007)
13. Pedersen, J.: Wave Forces and Overtopping on Crown Walls of Rubble Mound Breakwaters: An Experimental Study (1996)
14. Di, E., Lara, J.L., Maza, M., Losada, I.J., Contestabile, P., Vicinanza, D.: Stability analysis of a non-conventional breakwater for wave energy conversion. Coast. Eng. **145**, 36–52 (2019). https://doi.org/10.1016/j.coastaleng.2018.12.008
15. Barbosa, D.V.E., Santos, A.L.G., Santos, E.D., Souza, J.A.: Overtopping device numerical study: openfoam solution verification and evaluation of curved ramps performances. Int. J. Heat Mass Transf. **131**, 411–423 (2019). https://doi.org/10.1016/j.ijheatmasstransfer.2018.11.071
16. Varghese, R.V., Shirlal, K.G., Hari, B., Mohanan, P.S.: Review of developments in estimation of wave reflection from coastal structures. IOSR J. Mech. Civil Eng. **30**, 32 (2016). https://www.iosrjournals.org/iosr-jmce/ppers/ICETEM/Vol.%201%20Issue%202/19-30-32
17. Yoshimi Goda, T.S.: Estimation of incident and reflected waves in regular wave experiments. Ocean Eng. **22**, 77–86 (1995). https://doi.org/10.1016/0029-8018(93)E0011-G
18. Klopman, G., van der Meer, J.W.: Random wave measurements in front of reflective structures. J. Waterw. Port Coast. Ocean Eng. **125**(1), 39–45 (1999). https://doi.org/10.1061/(ASCE)0733-950X(1999)125:1(39)
19. Maliki, A.Y., Musa, M.A., Ahmad, M.F., Zamri, I., Omar, Y.: Comparison of numerical and experimental results for overtopping discharge of the OBREC wave energy converter **12**, 1337–1353 (2017). http://jestec.taylors.edu.my/
20. Zanuttigh, B., van der Meer, J.W.: Wave reflection from coastal structures in design conditions. Coastal Eng. **55**(10), 771–779 (2008). https://doi.org/10.1016/j.coastaleng.2008.02.009
21. Zanuttigh, B., van der Meer, J.W., Andersen, T.L., Lara, J.L., Losada, I.J.: Analysis of wave reflection from structures with berms through an extensive database and 2DV numerical modelling. In: Coastal Engineering 2008: (In 5 Volumes), pp. 3285–3297 (2009). https://doi.org/10.1142/9789814277426_0272
22. Barkhudarov, M.: Lagrangian VOF Advection Method for FLOW-3D, pp. 1–11. Flow Science, Inc. FSI-03-TN6 (2004). https://flow3d.co.kr/wp-content/uploads/FSI-03-TN63-R_lagrangian-vof
23. Hebsgaard, M., Sloth, P., Juhl, J.: Wave overtopping of rouble mound breakwaters. In: International Conference in Coastal Engineering, pp. 2235–2248 (1998). https://doi.org/10.1061/9780784404119.167
24. Musa, M.A., Roslan, M.F., Ahmad, M.F., Muzathik, A.M., Mustapa, M.A., Fitriadhy, A., Mohd, M.H., Rahman, M.A.A.: The influence of ramp shape parameters on performance of overtopping breakwater for energy conversion. J. Mar. Sci. Eng. **8**(11), 875 (2020). https://doi.org/10.3390/jmse8110875
25. Ahmad, M.F., Ramli, M.I., Musa, M.A., Goh, S.E.G., Che Wan Othman, C.W.M.N., Ariffin, E.H., Mokhtar, N.A.: Numerical simulation for overtopping discharge on tetrapod breakwater. In: AIP Conference Proceedings, vol. 2746, issue 1 (2023). https://doi.org/10.1063/5.0153371

26 Musa, M.A., Roslan, M.F., Fitriadhy, A., Eissa, Y.W., Ahmad, S.Z.A.S., Ahmad, M.F.: Numerical simulation of geometrical parameters effect on wavecat energy converter (May 13, 2023). In: Proceedings of the 13th International Conference on Marine Technology (MARTEC 2022). SSRN. https://doi.org/10.2139/ssrn.4447485
27. Musa, M.A., Roslan, M.F., Fitriadhy, A., Albani, A., Mohd, M.H., Rahman, M.A.A., Ahmad, M.F., Nazri, M.N.: Experimental comparison of linear and cubic ramps for hybrid overtopping breakwater energy conversion. J. Sustain. Sci. Manag. **18**(1), 215–227 (2023). https://doi.org/10.46754/jssm.2023.01.013

Chapter 16
Experimental Works on Pit Mining Migration and River Bed Changes

Nur Sara Hazeqqa Johari, Mohd Sofiyan Sulaiman, Nur Ayuni Zaiddin, Muhammad Hafizzudin Muhammad Ali, and Ahmad Fitriadhy

Abstract Sand and gravel are important resources in infrastructure development and construction. Excessive instream sand mining will have an impact on the river ecosystem, particularly the morphology and hydraulic regime. The dynamic nature of the river system has a significant impact on pit migration and bed profile changes. An experimental study in a laboratory flume using a mobile channel bed was carried out to explain the changes in bed profile in the presence of instream pit mining. In the flume, four different pit sizes ranging from 50 to 70 cm in length and 30–50 cm in width were created. Flow velocities were set to range from 0.6 to 1.5 m/s. Three major findings were observed: (i) the nick point at the downstream were eroded and further migrated before equaling the original bed profile; (ii) the deposition occurred in the middle of the pit, indicating the filling mechanism of pit mining; and (iii) the headcutting at the upstream of pit occurred due to over-steepened nick point, which can migrate further upstream from the original designated pit area. These findings were critical for future instream sand mining guidelines, which should account for the possibility of pit migration in the risk analysis.

Keywords Sand mining · Pit migration · Bed changes · Nick point · Deposition

---

N. S. H. Johari · M. S. Sulaiman (✉) · N. A. Zaiddin · M. H. M. Ali · A. Fitriadhy
Faculty of Ocean Engineering Technology, Universiti Malaysia Terengganu, 21030 Kuala Nerus, Terengganu Darul Iman, Malaysia
e-mail: sofiyan@umt.edu.my

N. S. H. Johari
e-mail: nursarahazeqqajohari@gmail.com

N. A. Zaiddin
e-mail: ayunizaiddin@gmail.com

M. H. M. Ali
e-mail: muhdhafizzudinali@gmail.com

A. Fitriadhy
e-mail: a.fitriadhy@umt.edu.my

A. Ismail et al. (eds.), *Exploring Diversity in Engineering and Technology for Knowledge and Innovation*, Advanced Structured Materials 215,
https://doi.org/10.1007/978-3-031-64330-9_16

16.1 Introduction

River sand mining, reservoir construction, and excessive groundwater extraction are a few examples of anthropogenic stresses which might alter river sustainability in the future. These processes are believed to disrupt freshwater and sediment discharge in the river system. The repercussion from those activities will lead to sedimentation, river bank erosion, high concentration of turbid water, saltwater intrusion, and sea-level rise [1, 2]. Sand mining can be associated with removal of sand from the river bank and bed [3]. The mining activities as a filling material for construction are prevalent in most developing and developed countries [4]. In order to fulfill the demands, the mining of sand and gravel were performed using various techniques such as instream mining, dry mining, and bar skimming. Koehnken et al. [5] claim that rivers are a large source of sand and gravel since transportation costs are typically lower and the materials are strong minerals with angular shapes, which are desirable for construction. Kondolf et al. [6] argued that sand mining activities will create the hungry water phenomenon. The phenomenon can be defined as the river flow which has excess transport capacity and more stream power to transport than available sediment in the river [7]. The sediment supply is more limited than the transport capacity which might induce the erosion of river bed and banks.

A physical model was constructed in the hydraulic laboratory to estimate maximum scour depth, effective length of pit migration, and migration speed of river bed. To test those behavior, the mining pit was defined in the model. Sand mining has a variety of effects on the ecosystem at the global level. If hydraulic structures are present close to the mining area, the nick point migration could result in scouring.

The objectives of this study have been set as follows: (i) to perform a physical simulation of a mobile bed with the presence of artificial mining pit and (ii) to observe the nick point migration, deposition pattern, and headcutting in the artificial pit.

16.2 Methodology

Physical modeling using the pre-defined flow was simulated at the National Hydraulic and Instrumentation Laboratory, NAHRIM, which has a length of 100 m, width 1.5 m, and depth 1.5 m. The flume walls are made of glass wall as shown in Fig. 16.1. Bed channel has been constructed using mobile bed sediment with the length of 6 m, width of 1.5 m, and depth of 0.7 m using 100-micron meter grain size. These parameters were made based on previous experimental studies which are using 100–140-micron meter grain size and average size for channel bed that is suitable in the flume [8–11]. The regular and irregular pit has been constructed at 2.5 m from 0.0 m of the channel bed (see Fig. 16.2).

In each set of experiment, the channel bed was leveled, and pit was excavated from the bed with four different pit sizes. The experimental set is illustrated in Table 16.1.

Fig. 16.1 Physical model in the laboratory

Fig. 16.2 Artificial pit mining in the model

Table 16.1 Experimental setup with dimensions of the pits

Pit no	Rotational speed (RPM)	Flow velocity (m/s)	Length (cm)	Width (cm)	Depth (cm)
A	240	0.4–0.94	60	40	10
B	240	0.66–1.49	70	50	10
C	240	0.46–1.57	60	40 30	10
D	240	0.76–1.49	70	40 30	10

During each experiment, the flows were routed for approximately 5 min and then water was drained to ensure the channel bottom is in accordance to the actual natural stream. Initial channel bed elevation was measured using laser distance at a predefined datum elevation. A second phase of the experiment was performed to capture the bed changes in the presence of an artificial mining pit. Water discharge during the experiment was controlled by RPM moderator/machine at 220 RPM about 15 min, and then, the flume was drained to measure the bed elevation after the experiment.

16.3 Results and Discussion

In this study, the elevation for mobile bed was mapped out using the ArcGIS software. The digital elevation model (DEM), 3D plot and contour lines were compared at pre- and post-simulation to observe the difference between them. Apart from those plots, the longitudinal profiles and cross-sectional plots were visualized to assess the pattern of migration and bed change in this study.

It can be observed from Fig. 16.3 that that steep contour lines were slowly vanished at post-simulation plot. The filling processes in the pit are predominant in this plot. Since two artificial pits were created in thin simulation run, the contour lines slowly migrate to the downstream, which are similar to previous study that hydraulic condition of the flow end pit location effect downstream pit migration [12].

The digital elevation model (DEM) in Fig. 16.4 revealed the decreased in bed level as depicted in the min–max legend of elevation. This hypothesis is supported by looking at the longitudinal profile in Fig. 16.5. The maximum elevation is dropped as shown in y-axis. Another important observation in Fig. 16.5 is the migration of the pit downstream from the original pit. Although the filling process is dominant, the elevation at downstream from the pit is decreased due to erosion of the channel bed and hence the migration of the pit. The cross-sectional profiles in Fig. 16.6 depict the same; the filling process occurred inside the mining pit while erosion occurred at the head of the pit.

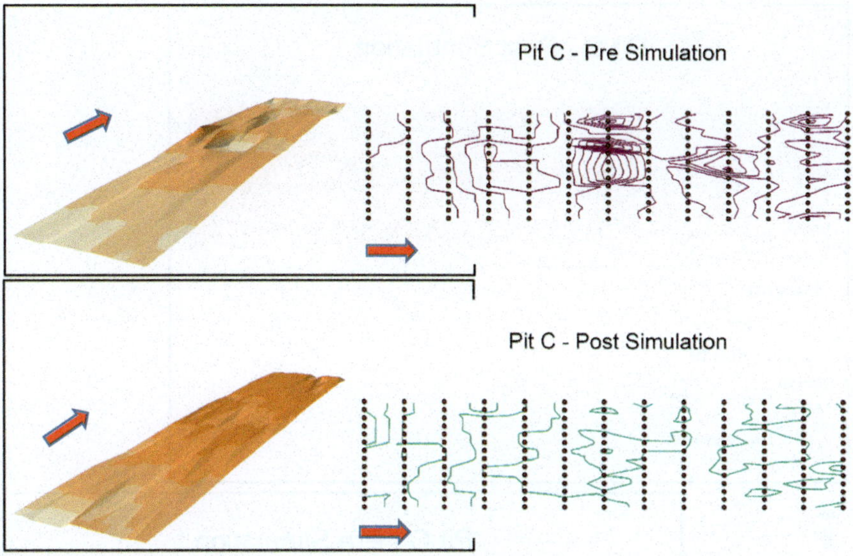

Fig. 16.3 3D and contour plots at pre- and post-simulation

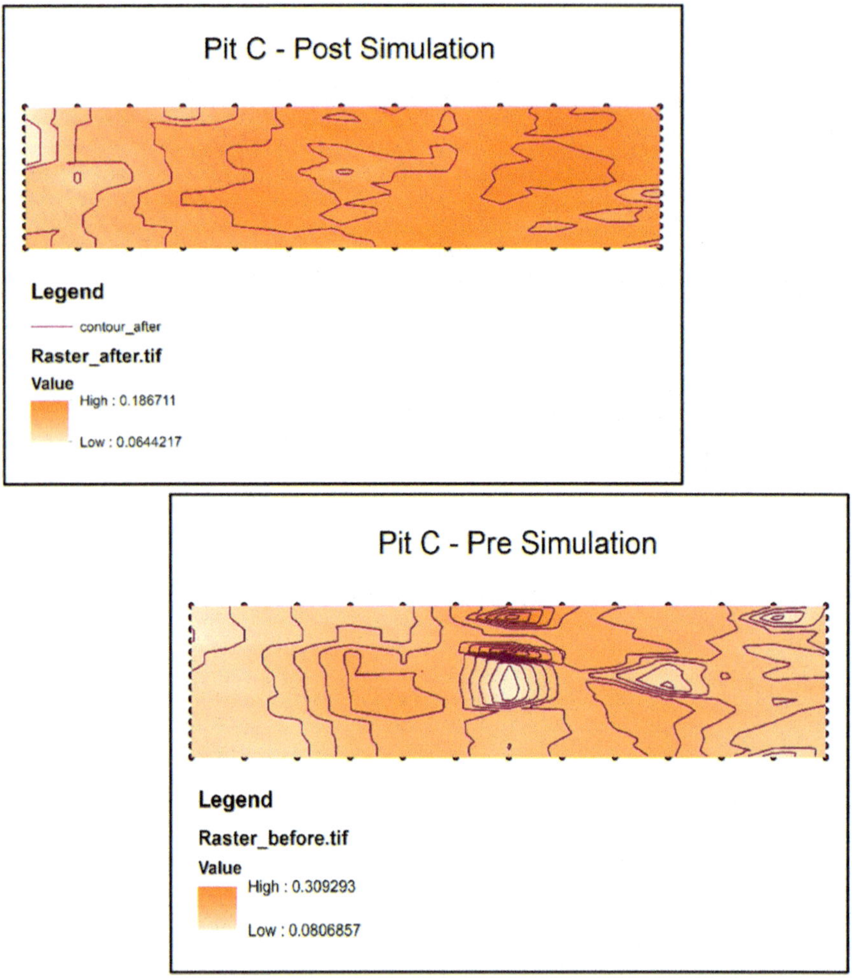

Fig. 16.4 Digital elevation model depicting decreasing of elevation

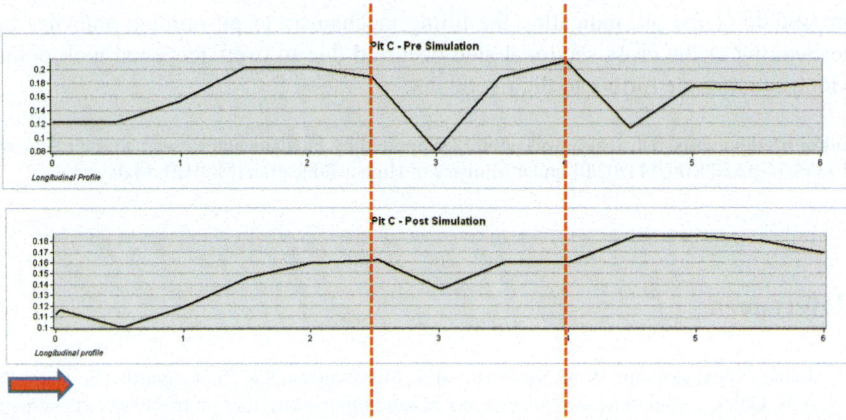

Fig. 16.5 Longitudinal profiles of pit *C*

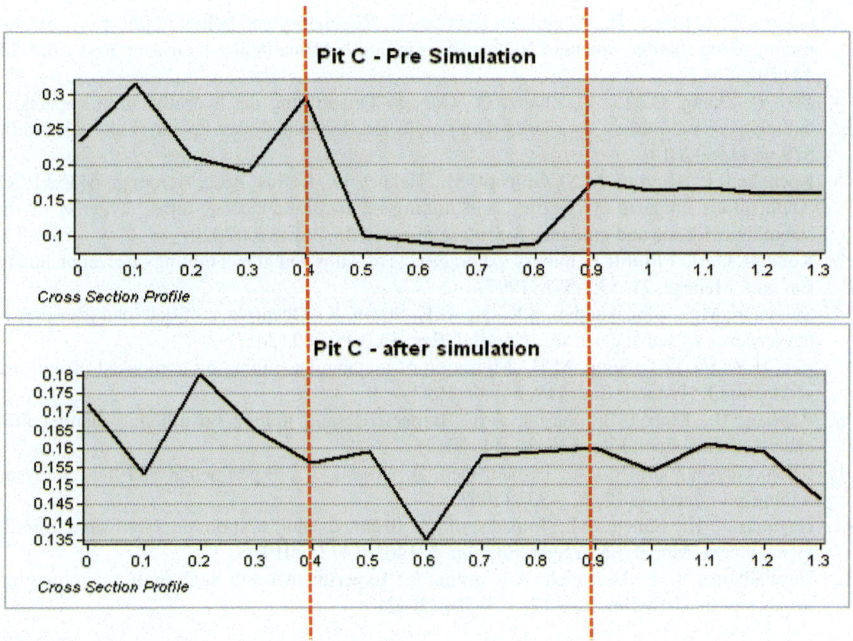

Fig. 16.6 Cross-sectional profiles of pit *C*

16.4 Conclusion

It can be concluded that sand mining does effect the channel bottom. Three major findings were observed: (i) the nick point at the downstream was eroded and further migrated before equaling the original bed profile; (ii) the deposition occurred in the middle of the pit, indicating the filling mechanism of pit mining; and (iii) the headcutting at the cross-sectional side occurred due to over-steepened nick point, which can migrate further to the river bank.

Acknowledgements This study was entirely supported by Fundamental Research Grant Scheme [FRGS/1/2020/TK0/UMT/02/9] under Ministry of Higher Education (MOHE), Malaysia.

References

1. Johari, N.S.H., Zaiddin, N.A., Sulaiman, M.S., Sinnakaudan, S.K., Sathiamurthy, E., Hamzah, A.F.: Quick scoping review on the impact of sand mining activities on freshwater ecosystem, hydraulic regime and river morphology. In: AIP Conference Proceedings, vol. 2484 (2023)
2. Haron, N.A., Yusuf, B., Sulaiman, M.S., Ab Razak, M.S., Abu Bakar, S.N.: Assessing river stability and hydraulic geometry of fluvial river in Malaysia. Int. J. Integr. Eng. **11**, 214–223 (2019)
3. Gondo, T., Mathada, H., Amponsah-Dacosta, F.: Regulatory and policy implications of sand mining along shallow waters of Njelele River in South Africa Jamba-J. Disaster Risk Stud. **11**, 12 (2019)
4. Yao, J., Zhang, D., Li, Y., Zhang, Q., Gao, J.: Quantifying the hydrodynamic impacts of cumulative sand mining on a large river-connected floodplain lake: Poyang Lake. J. Hydrol. **579**, 124156 (2019)
5. Koehnken, L., Rintoul, M.S., Goichot, M., Tickner, D., Loftus, A.C., Acreman, M.C.J.R.R.: Applications, impacts of riverine sand mining on freshwater ecosystems: a review of the scientific evidence and guidance for future research **36**, 362–370 (2020)
6. Kondolf, G.M.: PROFILE: hungry water: effects of dams and gravel mining on river channels. Environ. Manage. **21**, 533–551 (1997)
7. Sulaiman, M.S., Sinnakaudan, S.K., Ng, S.F., Strom, K.: Occurrence of bed load transport in the presence of stable clast. Int. J. Sedim. Res. **32**, 195–209 (2017)
8. Lee, H.Y., Fu, D.T., Song, M.H.: Migration of rectangular mining pit composed of uniform sediments. J. Hydraul. Eng. **119**, 64–80 (1993)
9. Barman, B., Kumar, B., Sarma, A.K.: Impact of sand mining on alluvial channel flow characteristics. Ecol. Eng. **135**, 36–44 (2019)
10. Salehi Neyshabouri, S.A.A., Farhadzadeh, A., Amini, A.: Experimental and field study on mining-pit migration **17**, 323–331 (2002)
11. Haghnazar, H., Saneie, M.: Impacts of pit distance and location on river sand mining management. Model. Earth Syst. Environ. **5**, 1463–1472 (2019)
12. Neyshabouri, S., Farhadzadeh, A., Amini, A.: Experimental and field study on mining-pit migration. Int. J. Sedim. Res. **17**, 323–331 (2002)

Chapter 17
Wheel Over Point Calculation Using Constant Rate of Turn Technique: Assessing Traditional Technique Novelty Through a Simulation Study

Amir Syawal Kamis, Ahmad Faizal Ahmad Fuad, Azmirul Ashaari, and Che Wan Mohd Noor

Abstract This research intends to evaluate the constant rate of turn approach in determining the wheel over point. While mariners navigating the ships are familiar with the aforementioned practice, there is a lack of data in scientific papers supporting the approach's efficacy; thus, its novelty is frequently questioned. This research next conducts manoeuvring analysis using a ship simulator to establish the technique's efficiency. The data obtained through the simulation study, specifically the cross-track distance data, was validated by comparing it to the cross-track limit in three different navigation areas. The approach validated in this study is planned to be used aboard the ship to assist navigators in making more effective course changes to maximise navigational safety.

Keywords Wheel over point · Constant rate of turn · Passage planning · Course optimisation · Course alteration

A. S. Kamis
Rating Programs, Akademi Laut Malaysia, BT 30 Kg Tg Dahan, 78200, Kuala Sungai Baru, Melaka, Malaysia
e-mail: amirsyawal87@gmail.com

A. S. Kamis · A. F. A. Fuad (✉)
Faculty of Maritime Studies, Universiti Malaysia Terengganu, 21030 Kuala Terengganu, Terengganu, Malaysia
e-mail: faizalfuad75@gmail.com

A. Ashaari
Azman Hashim International Business School, Universiti Teknologi Malaysia, Jalan Bertingkat Skudai, 81310 Johor Bahru, Johor, Malaysia
e-mail: azmirul@utm.my

C. W. M. Noor
Faculty of Ocean Engineering Technology and Informatics, Universiti Malaysia Terengganu, 21030 Kuala Terengganu, Terengganu, Malaysia
e-mail: che.wan@umt.edu.my

A. Ismail et al. (eds.), *Exploring Diversity in Engineering and Technology for Knowledge and Innovation*, Advanced Structured Materials 215,
https://doi.org/10.1007/978-3-031-64330-9_17

17.1 Introduction

Figure 17.1 illustrates a recent accident involving the Hong Kong-based containership OOCL Durban and a berthed vessel, YM Constancy, while manoeuvring into its berth at the port of Kaohsiung, which occurred on 3 June 2021. Although the root cause of the turn is still being investigated, the incident can be understood as a direct cause of insufficient turning in confined water, as depicted in Fig. 17.1 [1].

An insufficient turning can occur due to one or a combination of the following reasons, namely (1) late course alteration, (2) insufficient rate of turn (ROT), (3) incorrect identification of wheel over point (WOP), and (4) inadequate knowledge regarding natural effects such as current, wind, and hydrodynamic effects [2].

Insufficient turning can be avoided by good passage planning. A comprehensive passage plan consists of four stages: ‘appraisal’, ‘planning’, ‘execution’, and ‘monitoring’ [3]. The term ‘appraisal’ refers to the procedure of compiling vital information about a voyage from its originating port to its destination port. The compiled data is then translated into a form of a charted route on paper or electronic navigation charts in the ‘planning’ stage, prior to the ‘execution’ and ‘monitoring’ of the passage plan. As seen in Fig. 17.2, the charted route comprises several lines with divergent courses connected by waypoints (WPT).

When altering a course, it is necessary to do it at a sufficient distance; otherwise, the ship may overshoot off the track, which results in a cross-track distance (XTD) [4]. Due to this reason, an accurate WOP is required to be presented for each of the course lines as an indication of the point of alteration [5]. XTD is the perpendicular distance between a ship’s centreline and its original intended track [6]. Although it

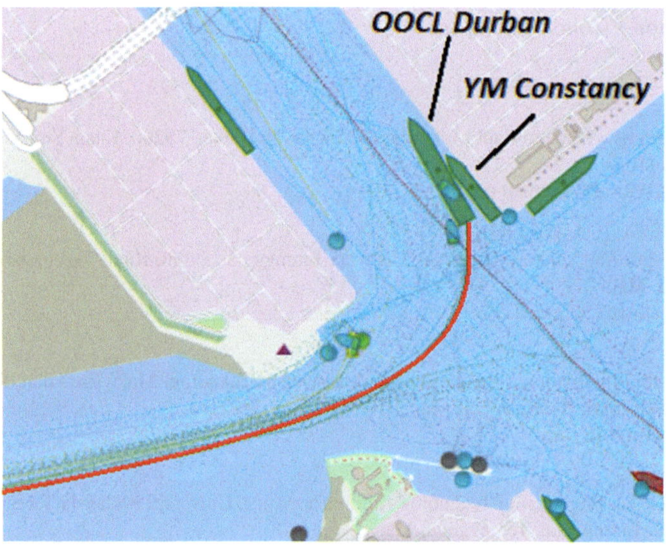

Fig. 17.1 OOCL Durban track prior to collision with YM constancy [1]

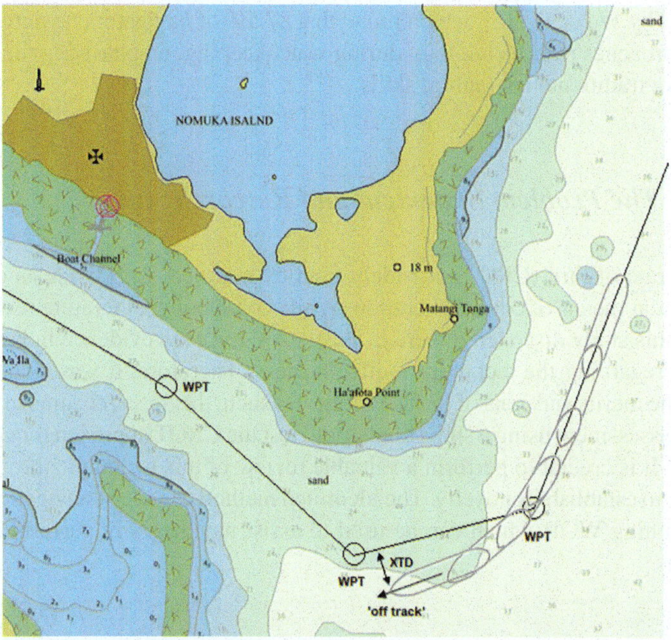

Fig. 17.2 Vessel overshooting a charted course and producing cross-track distance

increases fuel consumption [7], navigating off-track in the open sea is not a threat to navigation safety because there is sufficient sea room to manoeuvre. However, if the ship is in a confined channel, such as a river, a deviation from the intended route might result in incidents such as bottom contact or grounding [2]. Additionally, shallow water can cause a ship to be harder to steer, aggravating the situation [5, 8–11]. As a result, looking from the vessel's safety point of view, the cross-track distance can be considered a critical aspect of navigation, particularly in restricted waterways [4].

17.1.1 Electronic Chart Display Information System for Route Monitoring

The electronic chart display information system (ECDIS), paired with a sophisticated interface through the integrated bridge navigation system (IBNS), has greatly benefited navigation officers in route monitoring [12]. Nevertheless, according to the International Maritime Organisation (IMO), the technology is only required for cargo ships over 3000 gross tonnage and passenger ships over 500 gross tonnage [13]. Therefore, the remaining seafarers are still relying on their repertoire of traditional navigation skills while manoeuvring [14]. This assertion is aligned with research conducted by Wu et al. [15], which found that 87.5% of navigators favoured conventional approaches over technology during watchkeeping, emphasising the necessity to preserve traditional navigation skills.

17.1.2 The Problem Statement and Research Aim

Constant rate of turn (CROT) is widely used by mariners worldwide as one of the most popular means of course alteration. While conducting a literature review of the strategy, this study discovered there is a lack of significant evidence in high-quality literature regarding the technique's effectiveness. Therefore, it was challenging to identify the merits and cons of the technique. As such, this research aims to conduct a practical assessment using a ship simulator to test the CROT approach for calculating the WOP. It is crucial to perform a valuable review of this technique through a ship simulator to establish its novelty. The identified method can assist navigation officers in determining WOP, which can be used to assist navigators in performing course alterations.

17.2 Literature Review

17.2.1 Constant Rate of Turn

CROT is a technique that uses the rate of turn indicator (ROTI), to calculate WOP. A ROTI is a navigational device that indicates the ship's ROT. Excessive ROT might affect ship stability and produce significant rudder torque pressure [16]. For this reason, the CROT technique requires the ship's speed and the ROT to be agreed upon and discussed among the bridge team before it can be executed [5]. If the needle is pointing in the green area, it means that the ship is turning towards the starboard direction; if the needle is pointing in the red area, it means that the ship is turning towards the port direction. The numerical value indicates the rate of the turning.

This method also demands that the helmsman is competent at managing the rudder angle so that the rate of turn does not fluctuate considerably during the turning period [1, 5]. A more extensive application of the rudder angle will cause the ship to turn faster and vice versa.

As seen in Fig. 17.3, the formula of CROT is

$$d_{\text{WOP}} = \frac{\theta.V.0.5}{60 \times \text{ROT}} + F, \tag{17.1}$$

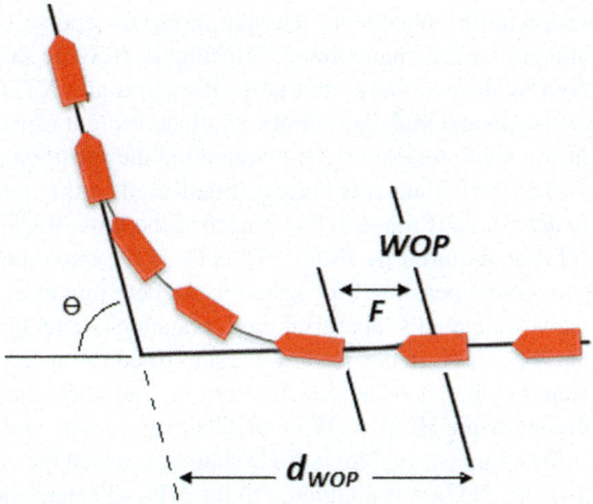

Fig. 17.3 Constant rate of turn

where ROT is the rate of turn in degree (°)/min, V is the ship's speed in knots, θ is the different between initial course and final course in degree (°), and F is the vessel's length overall (LOA) in nautical miles.

The result of the calculation is the distance between WOP and WPT, which must be noted on the course line, as seen in Fig. 17.3.

17.3 Methodology

The research began with a review of the germane literature pertaining to the constant rate of turn method, including how the method should be planned and executed. Prior to commencing the study, it is imperative for all authors to comprehend the notion of constant rate of turn. It is established that the speed (V) and rate of turn (ROT) are required to be discussed as per Eq. (17.1) before the manoeuvring analysis can begin. Subsequently, after a discussion, all authors had agreed collectively that the manoeuvring analysis would be conducted at a speed of 10 knots and a ROT of 10°/min. Following that, the WOP was determined and logged for each 10° turn. As a result, a sequence of courses was charted in the ECDIS simulator for each 10° turn. For this study, an LNG-type ship was selected for the analysis, and it was steered manually along a predetermined route. The course alterations were carried out as the vessel reached the prescribed WOP. The vessel's RPM was regulated appropriately to match a steady speed of 10 knots. As a result, the rudder was controlled to keep the rate of turn at 10°/min in accordance with the agreed ROT. Two designated helmsmen were invited to conduct the exercise. The first helmsman was an Able Body (AB) seaman with nine years of sailing expertise, whereas the second helmsman was a cadet with no prior experience steering an actual vessel. The cadet, however, had completed a formal steering course at a maritime academy. Each helmsman was instructed to manoeuvre the ship along the course line and then begin the course alteration while maintaining a 10°/min ROT. If the ship strayed from the course line, both helmsmen were ordered to disregard any XTD developed during the course alteration and make no further efforts to bring it nearer to the course line; they were instructed to focus only on maintaining the specified rate of turn.

The XTD data was collected and used in this analysis. The research was able to determine if the ship had breached the cross-track limit (XTL) by observing the XTD as required by IMO [17]. XTL is the maximum deviation from its intended course line permitted for a vessel. As determined by the navigation officer during the voyage plan's 'appraisal' and 'planning' stage, the safe zone is contained inside the XTL. XTL in this study was established following the value given in research by Kamis et al. [1] as the selected ship has the same characteristics. The XTL used for this study are HCW = 145.3 m, CN = 1022.3 m, and OS = 1955.8 m.

In summary, the ship's XTD shall not exceed the XTL specified. After collecting data, the XTD was compared to the XTL to determine the degree of efficacy of the constant rate of turn technique.

17.4 Result and Discussion

Throughout the simulation, a total of fifty-four manoeuvrings were performed. Twenty-seven turns were completed by the AB seaman, whereas the other twenty-seven turns were completed by the cadet, as given in Table 17.1.

Table 17.1 TD results versus compliance with XTL

Location	θ	WOP	AB seaman				Cadet			
			XTD result	Comply with XTL?			XTD result	Comply with XTL?		
				HCW	CN	OS		HCW	CN	OS
Kemaman, Malaysia	10	0.24	5	Yes	Yes	Yes	22	Yes	Yes	Yes
	20	0.33	4	Yes	Yes	Yes	39	Yes	Yes	Yes
	30	0.41	9	Yes	Yes	Yes	66	Yes	Yes	Yes
	40	0.49	36	Yes	Yes	Yes	111	Yes	Yes	Yes
	50	0.58	46	Yes	Yes	Yes	135	Yes	Yes	Yes
	60	0.66	106	Yes	Yes	Yes	185	*No*	Yes	Yes
	70	0.74	204	*No*	Yes	Yes	296	*No*	Yes	Yes
	80	0.83	314	*No*	Yes	Yes	389	*No*	Yes	Yes
	90	0.91	463	*No*	Yes	Yes	537	*No*	Yes	Yes
Cumulative XTD and compliances to XTL in %			1187	*67%*	100%	100%	1780	*55%*	100%	100%

17.4.1 *Findings*

17.4.1.1 Expertise Matter

The findings indicate that the helmsman's expertise is crucial while performing this sort of turn by controlling the rudder angle to maintain the ROT. The XTD findings from the manoeuvring study indicated that the AB seaman's manoeuvring complied with 67% of the established XTL in the harbour and confined water. In comparison, the cadet's XTD result complied with the XTL by 55%. In terms of XTD value, it can be observed that the cadet produced a higher XTD value than the AB seaman. This demonstrates that the AB seaman is able to maintain closer proximity to the course line. During the manoeuvring study, it was determined that the cadet struggled to maintain a 10°/min rate of turn while steering the ship owing to inefficient rudder application.

17.4.1.2 Planning of WOP for Harbour and Confined Water, Coastal Water, and Open Sea

During the first phase of this study, it was determined that before the simulation research was carried out, the speed and ROT must be specified. Therefore, it can be inferred that this methodology is appropriate for calculating WOP for the open sea and coastal water since the ship's speed in these areas can be estimated using the ship's typical service speed from past voyages. Similarly, before the voyage begins, the bridge team will review and agree upon the ROT based on their previous experience in steering the ship. With that being said, the ROT in the open sea and coastal navigation is possible to be decided upon by the bridge team; thus, this technique is possible. On the other hand, the authors considered and agreed that throughout pilotage in the HCW, due to the traffic density, manoeuvrability, and environmental factors, the ship's ROT and speed will often be modified corresponding to the pilot's preference. Due to this reason, without knowledge of the speed and ROT, it is impossible to determine the WOP for courses in the HCW while planning prior to voyage commencement. However, for small boats that do not require pilotage services, this technique can still be used using average speed and ROT that the navigators are comfortable with.

17.4.1.3 XTL Compliance in the Various Navigation Area

The XTD was checked to see if it complied with the XTL for each navigation region, as stated in Table 17.1. Even if a ship reacts differently at various water depths, the criterion that determines the success of this study is the helmsman's ability to maintain the ROT via rudder control. Although the turning circle of a ship is influenced by water depth, a 10°/min ROT in HCW, CW, or the OS produces a turning circle that

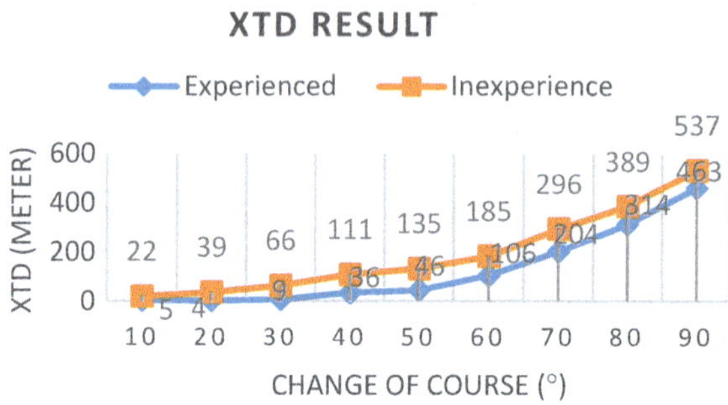

Fig. 17.4 XTD results for the AB and the cadet

is relatively comparable, resulting in a similar XTD. As given in Table 17.1, for both the AB seaman and the cadet, the XTD failed to comply with the XTL in the HCW. On the contrary, compliance with XTL was 100% in CW and the OS regardless of expertise. Thus, based on this complete compliance with the XTL, the result indicated that the CROT approach is acceptable for coastal water and open sea navigation.

17.4.1.4 A Significant Increase for Course Alteration of More Than 50°

Figure 17.4 shows the nonlinear graph of the XTD result for the AB seaman and the cadet.

It can be observed that the XTD increased significantly for an alteration angle of more than 50°. This observation proves that the technique has a gap which causes the XTD to increase drastically after a change of course of more than 50°. It may be possible to close the gap by studying the formula given. The process of bridging the gap will eventually create an improved mathematical model.

17.5 Conclusion

It is important to note that the purpose of this study was to determine the accuracy of a constant rate of turn (CROT) in computing WOP to improve ship navigation safety. It is one way to ensure the ship stays on the course when changing course. While ECDIS has reduced seafarers' reliance on this sort of conventional methodology, it is essential to remember that not all vessels are equipped with ECDIS; hence, the approach remains valid for the majority of mariners who navigate without the technology. The assessment of the CROT was conducted using a ship simulator. The procedure is necessary to demonstrate the technique's novelty since there is a minimal study on its effectiveness, despite being regularly utilised by mariners worldwide.

It can be concluded that the research established the CROT technique's efficacy. Interestingly, this study adds to the body of knowledge by indicating that the CROT methodology is more suited for coastal and open sea navigation than harbour and confined waterways. Additionally, it was discovered that helmsman expertise is critical for optimising CROT implementation.

A better version of the mathematical model can be generated using mathematical analysis to fill the gaps in the CROT formula, not only for safety but also to reduce fuel consumption, as a suggestion for future research.

References

1. Kamis, A.S., Ahmad Fuad, A.F., Anwar, A.Q., Ali, S.A.: Assessing the efficacy of the advance transfer technique in calculating the wheel over point through simulation studies. In: Advanced Structured Materials, pp. 129–144 (2022)
2. Kamis, A.S., Ahmad Fuad, A.F., Anwar, A.Q., Monir Hossain, M.: A systematic scoping review on ship accidents due to off-track manoeuvring. WMU J. Marit. Aff. **39** (2022). https://doi.org/10.1007/s13437-022-00274-2
3. Skora, K., Wolski, A.: Voyage planning. Sci. J. Silesian Univ. Technol. Ser. Transp. **92**, 123–128 (2016). https://doi.org/10.20858/sjsutst.2016.92.12
4. Vujičić, S., Mohović, R., Tomaš, I.Đ.: Methodology for controlling the ship's path during the turn in confined waterways. Pomorstvo **32**, 28–35 (2018). https://doi.org/10.31217/p.32.1.2
5. Kamis, A.S., Ahmad Fuad, A.F., Ashaari, A., Noor, C.W.M., Ali, S.A.: Development of WOP mathematical model for optimum track-keeping. A ship simulation study using VLCC, focusing on hard over rudder turning circle with three stages of validation analysis. Polish Marit. Res. **28**, 156–174 (2021). https://doi.org/10.2478/pomr-2021-0043
6. Kamis, A.S., Ahmad Fuad, A.F., Ashaari, A., Mohd Noor, C.W.: Wheel over point mathematical model. Ocean Syst. Eng. **11**, 203–216 (2021). https://doi.org/10.12989/ose.2021.11.3.203
7. Reid, R.E.: Improvement of Ship Steering Control for Merchant Ships—Phase IIA. Kings Point, New York (1978)
8. Briggs, M.J., Borgman, L.E., Bratteland, E.: Probability assessment for deep-draft navigation channel design. Coast. Eng. **48**, 29–50 (2003). https://doi.org/10.1016/S0378-3839(02)00159-X
9. Masumi, Y., Nikseresht, A.H.: 2DOF numerical investigation of a planing vessel in head sea waves in deep and shallow water conditions. Appl. Ocean Res. **82**, 41–51 (2019). https://doi.org/10.1016/j.apor.2018.10.017
10. Paffett, J.A.H.: Ship manoeuvring characteristics. J. Navig. **26**, 113–122 (1973). https://doi.org/10.1017/S0373463300022943
11. Mucha, P., Dettmann, T., Ferrari, V., el Moctar, O.: Experimental investigation of free-running ship manoeuvers under extreme shallow water conditions. Appl. Ocean Res. **83**, 155–162 (2019). https://doi.org/10.1016/j.apor.2018.09.008
12. Rutkowski, G.: ECDIS limitations, data reliability, alarm management and safety settings recommended for passage planning and route monitoring on VLCC tankers. TransNav, Int. J. Mar. Navig. Saf. Sea Transp. **12**, 483–490 (2018). https://doi.org/10.12716/1001.12.03.06
13. ISES: The majority of international fleet now fitted with ECDIS-ISES Association. https://www.isesassociation.com/the-majority-of-international-fleet-now-fitted-with-ecdis/
14. Marine Insight: Understanding Different Types of Manoeuvres of a Vessel. https://www.marineinsight.com/naval-architecture/different-types-of-manoeuvres-of-a-vessel/
15. Wu, J., Thorne-Large, J., Zhang, P.: Safety first: the risk of over-reliance on technology in navigation. J. Transp. Saf. Secur., 1–28 (2021). https://doi.org/10.1080/19439962.2021.1909681
16. Molland, A.F.: Chapter 8—manoeuvring. Marit. Eng. Ref. B. **19**, 578–635 (2008). https://doi.org/10.1016/B978-0-7506-8987-8.00008-1

17. IMO MSC: Adoption of the Revised Performance Standards for Electronic Chart Display and Information Systems (ECDIS) Msc 82/24/Add.2. MSC Resolution, vol. 82 (2006). https://wwwcdn.imo.org/localresources/en/KnowledgeCentre/IndexofIMOResolutions/MSCResolutions/MSC.232(82).pdf

Chapter 18
Drilling of Bone: Effect of Drill Bit Clearance on Thermal Damage

Mohd Faizal Ali Akhbar, Shahrizan Jamaludin, Razhan Hassan, and Akmal Wani Sulong

Abstract Bone drilling is a procedure in medical fields such as orthopedics, dentistry, and otology for installing screws and implants. However, it generates excess heat that can harm bone cells and jeopardize the stability of the implants. To address this issue, this research assesses the effect of the clearance part—on the drill body—on maximum temperature elevation during bone drilling. The research involves simulating bone drilling surgeries using DEFORM-3D, a commercial finite element software, and validating the results with physical experiments. The study examines drill bit clearances ranging from 0.05 to 0.25 mm, showing that larger clearances lead to more significant temperature increases in the bone, exceeding 20%. These experimental findings align closely with the simulations, demonstrating minor errors between 1.75 and 4.47%. This research sets the stage for future design optimization of the existing surgical drill to reduce thermal injury in these medical procedures.

Keywords FEM · Osteonecrosis · Surgical drill bit · Margin · Bone drilling

M. F. A. Akhbar (✉) · S. Jamaludin
Faculty of Ocean Engineering Technology (FTKK), University Malaysia Terengganu, 21030 Kuala Nerus, Terengganu, Malaysia
e-mail: mfaizalaa@umt.edu.my

S. Jamaludin
e-mail: shahrizanj@umt.edu.my

R. Hassan
Pusat Reka Bentuk & Inovasi Teknologi (PRInT), Universiti Malaysia Pahang, 26600 Pekan, Pahang, Malaysia
e-mail: jihan@ump.edu.my

A. W. Sulong
Terengganu Skill Development Center (TESDEC), Kawasan Perindustrian Bukit Kor, 21600 Marang, Terengganu, Malaysia
e-mail: akmalwani@gmx.com

A. Ismail et al. (eds.), *Exploring Diversity in Engineering and Technology for Knowledge and Innovation*, Advanced Structured Materials 215,
https://doi.org/10.1007/978-3-031-64330-9_18

18.1 Introduction

In many medical departments, e.g., orthopedics, neurosurgery, and dentistry, clinicians have been performing bone drilling to create holes for the implant and screw installation [1]. In bone drilling, intense heat is generated in the hole from two primary sources: (1) friction energy—from the drill bit and bone, and (2) shearing energy—from the bone-cutting process [2, 3]. This heat increases the bone temperature above the threshold level—47 °C [4]—for the permanent death of bone cells (thermal osteonecrosis). A bone temperature of more than 100 °C is not uncommon in bone drilling [4, 5]. Because of thermal osteonecrosis in the drilling hole, the implants and screws will be loosened, which could cause bone refracture [3, 6] and implant failure [7].

Researchers have determined the influences of drilling parameters [6, 8, 9] (rotational speed, diameter, initial temperature, hole depth, etc.) and drill design [10–13] (point angle, rake angle, clearance angle, etc.) on thermomechanical damage in bone. Several studies have invented novel tools to minimize thermomechanical damage in bone surgery [10, 14–16]. For example, Akhbar and Yusoff [10] optimized the existing drill bit and reduced the temperature rise and necrosis regions by up to 15.2%. Earlier, Saha et al. [14] unveiled that a narrower point angle (118°), a larger helix angle (36°), and split point web thickness produce smaller bone debris, enhance positioning ability, and reduce thermal damage by 41%. Others proposed a novel design drill bit with two cutting edges for a more efficient cutting process [16, 17]. However, none have considered the influence of clearance on thermal and mechanical damages in bone drilling.

The drill bit clearance is the height of the narrow part on the drill land, which is formed to stabilize the drill (Fig. 18.1). This part influences the friction energy during the rotation in the drilling hole [1, 18]. However, to date, the impact of drill bit clearance on bone temperature has never been reported in the literature, hence the impetus of this study.

This study aims to research the effect of clearance height on maximum bone temperature using the finite element method (FEM) and experimental bone drilling tests. The outcomes from these tests are crucial information that involves the optimal range of drill bit clearance size for future surgical drill bit optimization processes to decrease thermal damage in bone surgery.

18.2 Methodology

The FEM simulation setup was performed using the commercial FEM software DEFORM-3D (version 11, Scientific Forming Technologies Corporation, Columbus, Ohio). Figure 18.2 shows the 20.0 mm diameter of the human bone model with a thickness of 5.5 mm—adopted from the previous literature [3, 8, 19]. Whereas the drill bit diameter was 4.5 mm, the material was set as stainless steel AISI 316L due

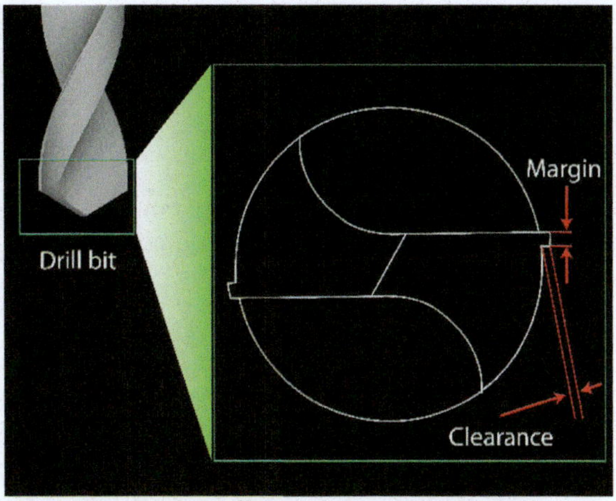

Fig. 18.1 Drill bit clearance

to its wide application in actual surgeries [20]. The drill bits with various clearance values [17, 21]—0.05, 0.09, 0.13, 0.17, 0.21, and 0.25 mm—were customized to execute the experimental drilling process. The surgical drill bit was assumed to be a rigid material, and the bone was assumed to be a plastic material [6, 22, 23]. The drill bit's initial temperature was established at 30 °C to replicate the ambient conditions in Malaysia, while the bone's initial temperature was maintained at 37 °C, representing the typical temperature of human bone. In this simulation, the properties of both models (Table 18.1) were extracted from the previous literature [10]. Furthermore, the machining parameters such as drilling speed, hole depth, and feed were fixed at 1000 rev/min, 3 mm, and 0.08 mm/rev, respectively.

As for the meshing process, the tetrahedral elements were selected, where the bone model was constructed with 109,725 elements, and the drill bit was designed with 38,417 elements, respectively (Fig. 18.2). The Cockroft-Latham damage criterion was employed to simulate the cutting process, as human bone can be considered subject to locally strain-controlled failure [24]. The shear friction of 0.3 was adopted in this study. This value has simulated accurate results in the previously published studies of bone drilling [3, 6, 10, 25].

For the experimental validation process, a conventional milling machine was used to execute the drilling process (Fig. 18.3). Bovine bone (femur) was selected as the replacement for the human bone [25]. The femur's diaphysis (middle part) was milled at the top surface with an end mill with a diameter of 20 mm. Figure 18.3 shows the scheme of the bone drilling experiment. For temperature measurement, four thermocouples and a data logger were used. The thermocouple's holes were drilled prior to the drilling process and filled with thermal paste to eliminate the air gap, which can affect the temperature variations. These holes were positioned at 0.5 mm from the circumference of the targeted location.

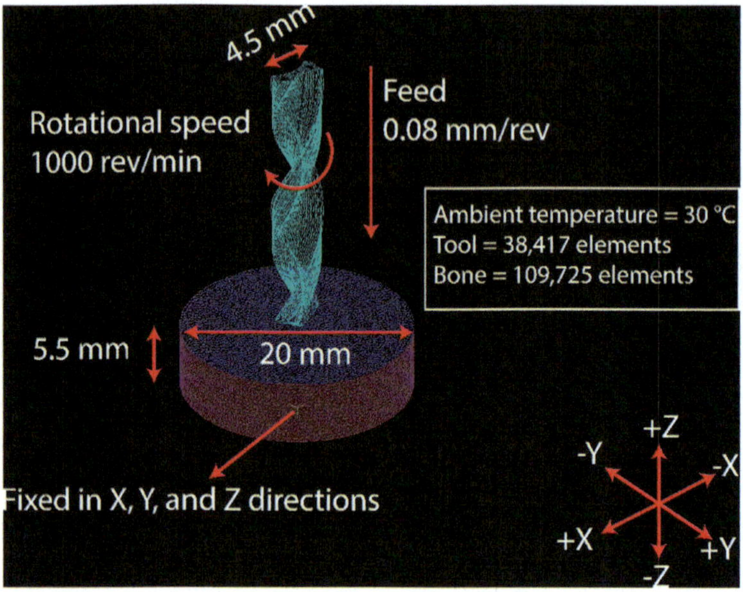

Fig. 18.2 Drill bit and bone models used for the drilling process

Table 18.1 Thermomechanical properties of simulation models

Properties	Bone	Drill bit
Density (kg/m^3)	2.10×10^3	7.99×10^3
Yield stress (Pa)	1.10×10^8	6.08×10^8
Young's modulus (Pa)	1.70×10^{10}	1.93×10^8
Poisson's ratio	0.40	0.30
Specific heat (J/kg K)	1.26×10^3	5.00×10^2
Thermal conductivity (W/m K)	3.80×10^{-1}	1.70×10
Shear modulus (Pa)	3.00×10^6	9.70×10^8
Heat expansion coefficient (mm/°C)	2.75×10^{-5}	1.21×10^{-5}
Ultimate compressive strength (Pa)	1.70×10^8	–
Thermal diffusivity (m^2/s)	1.44×10^{-7}	3.90×10^{-6}
Tensile strength (Pa)	2.00×10^8	5.79×10^8
Emissivity	9.70×10^{-1}	7.00×10^{-1}

18.3 Results and Discussion

Figure 18.4 shows the FEM and experimental test results of maximum bone temperature elevations with various drill bit clearance varying from 0.05 to 0.25 mm. The trends for FEM results agree with the experimental results, and the errors between

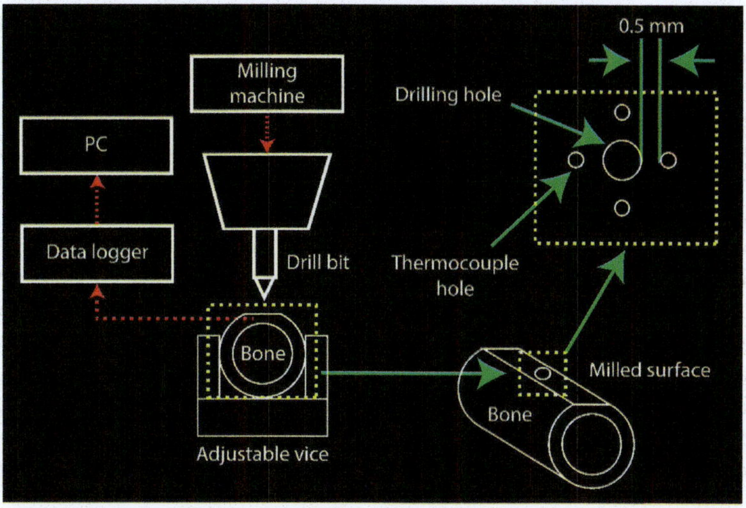

Fig. 18.3 Scheme diagram for the setup of experimental bone drilling

these methods vary from 1.75 to 4.5%. Therefore, we can confidently conclude that the bone drilling simulation used in this study is highly accurate in predicting outcomes that closely match those obtained in real bone drilling experiments.

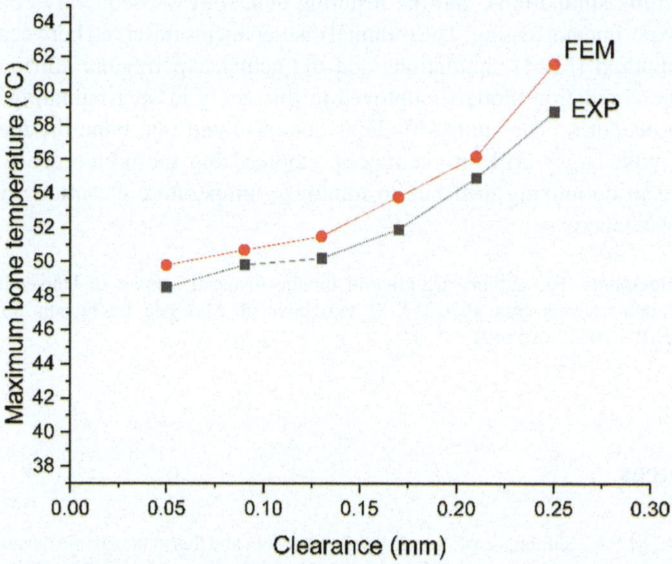

Fig. 18.4 Corresponding maximum bone temperature with clearance size of 0.05–0.25 mm using drilling simulations (FEM) and experiment test (EXP)

Based on this Fig. 18.4, the maximum bone temperature increases with the drill bit clearance. In every case of drill bit clearance, the maximum bone temperature surpassed the threshold for thermal osteonecrosis (47 °C [1]). The maximum bone temperature was enhanced by 23.7% when the drill bit clearance was increased from 0.05 to 0.25 mm. A larger drill bit clearance increases the contact between the drill bit and the bone, increasing the friction energy [1]. This friction energy induces excessive heat, subsequently increasing the bone temperature.

To date, no study has investigated the influence of drill bit clearance on temperature elevations in bone drilling. However, two studies by Soriano et al. [15, 17] revealed a direct relation between drill margin and bone temperature. They unveiled that a narrow margin width (0.1 mm) could reduce the bone temperature rise by 13% compared with the more significant margin (0.5 mm) [17]. In another study [15], they innovated a bone-cutting tool with two cutting edges (roughing and finishing) to reduce tissue damage. However, the bone suffered a temperature rise when using the new cutting tool because of the increase in margin number and, thus, friction energy.

18.4 Conclusion

In this study, a range of drill bit clearances were employed to investigate temperature profiles during the bone drilling process. This investigation involved the use of drilling simulations, and the resulting data were subsequently corroborated through experimental testing. The minimal discrepancies observed between the finite element method (FEM) simulations and the actual experiments affirm the accuracy of the simulation models employed in this study in approximating real bone drilling procedures. The study's findings demonstrated that bone thermal damage increases with larger drill bit clearances, emphasizing the preference for smaller clearances in optimizing drill bits to minimize temperature elevation and prevent thermal osteonecrosis.

Acknowledgements The authors are grateful for the financial support of Universiti Malaysia Pahang through research grant PDU213213 and Universiti Malaysia Terengganu through grant UMT/TAPE-RG 2022/Vot 55402.

References

1. Akhbar, M.F.A., Sulong, A.W.: Surgical drill bit design and thermomechanical damage in bone drilling: a review. Ann. Biomed. Eng. **49**, 29–56 (2021). https://doi.org/10.1007/s10439-020-02600-2
2. Lee, J., Chavez, C.L., Park, J.: Parameters affecting mechanical and thermal responses in bone drilling: a review. J. Biomech. **71**, 4–21 (2018). https://doi.org/10.1016/j.jbiomech.2018.02.025

3. Akhbar, M.F.A., Yusoff, A.R.: Fast & injurious: reducing thermal osteonecrosis regions in the drilling of human bone with multi-objective optimization. Measurement **152**, 107385 (2020). https://doi.org/10.1016/j.measurement.2019.107385
4. Benington, I.C., Biagioni, P.A., Crossey, P.J.J., Hussey, D.L.L., Sheridan, S., Lamey, P.J., Lameyt, P.: Temperature changes in bovine mandibular bone during implant site preparation: an assessment using infra-red thermography. J. Dent. **24**, 263–267 (1996). https://doi.org/10.1016/0300-5712(95)00072-0
5. Abouzgia, M.B., Symington, J.M.: Effect of drill speed on bone temperature. Int. J. Oral Maxillofac. Surg. **25**, 394–399 (1996). https://doi.org/10.1016/S0901-5027(06)80040-8
6. Akhbar, M.F.A., Yusoff, A.R.: Drilling of bone: thermal osteonecrosis regions induced by drilling parameters. Biomed. Phys. Eng. Express **5**, 065003 (2019). https://doi.org/10.1088/2057-1976/ab42f2
7. Christie, J.: Surgical heat injury of bone. Injury **13**, 188–190 (1981). https://doi.org/10.1016/0020-1383(81)90236-9
8. Akhbar, M.F.A., Malik, M., Yusoff, A.R.: Effects of drilling parameters in numerical simulation to the bone temperature elevation. In: AIP Conference Proceedings, p. 020002 (2018)
9. Augustin, G., Zigman, T., Davila, S., Udiljak, T., Staroveski, T., Brezak, D., Babic, S.: Cortical bone drilling and thermal osteonecrosis. Clin. Biomech. **27**, 313–325 (2012). https://doi.org/10.1016/j.clinbiomech.2011.10.010
10. Akhbar, M.F.A., Yusoff, A.R.: Multi-objective optimization of surgical drill bit to minimize thermal damage in bone-drilling. Appl. Therm. Eng. **157**, 113594 (2019). https://doi.org/10.1016/j.applthermaleng.2019.04.004
11. Ali Akhbar, M.F., Yusoff, A.R.: Drilling of bone: effect of drill bit geometries on thermal osteonecrosis risk regions. Proc. Inst. Mech. Eng. H **233**, 207–218 (2019). https://doi.org/10.1177/0954411918819113
12. Sui, J., Wang, C., Sugita, N.: Experimental study of temperature rise during bone drilling process. Med. Eng. Phys. **78**, 64–73 (2020). https://doi.org/10.1016/j.medengphy.2020.01.007
13. Li, S., Shu, L., Kizaki, T., Bai, W., Terashima, M., Sugita, N.: Cortical bone drilling: a time series experimental analysis of thermal characteristics. J. Manuf. Process. **64**, 606–619 (2021). https://doi.org/10.1016/j.jmapro.2021.01.046
14. Saha, S., Pal, S., Albright, J.A.: Surgical drilling: design and performance of an improved drill. J. Biomech. Eng. **104**, 245 (1982). https://doi.org/10.1115/1.3138356
15. Soriano, J., Garay, A., Ishii, K., Sugita, N., Arrazola, P.J., Mitsuishi, M.: A new surgical drill bit concept for bone drilling operations. Mater. Manuf. Process. **28**, 1065–1070 (2013). https://doi.org/10.1080/10426914.2013.811745
16. Sugita, N., Ishii, K., Sui, J., Terashima, M.: Multi-grooved cutting tool to reduce cutting force and temperature during bone machining. CIRP Ann. Manuf. Technol. **63**, 101–104 (2014). https://doi.org/10.1016/j.cirp.2014.03.069
17. Soriano, J., Garay, A., Aristimuño, P., Arrazola, P.J.: Study and improvement of surgical drill bit geometry for implant site preparation. Int. J. Adv. Manuf. Technol. **74**, 615–627 (2014). https://doi.org/10.1007/s00170-014-5998-x
18. Akhbar, M.F.A.: Thermomechanical damage in cortical bone caused by margins of surgical drill bit: a finite element analysis. Comput. Methods Programs Biomed. **231**, 107361 (2023). https://doi.org/10.1016/j.cmpb.2023.107361
19. Treece, G.M., Gee, A.H., Mayhew, P.M., Poole, K.E.S.: High resolution cortical bone thickness measurement from clinical CT data. Med. Image Anal. **14**, 276–290 (2010). https://doi.org/10.1016/j.media.2010.01.003
20. Bertollo, N., Walsh, W.R.: Biomechanics in Applications. InTech (2011)
21. Soriano, J., Garay, A., Aristimuño, P., Iriarte, L.M., Eguren, J.A., Arrazola, P.J.: Effects of rotational speed, feed rate and tool type on temperatures and cutting forces when drilling bovine cortical bone. Mach. Sci. Technol. **17**, 611–636 (2013). https://doi.org/10.1080/10910344.2013.837353
22. Qi, L., Wang, X.N., Meng, M.Q.: 3D finite element modeling and analysis of dynamic force in bone drilling for orthopedic surgery. Int. J. Numer. Methods Biomed. Eng. **30**, 845–856 (2014). https://doi.org/10.1002/cnm.2631

23. Akhbar, M.F.A., Yusoff, A.R.: Optimization of drilling parameters for thermal bone necrosis prevention. Technol. Health Care **26**, 621–635 (2018). https://doi.org/10.3233/THC-181221
24. Nalla, R.K., Kinney, J.H., Ritchie, R.O.: Mechanistic fracture criteria for the failure of human cortical bone. Nat. Mater. **2**, 164–168 (2003). https://doi.org/10.1038/nmat832
25. Ali Akhbar, M.F., Yusoff, A.R.: Comparison of bone temperature elevation in drilling of human, bovine and porcine bone. Procedia CIRP **82**, 411–414 (2019). https://doi.org/10.1016/j.procir.2019.03.220

Chapter 19
Synthesis and Morphological Studies of Cerium Oxide-Zinc Oxide as a Potential Heterostructured Material

Sharifah Zahratul Ashirah Syed Abdullah, Wan Rafizah Wan Abdullah, Fazilah Ariffin, and Mohd Sabri Mohd Ghazali

Abstract ZnO is a versatile metal oxide owing to its binding energy of 6 meV and direct band gap of 3.3 eV. The microstructure-electronic properties of ZnO have been altered to create an intentional effect that is beneficial for many engineering applications. New form of functional materials made of ZnO-based heterostructured materials have been explored for overcoming the limitations of conventional ZnO. The multi-phase ZnO heterostructured materials allow the fine-tuning of the band structure while extending the electron–hole separation. Such effects are required for performance enhancement of photocatalytic and photonic materials. This paper reported the synthesis of CeO_2–ZnO heterostructured materials containing different CeO_2 to ZnO molar ratios (0.1:100, 1:100, and 10:100) via a nanofabrication technique that combined the citrate gelation complexation reaction and solid-phase sintering up to 1200 °C for 5 consecutive hours. The microstructure evolutions of CeO_2–ZnO materials and establishment of the CeO_2–ZnO heterojunctions were investigated using SEM and FESEM-EDX analyses. The conversion of raw materials and the phase analysis were confirmed using XRD.

Keywords Nanofabrication · Heterostructure · Zinc oxide · Cerium oxide

S. Z. A. S. Abdullah · W. R. W. Abdullah (✉)
Faculty of Ocean Engineering Technology, Universiti Malaysia Terengganu, 21030 Kuala Nerus, Terengganu, Malaysia
e-mail: wanrafizah@umt.edu.my

S. Z. A. S. Abdullah
e-mail: p4967@pps.umt.edu.my

F. Ariffin · M. S. M. Ghazali
Faculty of Science and Marine Environment, Universiti Malaysia Terengganu, 21030 Kuala Nerus, Terengganu, Malaysia
e-mail: fazilah@umt.edu.my

M. S. M. Ghazali
e-mail: mohdsabri@umt.edu.my

A. Ismail et al. (eds.), *Exploring Diversity in Engineering and Technology for Knowledge and Innovation*, Advanced Structured Materials 215,
https://doi.org/10.1007/978-3-031-64330-9_19

19.1 Introduction

A heterostructured material is a dual or multi-phase material having functional heterogeneous zones. In order to obtain specific intended properties, these zones are manipulated and the synergistic effect between two phases in contact improves the quality of the composite in comparison with the conventional single-phase material. Ongoing research trends underscore the rapid advancement of sophisticated materials to cater to the evolving interests of the research community in the field of material engineering. Several known forms of heterostructured materials include typical heterostructured materials [1], lamella structures [2], gradient [3, 4], laminated materials [5, 6], dual or multi-phase [7], core–shell [8], and multi-modal structures [9]. Larger scale fabrication is possible as prevailing industrial equipment is able to manufacture this class of materials, thus lowering the production cost. The heterostructures are not only useful as structural materials [1], but they also offer qualities as advanced permanent magnet, thermoelectric, and new generation photo-induced catalytic materials [10].

In the area of photocatalysis study, more and more CeO_2–ZnO-based materials are being researched. In single form, semiconducting ZnO is incorporated in numerous consumer products as an additive such as paint, rubber, and even pharmaceutical products [11]. It is also an interesting component in electronic materials owing to its unique electrical stability and semiconductivity [12]. Equal qualities are offered by CeO_2 or better known as ceria. This rare-earth oxide shows a band gap of 3.2–3.9 eV, and it demonstrates excellent optical transparency in both visible and UV spectrum, high oxygen uptake capacity, and good refractive index [13]. CeO_2 can be found as an active component in solid fuel cells (SOFCs) [14, 15], gas sensors, and photocatalysts [16, 17]. Because of these good reputations, merging CeO_2 and ZnO as a new heterostructure combination is expected to result in more unique features.

Application of CeO_2–ZnO-based heterostructured materials for overcoming limitations in conventional photocatalysts has been reported previously. The interface of CeO_2–ZnO creates electronically active heterojunctions that could enhance light absorption spectrum and increases catalytic reactivity induced by visible light [18]. As a result, several CeO_2–ZnO photocatalysts developed so far exhibit enhanced quantum efficiency [19, 20], prolonged carrier lifespan, and better hole-electron charge separation [21, 22]. On the other hand, CeO_2–ZnO photocatalysts demonstrate superior catalytic efficiency for hydrogen recovery and water splitting processes to produce under visible light irradiation [23].

A proper materials selection, optimization of formulation, and suitable combination of synthesis parameters would ensure the production of CeO_2–ZnO heterostructured materials with unlimited potential. Driven by the above-mentioned vision, this paper highlights an alternative synthesis route for obtaining CeO_2–ZnO heterostructured material through implementation modified citrate gelation process and high temperature sintering in solid-phase.

19.2 Methodology

19.2.1 Nanofabrication of CeO_2–ZnO Heterostructured Materials

The wet-chemical mixing between nano-ZnO and Ce precursor was performed by using the modified citrate gelation technique. Pre-weighed cerium (III) nitrate ($Ce(NO_3)_3$) was reacted with citric acid ($C_6H_8O_7$) solution to form a viscous gel. The Ce citrate gel solution was subsequently combined with nano-ZnO powder (< 50 nm), and the mixture was constantly agitated at 300 rpm at the temperature of 70–80 °C. Excess organic compounds were removed in a calcination process at 500 °C for 4 consecutive hours. The precursor powder was oxidized in a solid-phase sintering process at 1200 °C and soaked for 5 h. Complete conversion of Ce precursor to CeO_2 and phase segregation was achieved at the end of the sintering stage. The sintered powder was eventually pulverized and sieved to collect fine particles with the size of less than 60 μm. The procedures were repeated to reproduce CeO_2–ZnO containing different molar ratios of CeO_2–ZnO such as 0.1:100, 1:100, and 10:100. A sintered purely ZnO was used as control. Phase analysis was carried out on sample by utilizing an X-ray diffractometer (Rigaku Miniflex II Model PW1710). Inspection of sample's microstructure and elemental distribution analysis were separately accomplished by using scanning electron microscopy (SEM) (JEOL JFC-1600) and field emission scanning microscopy (FESEM) (JEOL JSM 7600F) attached with energy dispersive X-ray spectroscopy (EDX), respectively.

19.3 Results and Discussion

19.3.1 Phase Analysis

Figure 19.1 compares the diffractograms of pure nano-ZnO, sintered ZnO, and CeO_2 and CeO_2–ZnO heterostructured with different molar ratios of 0.1:100, 1:100, and 10:100. The diffraction peaks from purely and sintered ZnO belong to the wurtzite hexagonal structured ZnO (PDF80-75). These peaks appeared in the XRD patterns of CeO_2–ZnO samples along with the diffraction peaks belong to CeO_2 in cubic-fluorite crystal structure (PDF 81-792). The intensity of peaks which matched the (111) and (200) planes of CeO_2 at 2θ of 29° and 33°, respectively, became more prominent with the increment in ratio of CeO_2 in sample. This finding indicates that the product collected from the synthesis process outlined in this work was a dual-phase system comprising combination of CeO_2 and ZnO.

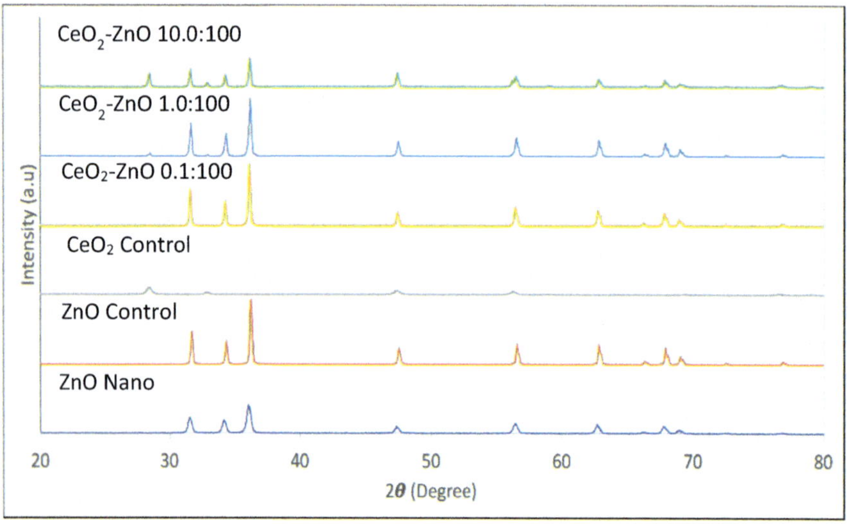

Fig. 19.1 XRD patterns of the synthesized CeO_2–ZnO heterostructured materials in comparison with ZnO nanoparticles, sintered ZnO, and CeO_2

19.3.2 Morphological Study of CeO_2–ZnO Heterostructured Materials

The surface microstructure of as-synthesized CeO_2–ZnO was inspected. SEM micrographs of CeO_2–ZnO samples in Fig. 19.2 were analyzed and they were compared to the micrographs of pure components. ZnO grain growth phenomenon was observed in the sintered pure ZnO samples due to calcination and sintering processes. The addition of CeO_2, on the other hand, caused pinning effect that slightly impeded the ZnO grains from expanding freely. As a result, the grains in CeO_2–ZnO materials were visibly smaller and more uniform compared to the sintered pure ZnO. Ce-rich precipitates can be seen as bright spots or nodules on the grain surface. Larger and randomly distributed Ce-rich nodules were observed on samples with higher CeO_2 molar ratio. FESEM-EDX analysis was performed on a CeO_2–ZnO sample containing CeO_2 to ZnO molar ratio of 1.0:100, and the EDX spectra were collected at three points on sample surface. A unique pattern of Ce distributions was observed. Ce-rich phase clustered up as nodules at the exterior of ZnO surface. It is believed that the exposure to high temperature sintering allowed migration of metal ions to preferential area. Insoluble Ce^{3+}/Ce^{4+} ions having ionic radii of 0.097–0.114 nm precipitated on the outer surface of ZnO crystals. Visibility of nodules and phase segregation became more prominent in sample containing high CeO_2 ration. Such microstructure contributed to establishment of CeO_2–ZnO heterojunctions. Comparable microstructures were reported [24] that worked with similar CeO_2–ZnO system.

The interphase between the two phases in contact creates complex interaction leading to distinctive physical, optical, mechanical, and electrical properties (Fig. 19.3).

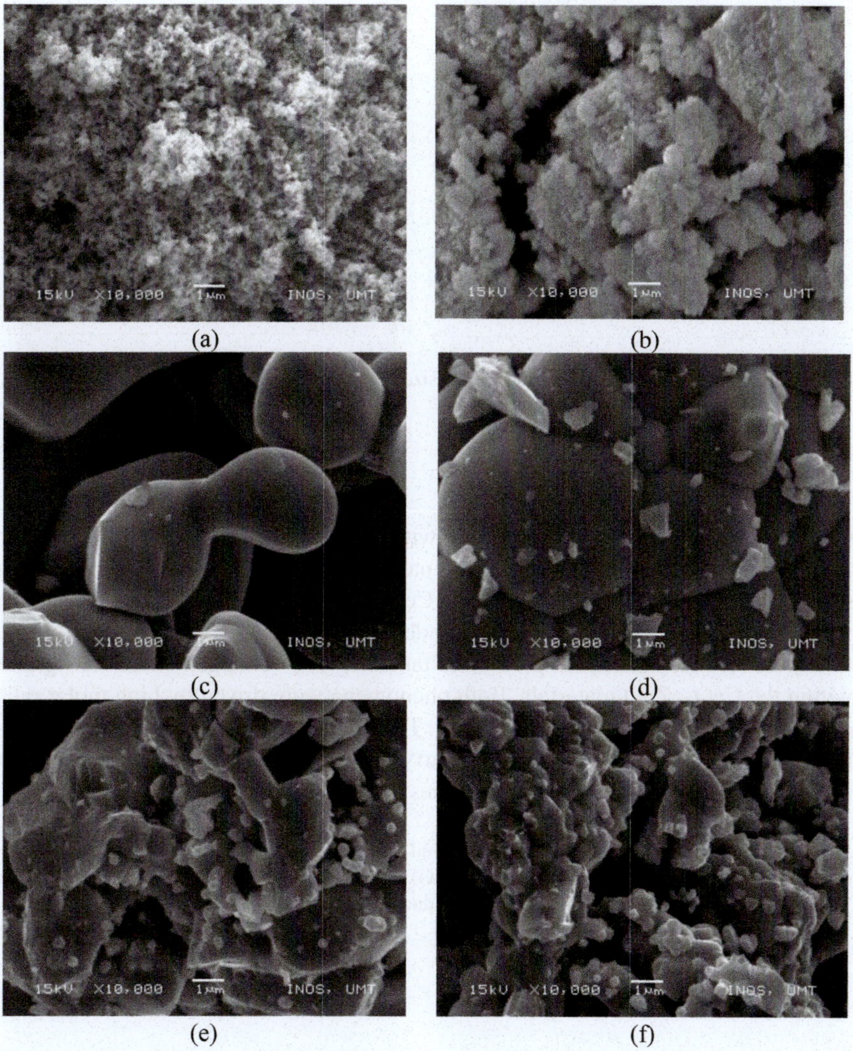

Fig. 19.2 SEM images of **a** ZnO nanoparticles, **b** CeO_2, **c** sintered ZnO and as-synthesized CeO_2–ZnO at molar ratios of **d** 0.1:100, **e** 1:100, and **f** 10:100

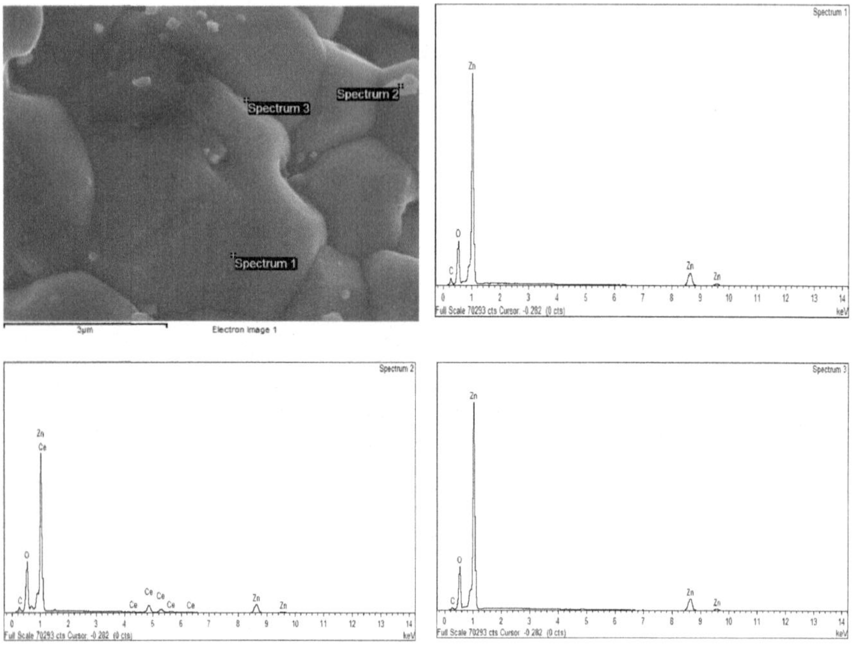

Fig. 19.3 FESEM-EDX spectrum of as-synthesized CeO_2–ZnO at molar ratio 1:100

19.4 Conclusion

CeO_2–ZnO heterostructured materials with molar ratios of 0.1:100, 1:100, and 10:100 were obtained from a sequential nanofabrication process as outlined in this study. Active heterojunctions formed as CeO_2 decorated the exterior of ZnO phase when Ce ions segregated and clustered under high temperature sintering. The formation of active heterojunctions grew simultaneously with enlargement of ZnO grains within the CeO_2–ZnO materials. Both phenomena were driven by high molar ratio of CeO_2 to ZnO in the initial formulation. The functionality CeO_2–ZnO heterostructured materials could be unlocked by carrying out more characterizations on their physicochemical-microstructure properties.

Acknowledgements The present study received financial support from the Talent and Publication Enhancement Research-Grant (TAPE-RG) (Vot 55215) provided by Universiti Malaysia Terengganu and Malaysia-Ministry of Higher Education.

References

1. Wu, X., Zhu, Y.: Heterogeneous materials: a new class of materials with unprecedented mechanical properties. Mater. Res. Lett. **5**(8), 527–532 (2017)
2. Wu, X., Yang, M., Yuan, F., Wu, G., Wei, Y., Huang, X., Zhu, Y.: Heterogeneous lamella structure unites ultrafine-grain strength with coarse-grain ductility. Proc. Natl. Acad. Sci. **112**, 14501–14505 (2015)
3. Wu, X., Jiang, P., Chen, L., Yuan, F., Zhu, Y.T.: Extraordinary strain hardening by gradient structure. Proc. Natl. Acad. Sci. U.S.A. **111**(20), 7197–7201 (2014)
4. Makeswaran, N., Orozco, C., Battu, A.K., Deemer, E., Ramana, C.V.: Structural, optical and mechanical properties of nanocrystalline molybdenum thin films deposited under variable substrate temperature. Materials **15**(3), 754 (2022)
5. Ma, X., Huang, C., Moering, J., Ruppert, M., Höppel, H.W., Göken, M., Narayan, J., Zhu, Y.: Mechanical properties of copper/bronze laminates: role of interfaces. Acta Mater. **116**, 43–52 (2016)
6. Bonadies, I.: Nanoscale materials in water purification. In: Thomas, S., Pasquini, D., Leu, S.-Y., Gopakumar, D.A. (eds.) Nanoscale Materials in Water Purification, pp. 231–246. Elsevier (2019)
7. Calcagnotto, M., Adachi, Y., Ponge, D., Raabe, D.: Deformation and fracture mechanisms in fine- and ultrafine-grained ferrite/martensite dual-phase steels and the effect of aging. Acta Mater. **59**(2), 658–670 (2011)
8. Shokry, A., Ahadi, A., Ståhle, P., Orlov, D.: Improvement of structural efficiency in metals by the control of topological arrangements in ultrafine and coarse grains. Sci. Rep. **11**(1), 1–14 (2021)
9. Han, B.Q., Huang, J.Y., Zhu, Y.T., Lavernia, E.J.: Strain rate dependence of properties of cryomilled bimodal 5083 Al alloys. Acta Mater. **54**(11), 3015–3024 (2006)
10. Ma, E., Zhu, T.: Towards strength–ductility synergy through the design of heterogeneous nanostructures in metals. Mater. Today **20**(6), 323–331 (2017)
11. Bahrami, M., Nezamzadeh-Ejhieh, A.: Effect of the supported ZnO on clinoptilolite nano-particles in the photodecolorization of semi-real sample bromothymol blue aqueous solution. Mater. Sci. Semicond. Process. **30**, 275–284 (2015)
12. Wibowo, A., Marsudi, M.A., Amal, M.I., Ananda, M.B., Stephanie, R., Ardy, H., Diguna, L.J.: ZnO nanostructured materials for emerging solar cell applications. RSC Adv. **10**(70), 42838–42859 (2020)
13. Cui, Z., Zhang, D., Hu, J., Fang, C.: CdS/CeO_2 heterostructures as visible-light photocatalysts for the reduction of nitro to amine organics. J. Alloys Compd. **885**, 160961 (2021)
14. He, Y., Yang, B., Cheng, G.: Controlled synthesis of CeO_2 nanoparticles from the coupling route of homogenous precipitation with microemulsion. Mater. Lett. **57**(13–14), 1880–1884 (2003)
15. Li, G., Nie, X., Chen, J., Jiang, Q., An, T., Wong, P.K., et al.: Enhanced visible-light-driven photocatalytic inactivation of *Escherichia coli* using g-C_3N_4/TiO_2 hybrid photocatalyst synthesized using a hydrothermal-calcination approach. Water Res. **86**, 17–24 (2015)
16. Izu, N., Itoh, T., Nishibori, M., Matsubara, I., Shin, W.: Effects of noble metal addition on response of ceria thick film CO sensors. Sens. Actuators B: Chem. **171–172**, 350–353 (2012)
17. Renganathan, B., Sastikumar, D., Gobi, G., Yogamalar, N.R., Bose, A.C.: Gas sensing properties of a clad modified fiber optic sensor with Ce, Li and Al doped nanocrystalline zinc oxides. Sens. Actuators B: Chem. **156**(1), 263–270 (2011)
18. Munawar, T., Mukhtar, F., Nadeem, M.S., Mahmood, K., Hussain, A., Ali, A., Arshad, M.I., Un Nabi, M.A., Iqbal, F.: Structural, optical, electrical, and morphological studies of rGO anchored direct dual-Z-scheme ZnO-Sm_2O_3–Y_2O_3 heterostructured nanocomposite: an efficient photocatalyst under sunlight. Solid State Sci. **106**, 106307 (2020)
19. Johra, F.T., Jung, W.G.: RGO–TiO_2–ZnO composites: synthesis, characterization, and application to photocatalysis. Appl. Catal. A Gen. **491**(1), 52–57 (2015)

20. Rani, M., Shanker, U.: Sun-light driven rapid photocatalytic degradation of methylene blue by poly(methyl methacrylate)/metal oxide nanocomposites. Colloids Surf A Physicochem Eng Asp **559**, 136–147 (2018)
21. Zhao, H., Ding, X., Zhang, B., Li, Y., Wang, C.: Enhanced photocatalytic hydrogen evolution along with byproducts suppressing over Z-scheme $Cd_xZn_{1-x}S/Au/g\text{-}C_3N_4$ photocatalysts under visible light. Sci. Bull. **62**(9), 602–609 (2017)
22. Zeng, D., Yang, K., Yu, C., Chen, F., Li, X.X., Wu, Z., Liu, H.: Phase transformation and microwave hydrothermal guided a novel double Z-scheme ternary vanadate heterojunction with highly efficient photocatalytic performance. Appl. Catal. B Environ. **237**, 449–463 (2018)
23. Hezam, A., Wang, J., Drmosh, Q.A., Karthik, P., Abdullah Bajiri, M., Namratha, K., Zare, M., Lakshmeesha, T.R., Shivanna, S., Cheng, C., Neppolian, B.: Rational construction of plasmonic Z-scheme Ag-ZnO-CeO_2 heterostructures for highly enhanced solar photocatalytic H_2 evolution. Appl. Surf. Sci. **541**, 148457 (2021)
24. Zhu, L., Li, H., Xia, P., Liu, Z., Xiong, D.: Hierarchical ZnO decorated with CeO_2 nanoparticles as the direct Z-scheme heterojunction for enhanced photocatalytic activity. ACS Appl. Mater. Interfaces **10**(46), 39679–39687 (2018)

Chapter 20
Structural Analysis of Marine Structures Using the Finite Element Method

Zulzamri Salleh, Muhammad. Akmal Afiq Zullastri, Mohd. Hafizi Zohari, Asmawi Abdul Malik, and Jaronie Mohd Jani

Abstract Highly demanding in the fast-growing marine construction industry, increasing revenue and profitability requires advances in technology for cost-effective production in key areas of ship production and ship production management. The absence of information, particularly in the mechanical damage behavior of marine structures, is addressed by structural health monitoring (SHM), which focuses on providing accurate and real-time information on the state and performance of structures. This research will focus on finite element analysis (FEA) formulation for plate and shell structures, with a focus on the first-order shear deformation theory. Then, validation for four cases of different ages of damage with different sizes, locations, and number of damages are considered as marine structures. To experimentally simulate this model in a real maritime environment, hydrodynamics will be used to quantify it, and a method will be developed utilizing newly introduced damage parameters and an isogeometric FEA formulation of the von Mises strain distribution with the PTC CREO software. All of this information is extremely beneficial to ship repair firms that have been highly recommended by the International Maritime Organization (IMO).

Z. Salleh (✉) · Muhammad. A. A. Zullastri · A. A. Malik
Universiti Kuala Lumpur, Malaysian Institute of Marine Engineering Technology, Lumut, Malaysia
e-mail: zulzamri@unikl.edu.my

Muhammad. A. A. Zullastri
e-mail: makmal.zullastri@s.unikl.edu.my

A. A. Malik
e-mail: asmawiam@unikl.edu.my

Mohd. H. Zohari
Faculty of Mechanical and Automotive Engineering Technology, Universiti Malaysia Pahang, Pekan, Malaysia
e-mail: hafizi@umpsa.edu.my

J. M. Jani
School of Science, Engineering and Technology, RMIT Vietnam, Ho Chi Minh City, Vietnam
e-mail: jaronie.mohdjani@rmit.edu.vn

A. Ismail et al. (eds.), *Exploring Diversity in Engineering and Technology for Knowledge and Innovation*, Advanced Structured Materials 215,
https://doi.org/10.1007/978-3-031-64330-9_20

Keywords Finite element analysis · Marine structure · Stress monitoring · Structure monitoring

20.1 Introduction

In the contemporary era, advancements in technology are multifaceted, benefiting from the integration of cutting-edge innovations. Emphasizing not only the expansion of the public sector but also the progress of the maritime industry, the ship production and management sector faces a pressing need for technical innovations to ensure economical manufacturing due to heightened demand. Consequently, breakthroughs in marine technology, particularly in the realm of structural health monitoring, are now accessible. Structural health monitoring (SHM) is an engineering implement which controls, verifies, and informs about the condition or changes in the condition of a structure so that the engineers are able to obtain trustworthy information for management and decision making [1].

In recent years, structural health monitoring (SHM) has gained widespread recognition and adoption as a prominent technology in structural engineering across various countries. The advent of abbreviated construction timelines, heightened traffic loads, the introduction of high-speed trains, and the emergence of dynamic and fatigue challenges pose new complexities. Additionally, the use of novel materials, innovative construction approaches, slender designs, budget constraints, time-saving imperatives, and other factors underscore the need for enhanced control. In this context, SHM has become an indispensable tool, facilitating effective management to ensure both quality and safety for users. As per Shashank J. Patil, "the execution of a strategy for identifying and characterizing damage in engineering structures" [2]. The swift advancements in sensor technology, data acquisition, communication, signal analysis, and data processing have significantly enriched structural health monitoring (SHM). While SHM frequently furnishes dependable data on a structure's current state, its integration into actual civil engineering structures is relatively nascent. Given its multidisciplinary nature, there is room for improvement in various aspects of SHM technology. A standard structural health monitoring system comprises three fundamental subsystems: diagnostics, damage prognosis (DP), and predictive maintenance (PM) [3].

This occurrence leads to the progression and amplification of existing damage, potentially culminating in catastrophic ship damage if thorough on-site structural integrity assessments are not conducted. To tackle this issue, structural health monitoring emerges as a multidisciplinary technological field, providing dependable real-time data from structures. The primary element of SHM, primarily sourced from the system, involves strain sensors to furnish information about the structural condition. It is alternatively referred to as displacement and stress monitoring or shape-stress-sensing [4]. The introduction of iFEM by [5] has noted that there is no need to install numerous sensors to capture the entire structural displacement [5]. Thus, it shares similarities with other structural applications like finite element analysis

(FEA), computational fluid dynamics (CFD), and various other simulation software currently available.

Moreover, finite element analysis serves as a numerical approach for addressing engineering and mathematical physics challenges. The finite element (FE) method proves to be an efficient and extensively employed numerical technique in simulating the responses of diverse biological tissues and structures across a wide range of biomechanical engineering applications. This is particularly valuable in cases where conducting physically based experiments is either impractical or unethical [6]. It proves beneficial for addressing issues involving intricate geometry, diverse loading conditions, and complex material properties that defy analytical solutions. Furthermore, it finds application in the analysis of structures or materials too intricate for estimating feedback through ordinary differential equations, enabling the study of the correlation between form and function in morphological structures. FEA is capable of simulating the distribution of stress, strain, and deformation within a solid object when subjected to applied forces and constraints [7]. The precision of finite element analysis relies significantly on the mesh's quality. Thin or skewed elements, in particular, can yield inaccurate results. Enhancement techniques focus on refining the shape of individual elements based on specific quality metrics [8]. In a separate investigation, Akin emphasizes that finite element analysis stands out as the predominant tool for stress and structural analysis, often interconnected with various fields of study. For instance, the presence of a non-uniform temperature distribution creates indeterminate loading conditions in solid structural members. As a result, it is customary to perform thermal FEA to derive temperature results, subsequently used as input data for stress FEA. Furthermore, FEA can seamlessly integrate input data from other tools, including motion analysis (kinetic) systems and computational fluid dynamics (CFD) systems [9].

One of the three fundamental equivalent single layer (ESL) networks is the first-order shear deformation theory (FSDT), a concept introduced by Reissner and Mindlin. FSDT incorporates the impact of shear deformation across the thickness by employing a linear variation of in-plane displacement. However, as FSDT violates equilibrium conditions on the plate's top and bottom surfaces, a shear correction factor is necessary to reconcile the disparity between the actual stress state and the assumed constant stress state. The first-order shear deformation theory, while widely applicable, presents certain contradictions that may become significant in relatively thick laminates or in sandwich structures with low transverse shear modulus in their cores [10]. In FSDT, the removal of the third part of Kirchhoff's hypothesis results in the transverse normal no longer remaining perpendicular to the central surface after deformation [11]. This study primarily focuses on analyzing the stress distribution when various forces are applied to the boat model. The obtained results will be crucial for understanding the stress patterns in the structure.

Table 20.1 Model boat dimensions

	Model boat, mm
Length	1800
Height	400
Width	716.78

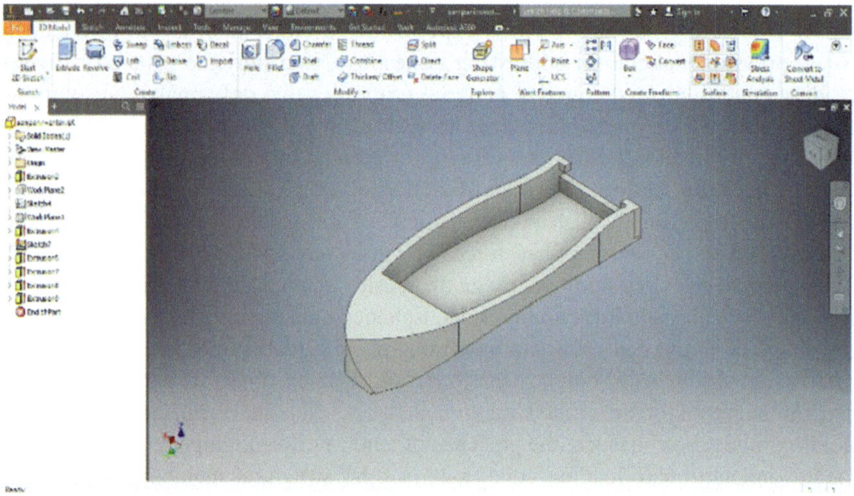

Fig. 20.1 Design of model boat using Autodesk inventor

20.2 Methodology

20.2.1 Design Phase

In the design phase, the CREO program serves as a medium to implement the design according to the provided references. This design process involves the creation of two sample models, specifically sample boats, each with a defined sample size as given in Table 20.1. The attachment for this phase can be referred to Figs. 20.1 and 20.2.

20.2.2 Analysis

Analytical work was completed and evaluated using application simulations facilitated by CREO software. The proposed model will undergo a systematic analysis procedure, with a number of designated areas selected as test subjects. This area

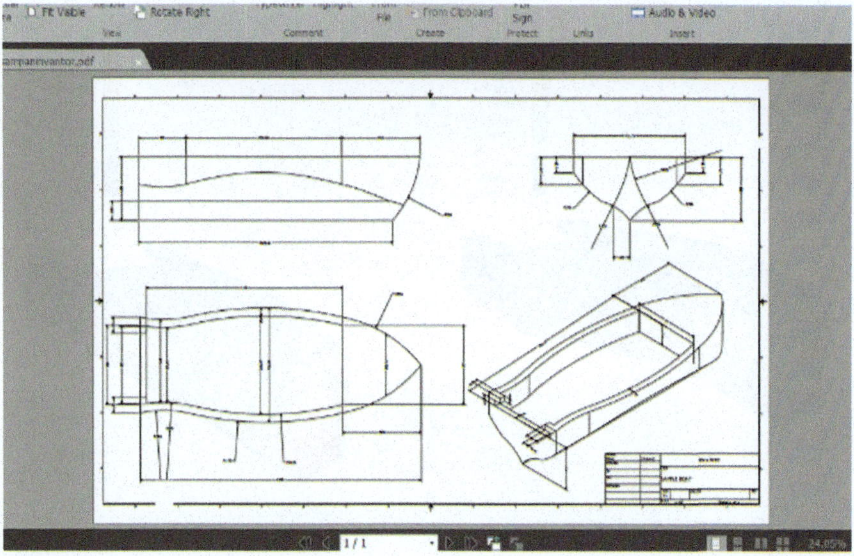

Fig. 20.2 Lines plan of the model boat

includes (a) the front of the boat and (b) the port and starboard side of the boat as indicated in Figs. 20.3 and 20.4.

20.2.3 Material

The data sources for this investigation are drawn from two independent outlets. This approach is adopted to account for potential variations in values and results, preventing a direct comparison between the two sources. The materials utilized in the study are fiberglass composite and silicon aluminum copper alloy, each with its respective specifications as given in Table 20.2.

20.3 Results and Discussion

The CREO simulation software yields two types of data or results: stress and displacement. Every set of specimen data was systematically collected and documented, guided by insights from prior experiments [10].

The stress and displacement values for various model boats are depicted in Figs. 20.3 and 20.4. Notably, the stress value demonstrates a decrease as the force rate increases, while the maximum displacement registers an increase. This observation indicates that, as the force rating for the boat model rises, its ability to

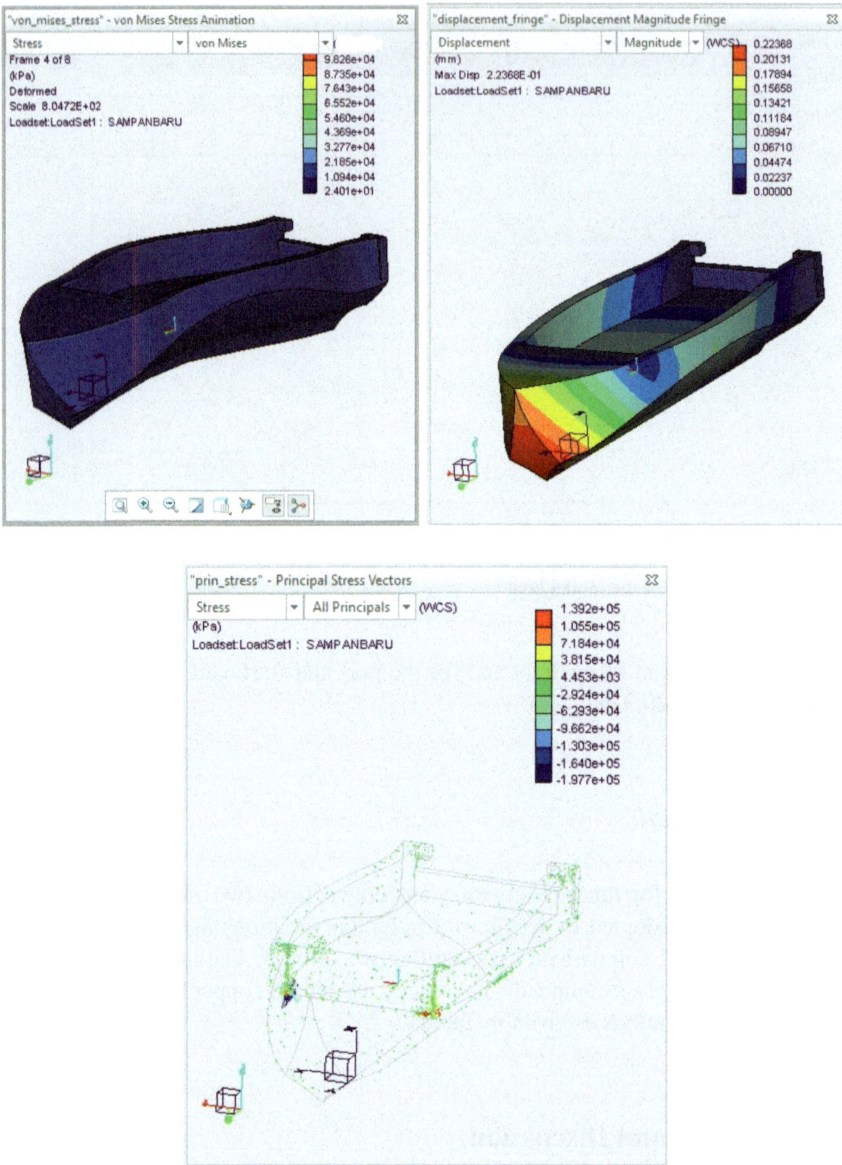

Fig. 20.3 Simulation result that been running using PTC CREO software (ahead side of model boat)

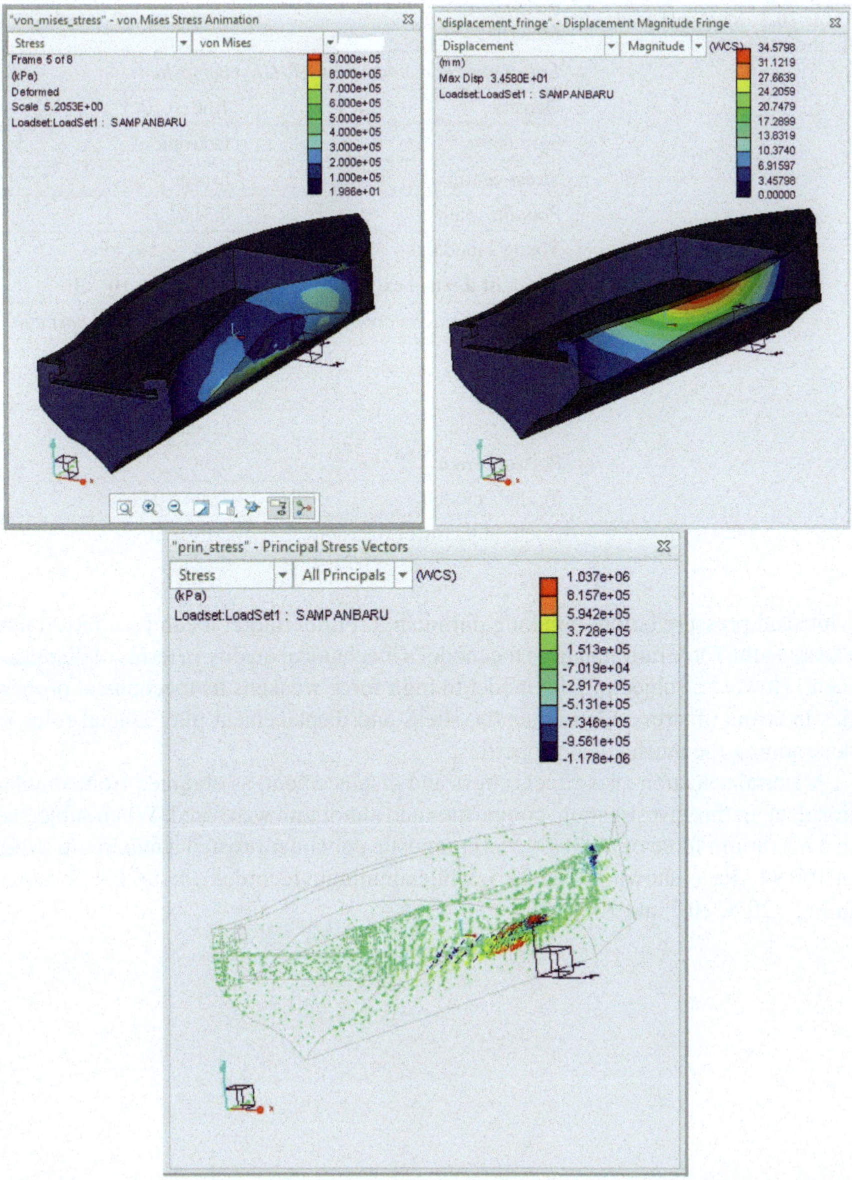

Fig. 20.4 Simulation result that been running using PTC CREO software (port side of the model boat)

Table 20.2 Materials used for this research

Materials	
Epoxy glassfibre composite EP-GF_composite	
Density	1.86×10^{-6}
Symmetry	Isotropic
Stress–strain	Linear
Poisson's ratio	0.314
Young's modulus	2.14×10^{7} kPa
Coeff. of thermal expansion	9.38889×10^{-6}/F
Aluminium silicon copper alloy cast Al–Si–Cu_alloy_cast.mtl	
Density	2680 kg/m^3
Symmetry	Isotropic
Stress–strain	Linear
Poisson's ratio	0.33
Young's modulus	7.3×10 Pa
Coeff. of thermal expansion	2.15×10^{-5}

withstand pressure before breaking diminishes. Furthermore, it can be inferred that elevating the force rate enhances the model's mechanical quality in terms of displacement. However, subjecting the model to high force weakens its mechanical properties in terms of stress. Consequently, stress and displacement play crucial roles in determining the mechanical properties.

Material selection also affects stress and displacement, as obtained from simulation data. In this investigation, composites and aluminum were used. When subjected to a maximum force of 5000 kN, the composite container resisted a maximum stress of 160.94 kPa as shown in Fig. 20.5, while aluminum recorded 548.96 kN as shown in Figs. 20.6, 20.7 and 20.8.

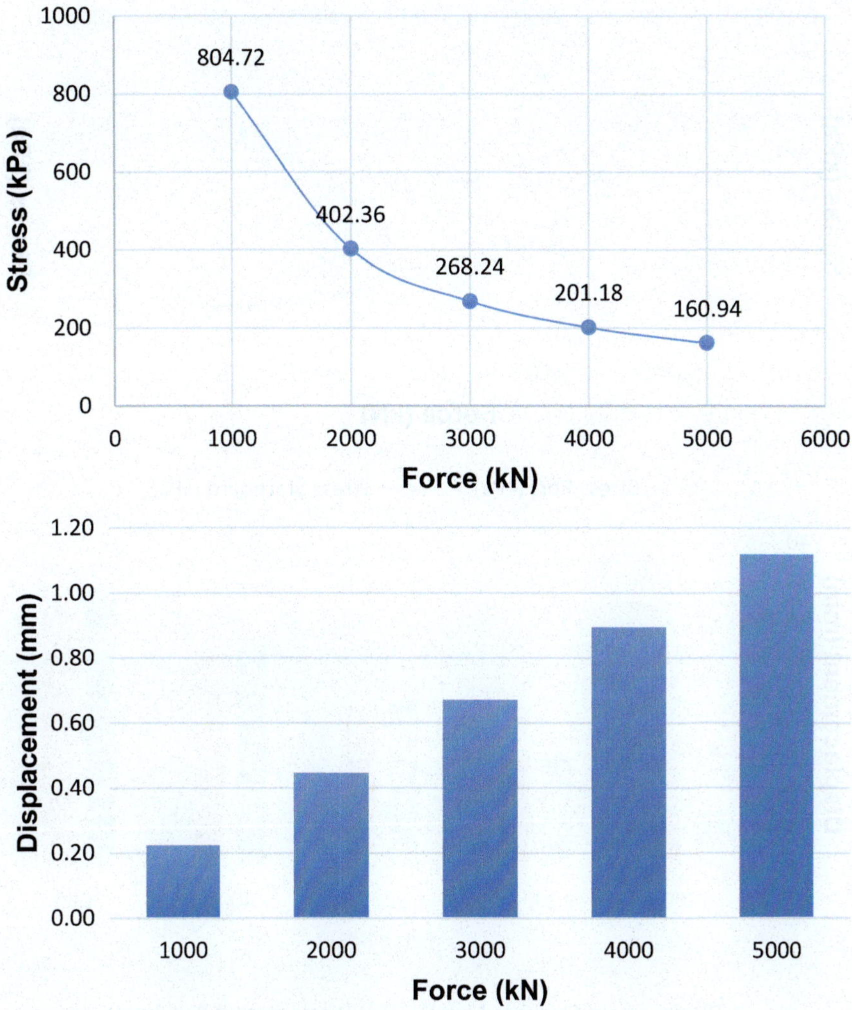

Fig. 20.5 Stress and displacement diagram for ahead

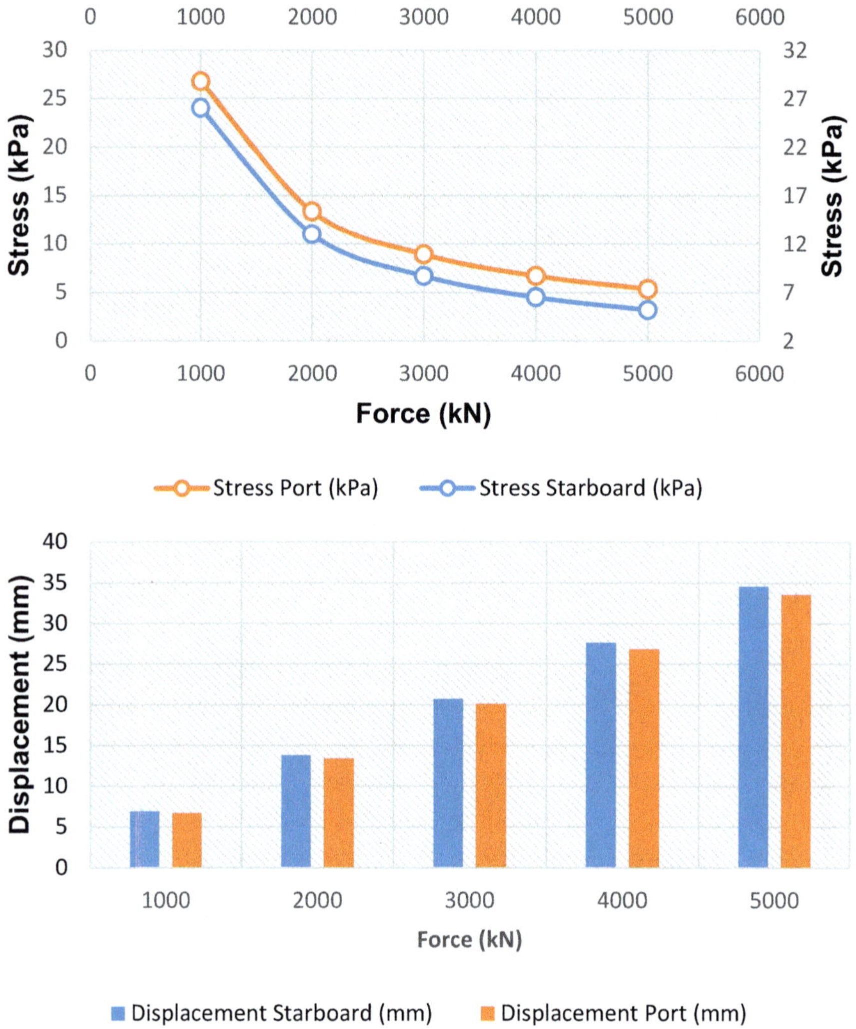

Fig. 20.6 Stress and displacement for port and starboard using composite material

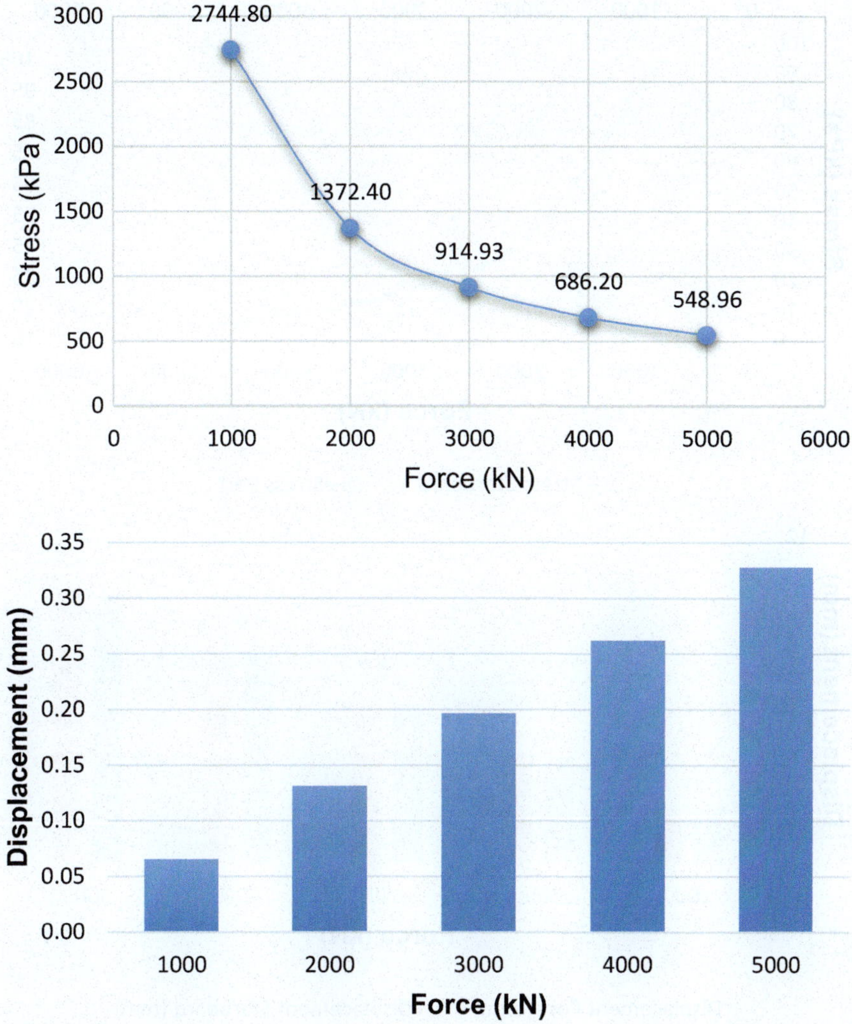

Fig. 20.7 Stress and displacement diagram for ahead

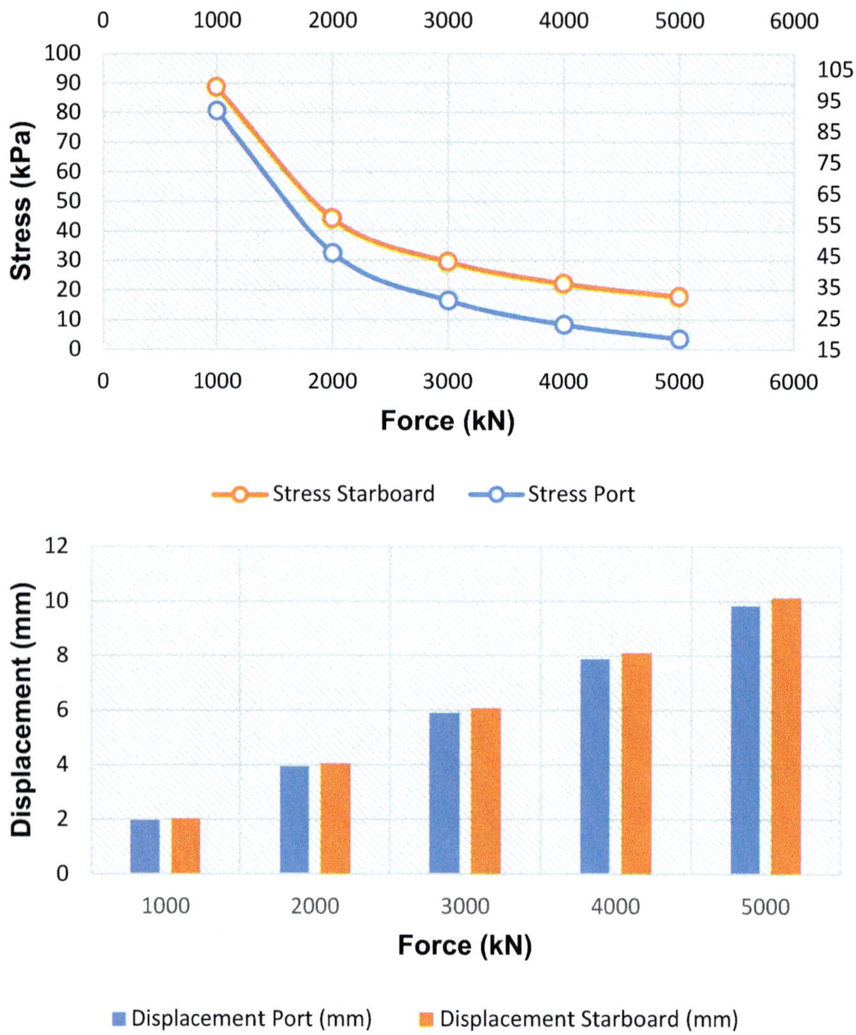

Fig. 20.8 Stress and displacement for port and starboard using aluminum material

20.4 Conclusion

In summary, this study utilizes finite element analysis to explore the correlation between stress and displacement. The acquired data streamlines the material quality assessment for the manufacturer before initiating the boat fabrication process. Overall, the project successfully achieved its objectives. Firstly, boat models were designed in response to specific requirements. Secondly, data from the FEA analysis was employed to determine optimal mechanical properties based on material selection, with composite and aluminum chosen for this investigation. When subjected to a 5000 kN force, the composite material exhibited a high stress value of 160.94 kPa, while aluminum registered a stress value of 548.96 kPa. Regarding displacement, the model with a 5000 kN force displayed measurements of 1.118 mm for the composite base and 0.328 mm for aluminum. The stress on the specimen varies according to the applied force rate, and these findings can be compared with a slight reduction in stress distribution across all participating components, as previously indicated [12]. The mechanical characteristics of the model exhibit changes as the applied force rate increases.

Acknowledgements The author would like to thank the Ministry of Higher Education (MoHE), Malaysia, for providing a grant (FRGS/1/2021/TK0/unikl/02/20) to the first author in carrying out and completing this work.

References

1. Chen, H.-P., Ni, Y.-Q.: Introduction to structural health monitoring. In: Structural Health Monitoring of Large Civil Engineering Structures, pp. 1–14. Wiley (2018). https://doi.org/10.1002/9781119166641.ch1
2. Bhosale, A.A., Patil, S.J.: Structural Health Monitoring: The Process of Implementing a Damage Detection (2019)
3. Abramovich, H.: 10. Structural health monitoring (SHM). Adv. Aerosp. Mater. **2019**, 272–305 (2019). https://doi.org/10.1515/9783110537574-010
4. Kefal, A., Oterkus, E.: Displacement and stress monitoring of a chemical tanker based on inverse finite element method. Ocean Eng. **119**(15), 2–15 (2016). https://doi.org/10.1016/j.oceaneng.2016.04.025
5. Kefal, A., et al.: Three dimension shape and stress monitoring of bulk carrier based on iFEM methodology. Ocean Eng. **147**(13), 256–267 (2018). https://doi.org/10.1016/j.oceaneng.2017.10.040
6. Ji, S., et al.: Automated subject-specific, hexahedral mesh generation via image registration. Finite Elem. Anal. Des. **47**(10), 1178–1185 (2011). https://doi.org/10.1016/j.finel.2011.05.007
7. David Polly, P.: G563 Quantitative Paleontology 2012 (2019) [online]. Available http://www.lusas.com/products/options/imdplus.html
8. Durand, R., Pantoja-Rosero, B.G., Oliveira, V.: A general mesh smoothing method for finite elements. Finite Elem. Anal. Des. **158**, 17–30 (2019). https://doi.org/10.1016/j.finel.2019.01.010
9. Finite Element Analysis Methods 1.1 Introduction (2009)
10. Birman, V., Genin, G.M.: Linear and nonlinear elastic behavior of multidirectional laminates. In: Comprehensive Composite Materials II, pp. 376–398. Elsevier (2017). https://doi.org/10.1016/B978-0-12-803581-8.09881-7
11. Carrera, E., Fazzolari, F., Cinefra, M.: Chapter 4—Fundamental of Mechanics of Beams, Plates and Shells. https://doi.org/10.1016/B978-0-12-849892-7.00006-6
12. Salleh, Z., Wei Kee, G.: Simulation of Three-Point Bending Sandwich Composite Panel Through Finite Element Analysis (2022)

Chapter 21
Chloride-Induced Damage Monitoring of Reinforced Concrete Using Ultrasonic Pulse Wave-Based Machine Learning

Julfikhsan Ahmad Mukhti and Seong-Hoon Kee

Abstract This research aims to utilize the machine learning approach for detecting the early stages of concrete corrosion within reinforced concrete based on ultrasonic pulse wave data. The study was conducted by performing accelerated corrosion on specimens with different rebar sizes and mix design strength, with each combination subjected to three levels of corrosion. The collected ultrasonic pulse data is then preprocessed, which includes smoothing, normalization, resampling, and shortening of the full-length signal to be used as the input for machine learning model training. Machine learning models developed in this study were based on the support vector machine (SVM), K-nearest neighbor (KNN), and subspace-KNN ensemble methods. We found that the machine learning models have a respectable performance for classifying concrete with a corrosion level of less and more than 3%. The SVM has the best-performing accuracy of 71.8%, F1-score of 0.715, and Cohen's kappa of 0.430.

Keywords Machine learning · Concrete corrosion · Ensemble · K-nearest neighbor · Support vector machine

21.1 Introduction

Concrete is the most commonly used material for modern infrastructures and buildings worldwide [1]. In most cases, the proportion of concrete used in buildings is twice as much as all other materials combined [2]. However, the overproduction of concrete creates pollution in the atmosphere [3] and global water supply [4]. As with

J. A. Mukhti · S.-H. Kee (✉)
Department of ICT Integrated Ocean Smart Cities Engineering, Dong-A University, Busan, South Korea
e-mail: shkee@dau.ac.kr

J. A. Mukhti
e-mail: julfikhsanam@donga.ac.kr

A. Ismail et al. (eds.), *Exploring Diversity in Engineering and Technology for Knowledge and Innovation*, Advanced Structured Materials 215,
https://doi.org/10.1007/978-3-031-64330-9_21

any construction material, concrete also deteriorates over time, making the material required to be monitored during the construction phase and on its service life. Therefore, improving the concrete lifespan on structures is necessary to reduce the negative impacts of concrete use and production, further supporting worldwide sustainability improvement.

The leading cause of concrete deterioration is reinforcement bar (rebar) corrosion [5]. The cost of repair and rehabilitation of deteriorated concrete on structures has been a significant financial impact on many countries. As an example, the cost of corrosion in the USA has been estimated to be approximately US$ 276 billion, with 8% of the cost from infrastructure [6].

The mechanism of chloride-induced deterioration occurs in three phases: initiation, rust propagation, and acceleration [7]. As the corrosion progresses, the damage level increases at an exponential rate. On the other hand, the repair cost is also expected to increase since more expensive repair methods (e.g., grouting) have to be used instead of more affordable solutions (e.g., recoating) that can be used to cover microcracks. Figure 21.1 shows the phases of concrete corrosion and its expected repair cost for each phase.

One of the most commonly used acoustic wave NDE methods for reinforced concrete is ultrasonic pulse wave. The method emits an ultrasonic pulse signal from one transducer to the receiving transducer through a medium. During the signal traveling period, the signal is diffused by the concrete due to its heterogenous composition [8]. The diffusion process includes signal scattering and reflection, which affects how the signal is received on the receiving transducer. Factors affecting the diffusion process include the aggregate composition, air voids, and microcracks [8].

The most commonly used property of the ultrasonic pulse wave method is its velocity, in which the measurement method has been standardized [9]. The pulse velocity method has been known to detect large defects such as delamination [10],

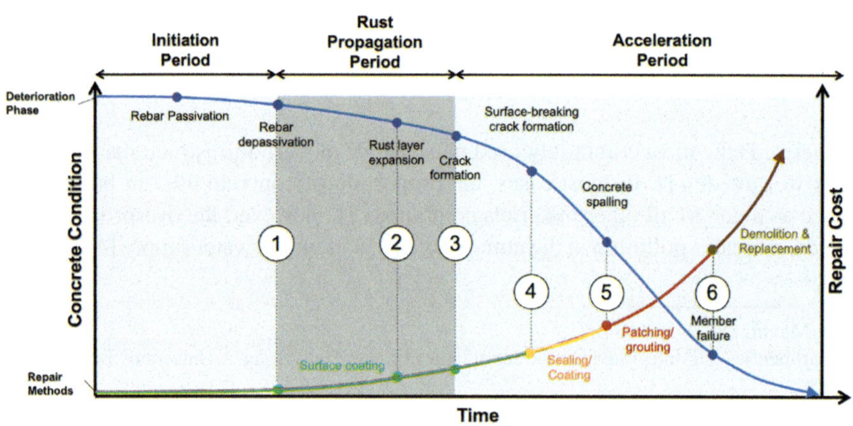

Fig. 21.1 Conceptual illustration of reinforced concrete deterioration process and its expected repair cost

surface-breaking cracks [11], and bottom-up cracks [12]. However, this method is not sensitive enough to detect microcracks where the gap created by the crack is not significant to delay the first arrival of the wave. Recent research explored nonlinear approaches to the analysis of the ultrasonic pulse wave. This approach analyzes the pulse wave in the frequency domain instead of the time domain. Recent research of nonlinear methods utilizing sideband peak count index (SPC-I) [13] and energy redistribution [14] has been proven to be able to detect corrosion initiation with a large increase in the parameter value around the cracking period. However, since the value parameter returns to ordinary value after some period, it is difficult to measure the severity of the damage based on nonlinear parameters.

A newer approach to signal analysis is by using machine learning to classify the concrete corrosion rate. The machine learning approach of ultrasonic pulse wave analysis has been thoroughly used in the past for the prediction of mechanical properties [15]. Using machine learning-based analysis, ultrasonic pulse wave has also been used for detecting cracks [16], voids [17], and corrosion [18]. However, there are currently no studies that utilize machine learning to identify the corrosion rate on heterogenous concrete mix designs in their initiation phase.

Based on the literature review above, it was found that existing NDE methods are not effective in detecting the early stages of concrete corrosion. Moreover, there are currently not enough studies regarding the use of machine learning to classify concrete rebar corrosion levels in the early stages with heterogeneous mix design, despite its potential. This study aims to investigate the effectiveness of the machine learning approach in analyzing the ultrasonic pulse wave data for the classification of concrete corrosion level in the early stage, which consists of the initiation period and early rust propagation period.

21.2 Methodology

21.2.1 General Methodology

The study was performed in the following order: (1) specimens were prepared with various mix designs and rebar sizes, (2) the specimens were subjected to accelerated corrosion with ultrasonic pulse wave data collected before, and after the process, (3) the collected data was preprocessed to be as effective as possible for training the machine learning models, (4) the preprocessed data was separated into datasets with the cracking condition and data collection points as the variables, (5) each dataset was used to train the prediction models, (6) the prediction model is tested using the testing dataset, and (7) the test result is evaluated based on established performance evaluation metrics. The following subchapters explain the detail of the methodology.

21.2.2 Specimen Preparation

In this study, reinforced concrete specimens in the form of 200 × 200 × 200 mm^3 cubes are prepared to be subjected to accelerated corrosion. The dimensions of the concrete specimens are shown in Fig. 2a. Variables of the specimens include three different compressive strength designs at 18, 24, and 40 MPa, referred to as mix 1, 2, and 3, respectively. Each specimen has a rebar with various diameters (10–13–19–22 mm, referred to as D10, D13, D19, and D22) planted in the middle of the specimen. The rebar is covered with 100 mm PVC on its middle part, so the corrosion is only expected to be focused on the concrete's center of gravity. The material composition of each mix design is given in Table 21.1. There were twelve combinations of specimen properties, with each combination subjected to different theoretical corrosion levels. The target theoretical corrosion level for D10 and D13 specimens were 3, 6, and 12%, while for D19 and D22 the targets were 5, 10, and 20%. Finally, there were three identical copies for each combination of mix design, rebar diameter size, and theoretical corrosion level.

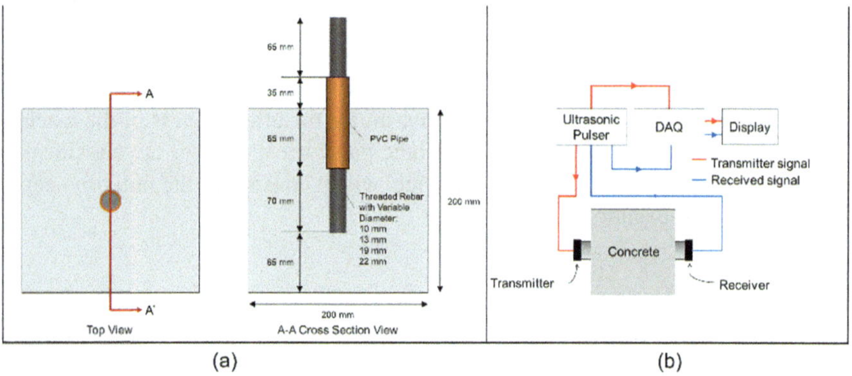

Fig. 21.2 Concrete specimen **a** design and **b** data acquisition setup

Table 21.1 Material composition of mix designs

Mix design	w/c ratio	Composition (kg/m^3)				
		Water	Cement	Gravel	Sand	Admixtures
Mix 1	0.585	168	287	898	957	2.58
Mix 2	0.507	170	335	956	870	2.50
Mix 3	0.346	166	480	993	720	4.32

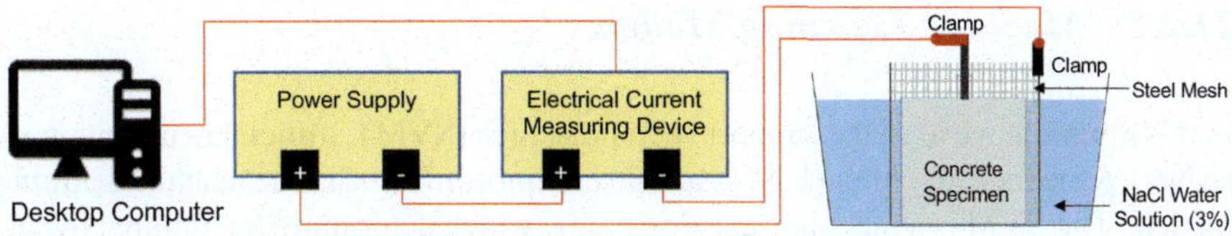

Fig. 21.3 Accelerated corrosion setup for the concrete specimens

21.2.3 Corrosion Acceleration

Each specimen was immersed in a 3% NaCl solution for five days during the accelerated corrosion process. Saturated surface dryness is expected to be reached within that period. After five days, the specimen was electrified with the setup, as shown in Fig. 21.3. The rebar became the anode in this setup, while the steel wire mesh became the cathode. The current was sent to the rebar through the NaCl solution. The theoretical mass loss was monitored using the current measuring device. The accelerated corrosion was stopped when the designated level had been reached.

Our experiment shows that surface-breaking cracks only appear on the D19 and D22 specimens when the designated theoretical mass loss has been reached. The specimens were then broken down, and the rebars were cleaned from the rust product to calculate the weight. The cleaned rebar was then weighed, in which the difference in weight between uncorroded and corroded rebar is the actual steel loss. This actual steel loss data was used as the corrosion level parameter in this study.

21.2.4 Ultrasonic Pulse Wave Measurement

The ultrasonic pulse data is collected from the specimens according to the ASTM C597M-16 ‘Standard Test Method for Pulse Velocity Through Concrete’ [9], with the setup shown in Fig. 2b. Equipment used in the experiment includes a 100 V square pulse generator (Olympus 5077PR), a data acquisition device (National Instruments PXIe-1073 Digital Oscilloscope), and a pair of transducers with a central frequency of 50 kHz as signal transmitter and receiver. During the data collection, the pulse generator generated a 10 ms square wave pulse that was emitted to the transmitter. The receiver then captures the signal on the opposite side of the specimen. The signal is then recorded by the acquisition device, which has a sampling rate of 10 MHz and captures 10 ms of signal data. In this study, the measurement was conducted before and after the accelerated corrosion process. On each measurement, data was collected with the transducers flanking the medium without any rebar (referred to as ‘Point 1’) and with rebar (referred to as ‘Point 2’) between them in a fully saturated condition. After the accelerated corrosion process ended, it was found that surface-breaking cracks were only found on specimens with D19 and D22 rebars. Pulse data was collected ten times on each Point 1 and Point 2.

21.2.5 Machine Learning Models

In this research, we used the support vector machine (SVM), artificial neural network (ANN), K-nearest neighbor (KNN), and an ensemble method as the machine learning models. The SVM is a machine learning model which determines the optimal hyperplane between data groups to classify the data. Most applications of the SVM model for concrete lean toward predicting compressive strength data through regression are based on a previously collected database [19]. The ANN, on the other hand, is a computational model that adapts the concept of the human brain. They consist of three main parts: input layer, inner or hidden layers, and output layer [20]. The inner layers consist of connected neurons, with each connection containing a specific weight w. Each neuron has its own threshold value and activation function $\boldsymbol{\Gamma}$ with its behavior depending on the input [20]. As with SVM, ANN has also been used to determine concrete compressive strength [21].

KNN is a classification method based on the proximity of the input to the existing data. The analyzed data (input) is classified either by counting the majority of the closest existing data ('neighbors') or weighted by distance. The ensemble method combines several machine learning classification models to produce the best prediction model. Several studies have shown that the ensemble method performs better than using only one machine learning model in some cases for determining concrete properties [22].

21.2.6 Data Preprocessing and Distribution

The original data has a 10 MHz sampling rate and 10 ms temporal length in its original form. To optimize the learning process, the data was (1) smoothed, (2) normalized, (3) resampled, and (4) shortened. The smoothing was done using a locally weighted scatter plot for every 30 data points to reduce the noise within the signal. Every sample was normalized, so each has the same amplitude. A previous study has pointed out that normalization improves machine learning performance [23]. The data was resampled from 10 to 1 MHz to reduce the training time without sacrificing signal clarity. Finally, for specimens only 20 cm thick, the signal is only significant in amplitude oscillation up to only 5 ms, which consists of 5000 data points.

Additionally, a dataset containing features in the time-frequency domain was also used. The feature used was spectral entropy, which is based on Shannon entropy in the field of information theory [24]. Using this feature, the number of data points in one sample is reduced to only 129.

The dataset prepared for training contained all specimens, with each specimen represented by two ultrasonic pulse data sampled on Point 1 and Point 2. The ultrasonic pulse data is then classified into two classes: specimens with less and more than 3% corrosion, represented as Class 1 and Class 2, respectively. Specimens in

Class 1 were assumed to be within the corrosion initiation phase, while specimens in Class 2 were assumed to be in the rust propagation and corrosion acceleration phase. For the development of machine learning models, 80% of the data was randomly allocated for training, while the remaining 20% was used for testing.

21.3 Result and Discussion

To evaluate the trained model, we used three statistical parameters: accuracy, F1-score, and Cohen's kappa. Accuracy is the ratio between correctly predicted data and the total number of predictions. Based on our preliminary test, the SVM and KNN methods perform best with cubic functions. On the other hand, the ensemble method has the best performance with the combination of subspace and KNN. This is in line with a previous study by Ma et al. [25], which presented that putting the data into subspace before the KNN can significantly improve modeling accuracy and reduce errors.

The performance evaluation of the classification result is given in Table 21.2. Based on the obtained result, the SVM and ensemble methods produce the highest accuracy of 71.8% using the normalized time series data. Using the entropy data, the obtained accuracy is slightly lower at 65.4% for SVM and significantly lower at 56.4% for the ensemble. However, using the KNN method, we found that the dataset with entropy input has a better accuracy at 67.9% compared to the same model with time series, which only obtained 60.3% accuracy.

The obtained macro-F1-score is very similar to the accuracy. However, the Cohen's kappa value shows a significant contrast between methods: only the SVM and ensemble methods with time series show input the Cohen's kappa value above 0.410. This means these methods reached a moderate agreement, which is typically in the range of 0.41 to 0.60 [26].

To further investigate the model performance, we analyzed the confusion matrices of each method using the best-performing input, as shown in Fig. 21.4. Based on each matrix, it can be seen that the true positive or TP for Class 1 was achieved by the

Table 21.2 Performance evaluation result of SVM, KNN, and ensemble machine learning models with time series and entropy input

Mix design	SVM—Cubic		KNN—Cubic		Ensemble—subspace-KNN	
	TS	EN	TS	EN	TS	EN
Accuracy (%)	71.8	65.4	60.3	67.9	71.8	56.4
F1-score	0.715	0.647	0.582	0.669	0.710	0.566
Cohen's kappa	0.430	0.294	0.149	0.333	0.414	0.113

TS = Time series input; EN = Entropy input

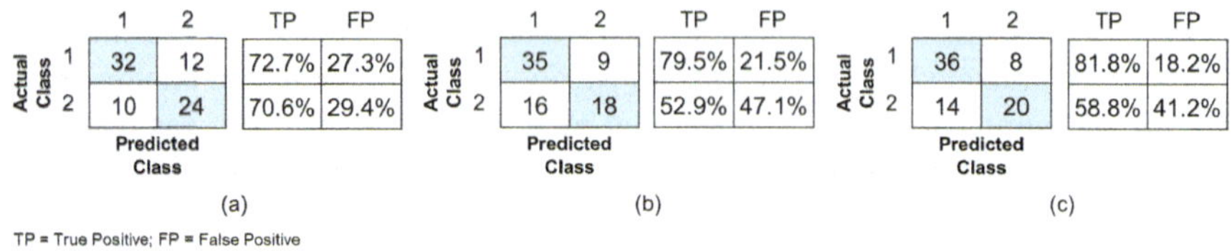

Fig. 21.4 Confusion matrices of the best-performing inputs from **a** SVM, **b** KNN, and **c** ensemble. Numbers within the each matrix represented the number of predictions

ensemble method, followed by the KNN and SVM. However, there is a significant difference in TP to Class 2 on the ensemble and KNN methods, where each only obtained 58.8 and 52.9%. The SVM, however, has a relatively balanced TP in each class, where it obtained 72.7% in Class 1 and 70.6% in Class 2.

Based on the above assessment, while both SVM and ensemble have very similar performance metrics, the SVM method is preferred for the following reasons: (1) the F1-score and Cohen's kappa were slightly higher on the SVM, and (2) the number of TP predictions on SVM was more balanced between classes. The balance is important because having a large gap of TP between classes will make the model more likely to predict one class over another, which makes the model ineffective as the signal profile starts shifting to the other class as the corrosion progresses.

21.4 Conclusion

Ultrasonic pulse wave data has been collected from specimens with four different rebar sizes, three different mix designs, and various corrosion levels. The collected data was then analyzed by three machine learning methods (support vector machine, K-nearest neighbor, and ensemble) to generate a classification model for concrete specimens with a corrosion level below and above 3%. Based on the obtained result, the following is the conclusion of this study:

(1) The machine learning method has been proven to be a viable method for the classification of concrete corrosion levels even for specimens without visible surface-breaking cracks, with the accuracy reaching above 70% and moderate agreement based on Cohen's kappa.
(2) The support vector machine method with time series input has the best performance with an accuracy of 71.8%, F1-score of 0.715, and Cohen's kappa of 0.430.
(3) Similar to the support vector machine, the ensemble method consisting of subspace and K-nearest neighbor using the time series input has an accuracy of 71.8%, F1-score of 0.710, and Cohen's kappa 0.414.

(4) The support vector machine model has a significantly more balanced true positive prediction between classes compared to the ensemble model. Therefore, we have more confidence in the support vector machine model over the ensemble method.

Acknowledgements This work was supported by the Korea Agency for Infrastructure Technology Advancement (KAIA) grant funded by the Ministry of Land, Infrastructure and Transport (Grant 21CTAP-C163815-01).

References

1. Venkatesan, P., Rajendran, A., Srinivasan, S., Kannan, S.: Corrosion of reinforced concrete exposed to marine atmosphere. Trans. SAEST **38**, 53–56 (2003)
2. Gagg, C.R.: Cement and concrete as an engineering material: an historic appraisal and case study analysis. Eng. Fail. Anal. **40**, 114–140 (2014). https://doi.org/10.1016/j.engfailanal.2014.02.004
3. Murugan, U.R.: Assessment condition of RC corroded column by non-destructive testing methods. Mater. Today Proc. **45**, 6645–6648 (2020). https://doi.org/10.1016/j.matpr.2020.11.866
4. Jose, A., Marquez, S., Nokken, M.R., Concordia, U.: Evaluation of saturation on surface electrical resistivity measurements evaluation of saturation on surface electrical resistivity measurements (2020)
5. Ranjith, A., Balaji Rao, K., Manjunath, K.: Evaluating the effect of corrosion on service life prediction of RC structures—a parametric study. Int. J. Sustain. Built Environ. **5**, 587–603 (2016). https://doi.org/10.1016/j.ijsbe.2016.07.001
6. Chen, F., Baji, H., Li, C.Q.: A comparative study on factors affecting time to cover cracking as a service life indicator. Constr. Build. Mater. **163**, 681–694 (2018). https://doi.org/10.1016/j.conbuildmat.2017.12.120
7. Bezuidenhout, S.R., van Zijl, G.P.A.G.: Corrosion propagation in cracked reinforced concrete, toward determining residual service life. Struct. Concr. **20**, 2183–2193 (2019). https://doi.org/10.1002/suco.201800275
8. Ahn, E., Shin, M., Popovics, J.S., Weaver, R.L.: Effectiveness of diffuse ultrasound for evaluation of micro-cracking damage in concrete. Cem. Concr. Res. **124**, 105862 (2019). https://doi.org/10.1016/j.cemconres.2019.105862
9. American Society for Testing and Material: ASTM C597-09 standard test method for pulse velocity through concrete (2009)
10. Petro, J.T., Kim, J.: Detection of delamination in concrete using ultrasonic pulse velocity test. Constr. Build. Mater. **26**, 574–582 (2012). https://doi.org/10.1016/j.conbuildmat.2011.06.060
11. Kee, S.H., Zhu, J.: Using piezoelectric sensors for ultrasonic pulse velocity measurements in concrete. Smart Mater. Struct. **22**(11), 115016 (2013). https://doi.org/10.1088/0964-1726/22/11/115016
12. Castellano, A., Fraddosio, A., Piccioni, M.D., Kundu, T.: Linear and nonlinear ultrasonic techniques for monitoring stress-induced damages in concrete. J. Nondestruct. Eval. Diagnostics Progn. Eng. Syst. **4**(4), 041001 (2021). https://doi.org/10.1115/1.4050354
13. Arumaikani, T., Sasmal, S., Kundu, T.: Detection of initiation of corrosion induced damage in concrete structures using nonlinear ultrasonic techniques. J. Acoust. Soc. Am. **151**, 1341–1352 (2022). https://doi.org/10.1121/10.0009621
14. Basu, S., Thirumalaiselvi, A., Sasmal, S., Kundu, T.: Nonlinear ultrasonics-based technique for monitoring damage progression in reinforced concrete structures. Ultrasonics **115**, 106472 (2021). https://doi.org/10.1016/j.ultras.2021.106472

15. Ben Chaabene, W., Flah, M., Nehdi, M.L.: Machine learning prediction of mechanical properties of concrete: critical review. Constr. Build. Mater. **260**, 119889 (2020). https://doi.org/10.1016/j.conbuildmat.2020.119889
16. Karimpouli, S., Tahmasebi, P., Saenger, E.H.: Ultrasonic prediction of crack density using machine learning: a numerical investigation. Geosci. Front. **13**, 101277 (2022). https://doi.org/10.1016/j.gsf.2021.101277
17. Saechai, S., Kongprawechnon, W., Sahamitmongkol, R.: Test system for defect detection in construction materials with ultrasonic waves by support vector machine and neural network. 6th Int. Conf. Soft Comput. Intell. Syst. 13th Int. Symp. Adv. Intell. Syst. SCIS/ISIS 2012, 1034–1039 (2012). https://doi.org/10.1109/SCIS-ISIS.2012.6505090
18. Zhang, J., Zhang, M., Dong, B., Ma, H.: Quantitative evaluation of steel corrosion induced deterioration in rubber concrete by integrating ultrasonic testing, machine learning and mesoscale simulation. Cem. Concr. Compos. **128**, 104426 (2022). https://doi.org/10.1016/j.cemconcomp.2022.104426
19. Muliauwan, H.N., Prayogo, D., Gaby, G., Harsono, K.: Prediction of concrete compressive strength using artificial intelligence methods. J. Phys. Conf. Ser. **1625**(1), 012018 (2020). https://doi.org/10.1088/1742-6596/1625/1/012018
20. Dastres, R., Soori, M.: Artificial neural network systems. Int. J. Imaging Robot. **21**, 13–25 (2021)
21. Tran, V.Q., Mai, H.V.T., Nguyen, T.A., Ly, H.B.: Investigation of ANN architecture for predicting the compressive strength of concrete containing GGBFS. PLoS ONE **16**, 1–21 (2021). https://doi.org/10.1371/journal.pone.0260847
22. Prayogo, D., Santoso, D.I., Wijaya, D., Gunawan, T., Widjaja, J.A.: Prediction of concrete properties using ensemble machine learning methods. J. Phys. Conf. Ser. **1625**(1), 012024 (2020). https://doi.org/10.1088/1742-6596/1625/1/012024
23. Akande, K.O., Owolabi, T.O., Twaha, S., Olatunji, S.O.: Performance comparison of SVM and ANN in predicting compressive strength of concrete. IOSR J. Comput. Eng. **16**, 88–94 (2014). https://doi.org/10.9790/0661-16518894
24. Asgharzadeh-Bonab, A., Amirani, M.C., Mehri, A.: Spectral entropy and deep convolutional neural network for ECG beat classification. Biocybern. Biomed. Eng. **40**, 691–700 (2020). https://doi.org/10.1016/j.bbe.2020.02.004
25. Ma, X., Yang, T., Chen, J., Liu, Z.: k-nearest neighbor algorithm based on feature subspace. Proc. 2021 Int. Conf. Big Data Anal. Comput. Sci. BDACS 2021, 225–228 (2021). https://doi.org/10.1109/BDACS53596.2021.00056
26. Ho, G.Y., Leonhard, M., Volk, G.F., Foerster, G., Pototschnig, C., Klinge, K., Granitzka, T., Zienau, A.K., Schneider-Stickler, B.: Inter-rater reliability of seven neurolaryngologists in laryngeal EMG signal interpretation. Eur. Arch. Oto-Rhino-Laryngology. **276**, 2849–2856 (2019). https://doi.org/10.1007/s00405-019-05553-y

Chapter 22
Interactive Model of a District Cooling System in a Cogeneration Plant

Adzuieen Nordin, Mohamad Asyraf Othoman, Didi Asmara Salim, and Shaharin Anwar Sulaiman

Abstract The interactive model of a district cooling system in a cogeneration plant is the model that allows researchers, students, and lecturers to know how the system of district cooling works. Basically, the method of teaching with chalk and talk from the lecturer to explain how the system run resulted in low understanding. This is because the real plant of the district cooling is huge and located far from the Institute of Politeknik Ungku Omar. In the program structure of Diploma in Mechanical Engineering (air conditioning and refrigeration), students in semester 4 must study the course of Air Conditioning and Refrigeration System which consists of the subtopic for district cooling system. Therefore, an interactive model of the district cooling system in cogeneration plants is developed for teaching and learning material. The most interesting part of this model is the synchronization between the models with the mobile application. Thus, the lecturer can explain with an attractive mode and the students can access and easily understand the function of each component of the district cooling system. The application is only available on android devices. From the pre- and post-test of using the interactive model, it is shown that the results of the theory test for Air Conditioning and Refrigeration System increased by 44%.

Keywords District cooling plant · Interactive model

A. Nordin (✉) · M. A. Othoman · D. A. Salim
Politeknik Ungku Omar, Ipoh, Perak, Malaysia
e-mail: adzuieen@puo.edu.my

M. A. Othoman
e-mail: masyraf@puo.edu.my

D. A. Salim
e-mail: didi@puo.edu.my

S. A. Sulaiman
Universiti Teknologi Petronas, Seri Iskandar, Perak, Malaysia
e-mail: shaharin@utp.edu.my

A. Ismail et al. (eds.), *Exploring Diversity in Engineering and Technology for Knowledge and Innovation*, Advanced Structured Materials 215,
https://doi.org/10.1007/978-3-031-64330-9_22

22.1 Introduction

In Politeknik Ungku Omar, the program structure of the Diploma in Mechanical Engineering (Air Conditioning and Refrigeration) requires students in the 4th semester to study courses in Air Conditioning and Refrigeration Systems, which consist of the subtopic cogeneration plant as a district cooling system. The cogeneration plant is defined as a simultaneous production of electricity and heat in a single source of input [1, 2]. The advantage of the cogeneration plant is the ability to improve the efficiency of fuel used in the production of electricity and heat [2, 3]. As a result, fuel energy is saved, and it is more cost-effective for users to implement the cogeneration system. In addition, the increase in efficiency, which corresponds with a decrease in fuel use by a cogeneration plant, results in large reductions in greenhouse gas emissions.

The use of the cogeneration plant is not limited to only electricity and steam generation, but also includes the production of chilled water using the steam generated [4]. The system that is used to generate electricity and chilled water consists of a gas turbine (GT), a heat recovery steam generator (HRSG), and a steam absorption chiller (SAC). This is called the district cooling (DC) system. The use of the cogeneration plant is not limited to only electricity and steam generation, but also includes the production of chilled water using the steam generated [5].

Gas-fueled DC systems in Malaysia have been installed since 1997. Kuala Lumpur City Centre (KLCC) is among one of the earliest developers of the Gas-fueled DC system in Malaysia, followed by Universiti Teknologi Petronas (UTP), Kuala Lumpur International Airport (KLIA), Putrajaya International Convention Centre (PICC), and Wisma Putrajaya in Kuala Lumpur.

However, the observation found that the students' level of understanding was quite limited when the lecturer gave a static diagram of the processes occurring in the centralized cooling plant system. To begin with, students will be described about the centralized cooling plant system. Due to the varying academic background of students, it is very difficult for students to understand the concept of a centralized cooling plant system. To obtain certainty, a set of pre-tests was carried out to find out the level of students' understanding of the operations taking place in the centralized cooling plant system. This pre-test contains only basic questions about the centralized cooling plant system. Pre-test results found that the average student score was in the range of 30–50%.

Therefore, a 3D-shaped teaching and learning innovation product and a smartphone application have been developed to explain the actual process in the cogeneration plant which consists of a district cooling system. In the early stages, this innovation product was only for android device users as most users used android devices.

22.2 Innovation Implementation Method

This interactive model was conducted for two semesters for the Diploma in Mechanical Engineering (air conditioning and refrigeration) (DPU) in semester 4. This interactive model was delivered through face-to-face teaching during session 1 2021/2022. Teaching and learning using this method can be implemented during or outside class hours by using 3D models of plants and android devices as key components of learning.

The 3D models resemble the main components of the cogeneration and district cooling plant and are produced using recycled materials as shown in Figs. 22.1 and 22.2, respectively. LED strip lights and display units are used to illustrate the cycle of the cogeneration and district cooling plant system. The lights and displays of this unit are controlled by an Arduino or CIP Microcontroller that has been programmed.

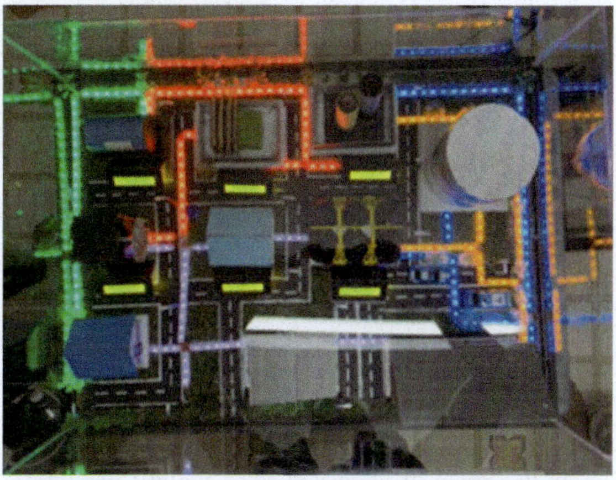

Fig. 22.1 3D interactive model from top view

Fig. 22.2 3D interactive model from side view

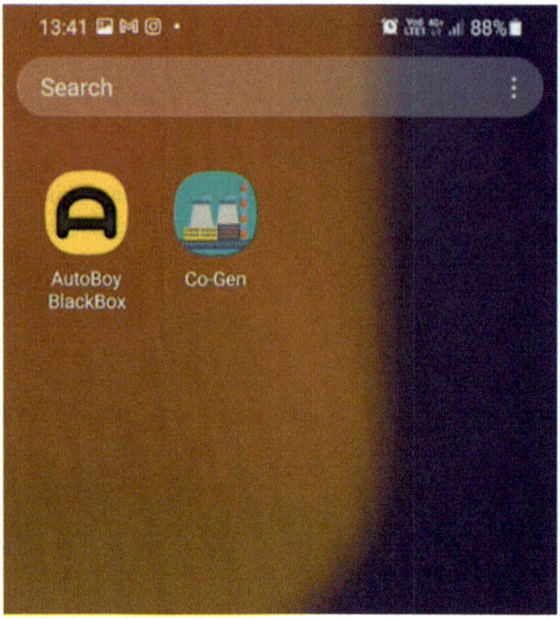

Fig. 22.3 Co-Gen DCS applications that can be downloaded through the Google Play store

Next, this Arduino is connected to an android device such as a smartphone or tablet with the Bluetooth method. Users need to download the DCS Co-Gen Application in the Google Play Store to explore the content of this interactive model. Users can choose the main components in the cogeneration and district cooling plant to find out the function of each main component such as gas turbine, auxiliary boiler, heat recovery steam generator, steam absorption chiller, thermal energy storage, electrical chiller, and others. Users can also control different colored LED lights to know the cycle of electrical flow, natural gas, chilled water, hot gases, and warm water (Fig. 22.3).

The differences in teaching and learning methods of the interactive model of cogeneration and district cooling plant are given in Table 22.1.

22.3 Results and Discussion

With the application of the interactive model of the cogeneration and district cooling plant teaching and learning improved by 44% with regard to the post-test scores. This shows a positive sign that this innovative product can achieve real teaching and learning objectives. Student achievement has increased as students have begun to be interested in what is being learned because they can see for themselves what is happening inside the centralized cooling plant. This also proves that interactive teaching can attract students [5]. Students gain more understanding to explain the

Table 22.1 Summary of differences in learning and learning implementation methods

Conventional teaching and learning	Teaching and learning using interactive model of co-generation and district cooling plant
• The lecturer explained the District Cooling Plant system using power point slides, static diagrams and animated videos	• Students and lecturers download the DCS Co-Gen App on the Google Play Store to explore the content of this interactive model • Students can choose the main components in the Co-Generation and District Cooling Plant to find out the function of each main component such as gas turbine, auxiliary boiler, heat recovery steam generator, steam absorption chiller, thermal energy storage, electrical chiller, and others • Students can also control different colored led lights to find out the cycle of electrical flow, natural gas, chilled water, hot gases, and warm water • Lecturer as a facilitator and explain the operation of the system in more depth

processes occurring in the centralized cooling plant system. This makes students more confident and motivated to master the field of air conditioning and cooling.

The success of the use of this interactive model was measured through the pre-test and post-test of DJU41012 subjects to students in session 1 2021/2022. One set of questions is used, and all the main components in the district cooling are tested in the question. The huge impact of student achievement can be seen through the pre-test and post-test process as shown in Fig. 22.4, when the average student scored 39% for pre-test and the increase in post-test achievement to 70%. This resulted in the score of the test increasing by 44%.

Apart from having a significant impact on student achievement, this innovation also saves lecturers time for the information process and manages to attract students to explore the topic. The students' confidence in explaining the district cooling system is also very convincing as well as enthusiastic as students are no longer learning in theory, but three-dimensional and interactive.

This innovation project is also very practical for lecturers to use. Lecturers do not need to sketch the components of the destring cooling by chalk and talk method. Lecturers only need to use the application on the phone to describe the system cycle.

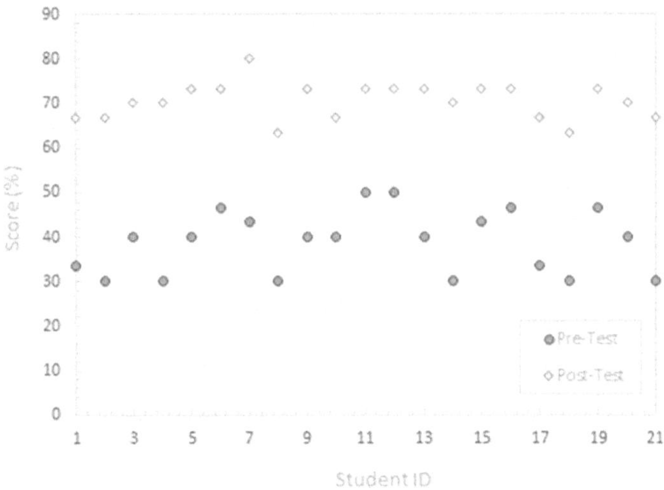

Fig. 22.4 Comparison of the students' results before (pre-test) and after (post-test) exposure to the newly developed interactive model

22.4 Conclusion

Overall, it can be concluded that the teaching materials developed can achieve the desired objective that students better understand the title for the subject DJU41012 Air Conditioning and Refrigeration System Components, District Cooling System through the 3D models and the interactive applications. In addition, students have a clearer understanding of the detail of the main components involved in a centralized cooling plant system. Therefore, interactive learning using the interactive model of cogeneration and district cooling plant highly impacts students' understanding and can expose students to the real situation in the industry.

Here it is also recommended that this Co-Gen application can be improved by including animations related to the functions of each main component. Several sets of questions in the form of quizzes can also be included to facilitate the process of assessing student understanding directly.

Acknowledgements The authors would like to express their appreciation to Politeknik Ungku Omar.

References

1. Rosen, M.A., Le, M.N., Dincer, I.: Efficiency analysis of a cogeneration and district energy system. Appl. Therm. Eng. **25**, 147–159 (2005)
2. Kanoglu, M., Dincer, I.: Performance assessment of cogeneration plants. Energy Convers. Manage. **50**, 76–81 (2009)

3. Martínez-Lera, S., Ballester, J.: A novel method for the design of CHCP (combined heat, cooling and power) systems for buildings. Energy **35**, 2972–2984 (2010)
4. Karim, Z.A.A., Yongo, P.W.: Analytical models for energy audit of cogeneration plant. J. Appl. Sci. **11**, 1519–1527 (2011)
5. Pradono, S., Astriani, M.S., Moniaga, J: A method for interactive learning. CommIT J. **7**(2), 46–48 (2013)

Chapter 23
Design of a Cold Room to Accommodate Excess Material for an External Warehouse

Woo Tze Keong, Adzuieen Nordin, and Mohd Nazrin Nahar

Abstract This project was conducted to design a cold room at an external warehouse, with the aim to increase the efficiency of inventory report cross-checking and to develop a standard operation procedure (SOP) for material movement. The data obtained is based on observation and a pilot test simulation before setup the actual design of the cold room. To execute this project, one design of the cold room has been designed to accommodate excess material at a minimum of 100 pallets. Therefore, material can be stored in the cold room with a specific temperature. The outcomes for the cold design can optimize space and accommodate more material in the cold room. This project also aims to increase the efficiency of the cross-checking inventory report between the internal and external warehouses and to develop a new SOP for movement of materials. To increase efficiency of the cross-checking report, one method has been suggested and some observations have been performed for movement of materials concept for SOP development in this project. The outcomes of this project are that one can store a higher volume of excess raw material to the cold room and a higher efficiency of the cross-checking inventory report between both warehouses.

Keywords Cold room · External warehouse · Inventory report · Standard Operation Process (SOP) · Warehouse

W. T. Keong (✉) · A. Nordin · M. N. Nahar
Politeknik Ungku Omar, Jalan Raja Musa Mahadi, 31400 Ipoh Perak, Malaysia
e-mail: wtkeong@puo.edu.my

A. Nordin
e-mail: adzuieen@puo.edu.my

M. N. Nahar
e-mail: nazrin@puo.edu.my

A. Ismail et al. (eds.), *Exploring Diversity in Engineering and Technology for Knowledge and Innovation*, Advanced Structured Materials 215,
https://doi.org/10.1007/978-3-031-64330-9_23

23.1 Introduction

Most cold rooms are able to handle temperatures ranging from – 2 to 10 °C [1]. The temperature range of a cold room is determined by the items being stored inside. A cool room temperature of 2–8 °C is appropriate for food items such as fresh produce. Pharmaceutical ICH storage will necessitate a temperature of 5 °C, with a 3 °C tolerance [2]. Freezer rooms, which may be set to – 30 °C for the storage of pharmaceuticals or chemicals, will be the preferable alternative when colder conditions are necessary [3]. This study explains how the range of the cold room temperature influences with their specification. For example, pharmaceuticals or chemical materials or products must be set at – 30 °C. So, that can maintain the quality of the products during storing in the cold room. However, for the cold room storage in an internal warehouse, for storage of chemical materials like DAF tape and mold compound, they need to be stored at 0 to – 10 °C for temperature control requirements.

The internal warehouse is the main warehouse for the company. Only one cold room has been setup in an internal warehouse to accommodate the chemical materials such as DAF tape, mold compound, and epoxy. The chemical materials need to be maintaining at the specific temperature requirement range in between 0 and – 10 °C. The capacity of the internal warehouse cold room only can accommodate a maximum of 28 pallets of chemical material. Therefore, the objective of this project is to design a cold room to accommodate excess of materials at an external warehouse, propose a new system of cross-checking report between the external warehouse, and develop a new standard operation procedure for the movement of materials.

23.2 Methodology

The main problem is that the cold room at the internal warehouse is unable to accommodate the excess of chemical materials. If the chemical material is not stored at the specific requirement temperature, the quality of the material will be affected and cannot be used for the production process. In order to accommodate all excess material, the flow process of the material movement from MMP internal warehouse to the external warehouse is included in this project. The inventory report is used to track the quantity of material transfer from the internal warehouse to the external warehouse. A lot of data for materials is available to verify in inventory report; thus, it may cause difficulties to cross-check between two locations which are the internal warehouse and the external warehouse.

The only extensively used method for large management of items between processing and distribution processing is the cold room. That is one of the strategies for preserving perishable items in a fresh and whole state for a longer length of time by regulating temperature and relative humidity within the storage system. The total capacity, number and size of cold rooms, refrigeration system, storage and handling equipment, and access facilities are all common features of a cold room operational system [4]. This study is mostly related to the design for storage of fruits and vegetables. The highlight of this study is the storage capacity and the design of the cold room. The author focuses more on the larger storage to install more chambers because the quantity of produce by unit of volume will increase the storage density. The design in rectangular shape is better than square shape.

Watching, listening, reading, and recording behavior and characteristics of phenomena are all examples of observation data gathering methods. As a data collection method, observation can be structured or unstructured. Data is collected using different methodologies and according to a pre-determined timetable in planned or dependent parameters. Observation is closely associated with ethnography, but is also common in action research, case study, and mixed methods as part of a bank of data collection methods to gather data on problems or issues or to answer a study's research questions. To be related in this project, the observation method can be completed by using simulation method and implementation of the process. By simulation method, the data can be collected and used as additional information for the process flow concept before to develop a standard operation procedure for movement of materials to the external warehouse. The first process is observation and from the observation one can deduct the problem statement of the project. Then, the next step is the design process and data collection.

A pilot testing will help alleviate some of these issues. A pilot test involves that the practitioner is run through a session or two ahead of the main scheduled study. All research methods, including the identification of eligible potential participants, their enrollment, recording of needed data, sample collection, supervision systems, quality control, and data processing, should be tested on a small scale. The pilot test should be designed to be as comparable as feasible to the techniques used in the main trial, and the participants should be representative of the trial population. In this project, the pilot test has been implemented for the design of the cold room for testing the capacity of the accommodated materials.

A standard operation procedure (SOP) guarantees that activities are carried out in a standardized and consistent manner, decreasing variation in how a process is carried out. SOPs placed around the premise that if of one does do a specific process regularly and in full compliance with the regulations, it is more likely to deliver a consistently high-quality finished product [5]. The SOP also can maintain a set of standard processes for the new standard process for any activities. For end-users of SOPs, they should have the chance to learn about the numerous issues involved with SOPs and also the benefits of following SOPs in order for SOPs to overcome obstacles and be recognized as practical guides toward reliable goods and services [2].

23.3 Results and Discussion

To solve all the issues on this inventory report cross-checking method, a Pivot table method has been created for a temporary solving of the cross-checking method between both reports. First step needs to eliminate the unnecessary information or contents on the external warehouse inventory report. As recommended, one only needs to select three important information on both reports as shown in Fig. 23.1. The information that needs to be stated for the cross-checking method are material, storage bin/pallet ID, and available stock for both inventory reports. So, it can be easier for the material specialist and user to cross-check between both inventory reports.

The result of the cross-checking method by using a Pivot table is shown in Fig. 23.2, i.e., Pivot table for the internal inventory report and a Pivot table for the external inventory report. Both Pivot tables have completely filtered the 542-XXX material part number. The total of the internal inventory report was tally compared to the external inventory report which is 32900 of total quantities available. This is a positive result by using this pivot table because it can increase the efficiency for cross-checking between both warehouses even for a temporary method.

In this study, a new system has been established for both side of warehouses that can directly provide and show all the detailed information in the report for automatically cross-checking. Based on Fig. 23.3, a recommended process flow is shown for the automatic cross-checking method for the inventory report. Shipping documents (Shipdoc) are the most important documents during transferring the material between both sides of warehouses. Shipdoc can keep all the details of the material including part number of the material, pallet ID, and quantity of the material. So, recommendation to implement this process will increase more efficiency for the cross-checking method because all of the process can be utilized by using the system.

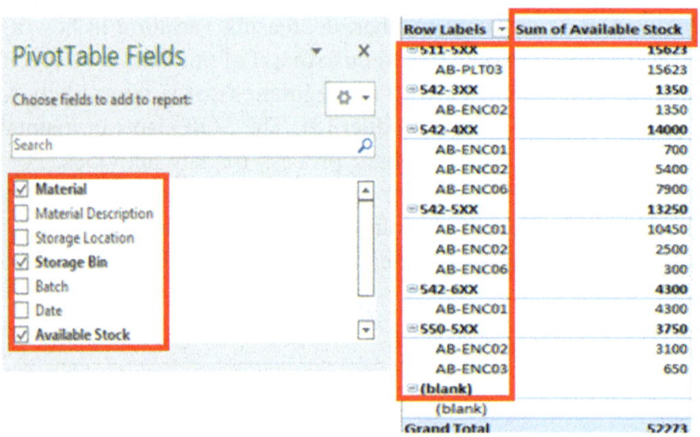

Fig. 23.1 Pivot table and selected information

Row Labels	Sum of Available Stock
542-3XX	1350
AB-ENC02	1350
542-4XX	14000
AB-ENC01	700
AB-ENC02	5400
AB-ENC06	7900
542-5XX	13250
AB-ENC01	10450
AB-ENC02	2500
AB-ENC06	300
542-6XX	4300
AB-ENC01	4300
Grand Total	32900

Row Labels	Sum of Available Qty
542-3XX	6400
AB-ENC01	6400
542-4XX	9250
AB-ENC02	9250
542-5XX	9050
AB-ENC01	9050
542-6XX	8200
AB-ENC06	8200
Grand Total	32900

Fig. 23.2 Cross-checking by using Pivot table (internal and external inventory report)

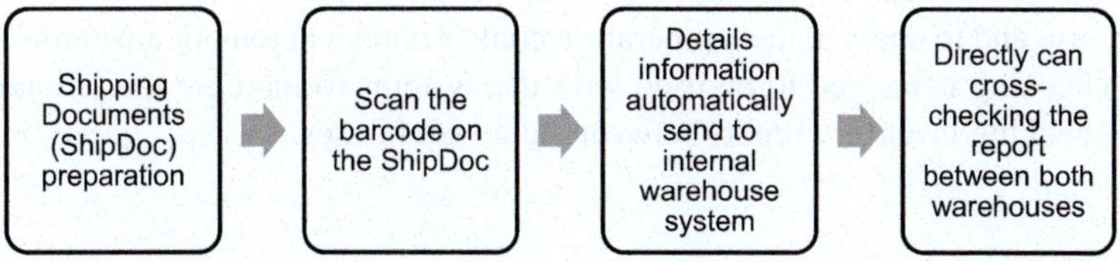

Fig. 23.3 Process flow for automatic cross-checking method

To develop and implement a SOP in an organization, one needs the have an observation and practice to complete some daily tasks that have been done by the material handler as a daily routine [6]. This procedure is to perform material sending/transfer from the internal warehouse to the external warehouse. The procedure also covers all types of materials from the internal warehouse. For the chemical material, cold room scope/requirements will be added in this procedure and freezer containers will be arranged to transfer chemical materials to the external warehouse. The freezer container shall ensure the temperature control range from 0 to – 10 °C while transferring chemicals to the external warehouse. The external warehouse shall manage all materials physically and systemically by using its system.

Table 23.1 shows the listing of standard operation procedures based on the observation result during the simulation process of movement of materials from the internal warehouse to the external warehouse. After implementation of the SOP, all the processes can be fully improved during the material transfer by using a freezer truck to the external warehouse. The internal warehouse management aims for zero incident during transferring materials to the external warehouse.

Table 23.1 List of standard operation procedure (SOP)

No	Standard operation procedure
1	Material transfer by using the freezer container
2	Shipment to external warehouse (MMP to EWH)
3	Shipment from external warehouse (EWH to MMP)
4	Chemical material pull-out from cold room procedure

23.4 Conclusion

As an overall conclusion for this project, the design of a cold room at the external warehouse completely can accommodate the excess material with more than 100 pallets of capacity. Moreover, the space can be optimized by using a two-layer racking system for the light material and utilizing the pallet space for storing heavy materials in the cold room. Other than that, the implementation of the standard operation procedure led to a smooth process during materials transfer from the internal warehouse to the external warehouse. As a rough recommendation for this project, one can utilize a new system for cross-checking inventory report which is more efficient compared to manually cross-checking by using the Microsoft Excel software. In the future study, this project can be continued to setup the designed cold room at the external warehouse and to analyze the temperature control and an automatic inventory report cross-checking is needed to be used with the systems to increase the efficiency of cross-checking inventory report between both warehouses.

References

1. Gopala Rao, C.: Design of small-capacity cold storages. In: Engineering for Storage of Fruits and Vegetables, pp. 389–425 (2015)
2. Hattemer-Apostel, R.: Standard operating procedures—a novel perspective. Qual. Assur. J. **5**(4), 207–219 (2001)
3. What is a cold room? Parameter. Retrieved June 12, 2022 (2021, September 9). From: https://humiditycontrol.com/what-is-a-cold-room/
4. Observation—Research-Methodology. Research (2019). Retrieved June 12, 2022, from: https://research-methodology.net/research-methods/qualitative-research/observation/
5. World Leaders in Research-Based User Experience. Pilot testing: Getting it Right (Before) the First Time. Nielsen Norman Group (2018)
6. Greg, A.: Reviewing the values of a standard operating procedure. Ethiop. J. Health Sci. (2020)

Chapter 24
Immersion Performance of Capsaicin Natural Antifouling Coating, Coated on Composite Fibreglass Marine Structures

Muhamad Sharin Izzuan Ramli, Mohamad Fahmi Shaharuddin, Syarmela Alaauldin, Puteri Zirwatul Nadila Megat, Norazlina Abd Nasir, Norazizah Che Mat, Sarah Nadiah Rashidi, Nor Aliya Hamizi, Mohd Rafie Johan, and Asmalina Mohamed Saat

Abstract The most successful antifouling (AF) paint which is tributyltin has been banned since 2008 by the International Maritime Organization (IMO). A new non-toxic and eco-friendly AF system is needed. In this project, a new AF paint was prepared with Capsaicin as a natural AF additive. Capsaicin is well known in the

---

M. S. I. Ramli · M. F. Shaharuddin · S. Alaauldin · P. Z. N. Megat · N. A. Nasir · N. C. Mat · S. N. Rashidi · A. M. Saat (✉)
Universiti Kuala Lumpur, Malaysian Institute of Marine Engineering Technology, 32200 Lumut Perak, Malaysia
e-mail: asmalina@unikl.edu.my

M. S. I. Ramli
e-mail: sharinizzuan.93@gmail.com

M. F. Shaharuddin
e-mail: mfahmi.shaharuddin@s.unikl.edu.my

S. Alaauldin
e-mail: syarmela@unikl.edu.my

P. Z. N. Megat
e-mail: puterizirwatul@unikl.edu.my

N. A. Nasir
e-mail: norazlinaan@unikl.edu.my

N. C. Mat
e-mail: norazizah@unikl.edu.my

S. N. Rashidi
e-mail: sarahnadiah@unikl.edu.my

N. A. Hamizi · M. R. Johan
Nanotechology and Catalyst Research Center, Universiti Malaya, Kuala Lumpur, Malaysia
e-mail: aliyahamizi@um.edu.my

A. Ismail et al. (eds.), *Exploring Diversity in Engineering and Technology for Knowledge and Innovation*, Advanced Structured Materials 215,
https://doi.org/10.1007/978-3-031-64330-9_24

pharmaceutical field to be good antimicrobial and environmentally friendly. Fibreglass panels with various percentages of Capsaicin were mixed in a rosin-modified base. The immersion performance of Capsaicin natural AF additives was conducted, and the percentage of weight change and marine growth was recorded. The increasing ratio of Capsaicin shows a decrease in the percentage of marine growth on fibreglass surfaces. The percentage of 6 and 8 wt% of Capsaicin additives has the lowest percentage of marine growth and weight change after being immersed in seawater. The regression analysis shows that uncoated obtained the highest R^2 value due to no AF protection mechanisms. The regression analysis also supports the finding that the optimum ratios of Capsaicin were in samples 6 and 8 wt% with the lowest R^2 due to the inhibited marine growth mechanism by Capsaicin.

Keyword Capsaicin antifouling · Composite fibreglass boat · Immersion · Marine growth

24.1 Introduction

The ship hull is one of the marine structures that is tremendously affected by marine growth. Marine growth began after immersion with the adhesion of organic particles. Within 24 h, bacteria diatom (microfouling) started to develop and in a week, macrofouling started to occur depending on the immersion location [1]. Marine growth can be divided into microfouling and macrofouling which is also known as animal and plant marine growth material. Examples of plant types are microalgae and macroalgae which are slime and weed. Animal marine growth occurs as soft bodies (tube worms) and hard shells (barnacles and mussels). The growth causes a critical challenge in the weight, resistance and performance of marine structures [2]. The colonization of marine growth, even in the form of slime, can increase in drag, weight, fuel consumption and cost of the total operation. Marine growth is reported to increase the hull surface roughness and lead to an increase in power consumption by 10–16% with just a slime attached to the surface.

Meanwhile, animal-type marine growth leads to an increase in weight reported to increase power consumption by up to 86% [3]. The cost of maintenance increases tremendously for the process of cleaning or hull scrubbing and recoating the hull surface. In the near coastal voyage, the hull is made from composite fibreglass, and hull scrubbing is not a favoured option. The use of effective protection strategies is crucial to ensure fuel saving and cleaner near coastal seawater [4]. Antifouling (AF) is a structure protection strategy commonly applied to marine structures. AF coatings are protective coating types of marine paints used to coat a ship's hull to prevent marine growth. The history of AF technology started with the use of lead and copper sheets attached to wooden boats. The usage of metal continued until tributyltin (TBT)

M. R. Johan
e-mail: mrafiej@unikl.edu.my

was introduced and widely used as AF. However, despite previously effective AF, TBT was banned by the IMO due to its toxicity. The development of new AF material needs to ensure an effective prevention mechanism and consider the level of toxicity, resistance and marine environment conditions [5]. In the case of near-voyage ships, it is much more detrimental to develop an effective AF mechanism as it will also affect marine food resources. Other challenges in marine AF are the various location, types of vessels and operation profiles [4].

Natural AF additives are reported to enhance the coating performance and meet the requirement by IMO. The usage of natural sources raised researcher interest due to their antibacterial properties, being environmentally friendly and easily available, and does not contain heavy metals. However, knowledge of the properties and various other types of natural sources in the marine field has still not been discovered. The development of effective natural AF requires an understanding of the interaction between marine growth and coated surfaces. Natural biocides are extracted from mangroves, seaweed, sponges, soft corals, fishes [6], algae [7], lemon [8], henna, Capsaicin [9] and others. Common polymer bases used in natural AF (fish and algae) were epoxy and vinyl chloride-vinyl acetate copolymer matrix [6]. The rosin modified which was extracted from the pine tree is a natural base of polymer matrix commonly used other than latex. The rosin-modified base is also reported to have good interaction with lemon natural biocide coated on fibreglass [8] and wood surface [10] as compared to commercial antifouling during the immersed period in seawater.

Capsaicin (Cap) has been established as one of the most effective natural antifoulants and is known as an alternative biocide for marine AF applications [11]. Cap is one of the main natural additives that have gained interest in AF due to its good antibacterial behaviour not only in the medical and biological fields but also in the application of marine technology [13]. Cap (trans-8-Methyl-N-vanillyl-6-nonenamide) is a steady, compelling and dominant component of the pepper with approximately about 69% of the total content of pepper. The component reacts as the active pungent component of chilli peppers, which effectively prevents and strongly inhibits various types of marine growth. A recent study involved Cap in the development of a pH-responsive smart coating in a multilayer film consisting of Cap, chitosan with polymer bases such as polyethylene glycol (PEG), polyvinyl alcohol (PVA) and alginate (ALG) and observed outstanding controlled release of Cap occurs in pH 8.5 [12].

Gu et al. summarized that the fouling release type coatings are suitable for ships with high speed and fast travel, and they are less effective for ships that are moored for a long time or travelling at low speed. Thus, to consider both moored and speed sailing, the coating needs to have multiple methods. The change in the surface performance of the coating or the release efficiency of AF additives is a combination of methods that can be solved by using both nanotechnology and natural resources, especially Cap additives [14]. Cap has been reported to have excellent AF activity and low environmental toxicity in the acrylate coating [15]. The mechanism of action of Cap in preventing marine fouling is not fully understood, but several proposed mechanisms suggest that the rapid increase in intracellular Ca^{2+} disrupts the cellular processes and inhibits fouling [16]. Cap derivatives have proven to inhibit the growth of various algae species, such as *P. tricornutum*, *Skeletonema costatum* and *Chaetoceros curvisetus* however no specific mechanism was discussed in this finding [17]. Cap is discussed as a toxic substance to microbe cells and contributes to preventing marine growth. Cap is also discussed as having pH-responsive properties that give antibacterial properties and prevent the attachment of marine growth [12].

The stable and controlled slow release of Cap from the coating is a challenge that researchers are working to solve [18]. It is crucial to understand the exact mechanism of action of Cap in preventing marine growth. The mechanism may vary, depending on the specific organism and environmental conditions. Further research is needed to fully understand the mechanism involved. Thus, this research focuses on modified commercial marine coatings with Cap natural additive. The modified Cap AF coating was immersed to evaluate the modified AF in the actual marine repairing and construction environment immersion.

24.2 Methodology

Composite fibreglass substrate panels with dimensions 150 mm × 400 mm are prepared using three layers of chopped strand mat (CSM) by the hand layup method. The preparation of composite fibreglass panels is benchmarking the fabrication process described elsewhere [19]. The coating formulation is prepared by using the rosin-modified base brand Binapaint obtained from Bina Integrated Sdn. Bhd. The various ratios of natural AF Cap additives (0, 2, 4, 6 and 8 wt%) were mixed with rosin and stirred to ensure the mixture was homogeneous. The composite fibreglass panels were coated with two layers of AF by using a brush. The method of coating application is similar to our previous paper [8]. The panels were immersed in seawater for 30 days to observe the marine growth activities following the ASTM D3623. The visual inspection, percentage marine growth and percentage weight change were recorded every 5 days. The observation data later were evaluated by using the regression analysis method to evaluate the performance of the natural Cap additives AF coating. Table 24.1 summarizes the natural Cap additives ratios and characterization conducted.

Table 24.1 Natural cap additives ratios and characterization

Specimen	Coating systems	Characterization
Uncoated	Fibreglass + gel coat	• Visual observation • Percentage weight change • Percentage of marine growth • Regression analysis
0 wt% cap	Fibreglass + gel coat + 0% AF	
2 wt% cap	Fibreglass + gel coat + AF + 2% cap	
4 wt% cap	Fibreglass + gel coat + AF + 4% cap	
6 wt% cap	Fibreglass + gel coat + AF + 6% cap	
8 wt% cap	Fibreglass + gel coat + AF + 8% cap	

24.3 Results and Discussion

Figure 24.1 shows the visual observation of the percentage marine growth for composite fibreglass panels for uncoated, 0% Cap, 2% Cap, 4% Cap, 6% Cap and 8% Cap. The uncoated samples observe an increase of 4.22% in marine growth. Visual observation shows that on day 0 samples are ready for immersion with no marine growth observed. The marine growth starts to grow with plant types (slime). The marine growth for uncoated samples shows the occurrence of animal-type marine growth at 4.22% on day 15 and significant growth to 30.84% on day 30. The major type of marine growth observed in all samples is slime. Samples with 6 and 8 wt% observed the lowest attachment of marine growth with 0.64% values for both panels.

Figure 24.2 shows the percentage marine growth (barnacles) for composite fibreglass panels, (a) uncoated; (b) 0% Cap; (c) 2% Cap; (d) 4% Cap; (e) 6% Cap and (f) 8% Cap. The percentage of marine growth (animal types) increases as the number of days increases. The uncoated fibreglass sample observed significant marine growth with a 30.84% increase after 30 days of immersion. Total % marine growth after 30 days of immersion for Cap natural AF samples with 0%, 2%, 4%, 6% and 8% ratio observed values of 1.94%, 1.94%, 0.97% and 0.64% weight increased, respectively. The percentage of marine growth is increasing due to the change from slime (plant type) into animal types of marine growth.

Figure 24.3 shows the percentage weight change for composite fibreglass panels, (a) uncoated; (b) 0% Cap; (c) 2% Cap; (d) 4% Cap; (e) 6% Cap and (f) 8% Cap. The percentage of weight change on composite fibreglass shows an increase in weight as the number of days increases. This finding is similar to the percentage of marine growth. The uncoated samples show the highest weight change after 30 days with 23.24% values. Meanwhile, the coated specimen observed with no Cap (0 wt%) shows the highest value (2.24%) and panels with 8 wt.% Cap show the lowest weight change (1.88%) after 30 days. This shows the addition of Cap leads to a lower % of weight change. A higher Cap ratio shows a lower value of % weight change. This finding shows the effectiveness of Cap in inhibiting marine growth. An increased percentage of marine growth (animal type) leads to the increasing weight of fibreglass panels. The weight change result supports the finding in visual and percentage marine growth.

The finding in the percentage of marine growth and weight change on fibreglass surfaces shows similar findings as in lemon additives AF embedded in a rosin-modified base coated on fibreglass surface. The effective ratio for Cap also observed a similar 6 wt% ratio reported using lemon natural additive [8].

Table 24.2 shows the regression analysis summary for composite fibreglass panels, (a) uncoated; (b) 0% Cap; (c) 2% Cap; (d) 4% Cap; (e) 6% Cap and (f) 8% Cap. All the samples show a positive relation of regression with increased slope of linear equation. The regression analysis produces an R^2 significant value for the uncoated sample in percentage marine growth (0.8707) and weight change (0.9827) showing its relationship with the number of days is highly significant. Observation on percentage marine growth and weight change shows a high percentage (> 80%) of R^2 fit the

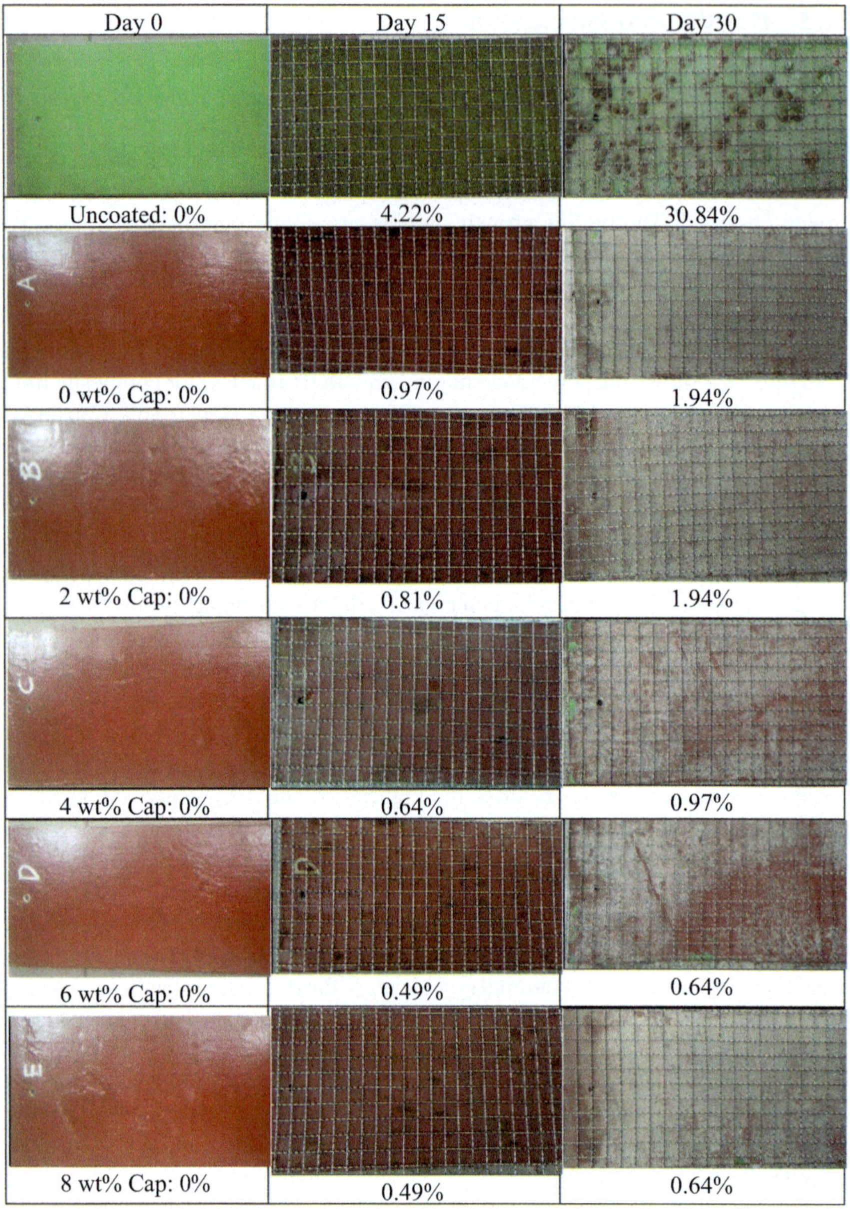

Fig. 24.1 Visual observation of percentage marine growth for composite fibreglass panels for uncoated, 0% cap, 2% cap, 4% cap, 6% cap and 8% cap

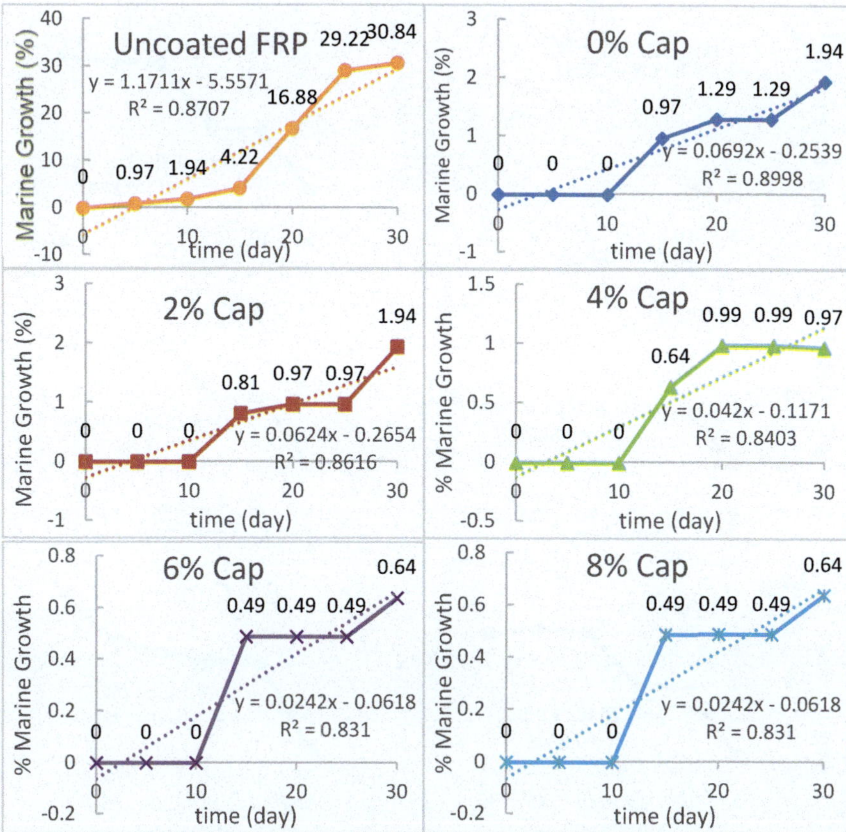

Fig. 24.2 Percentage marine growth (barnacles) for composite fibreglass panels, **a** uncoated; **b** 0% cap; **c** 2% cap; **d** 4% cap; **e** 6% cap and **f** 8% cap

linear equation except for sample 4 wt% Cap. An increase in Cap leads to a decrease in R^2 values. This confirms that Cap improves the performance of the AF coating by inhibiting marine growth. It shows that the immersion performance is affected by other factors such as seawater conditions, location of immersion and types of vessels [2]. The AF coating performance is also affected by the water pH, organic content, salinity, current flow, and coating surface [5]. The regression analysis and experimental data observed similar findings that immersion AF coating performance is influenced by the location of immersion, water quality and position of immersion as discussed earlier [20].

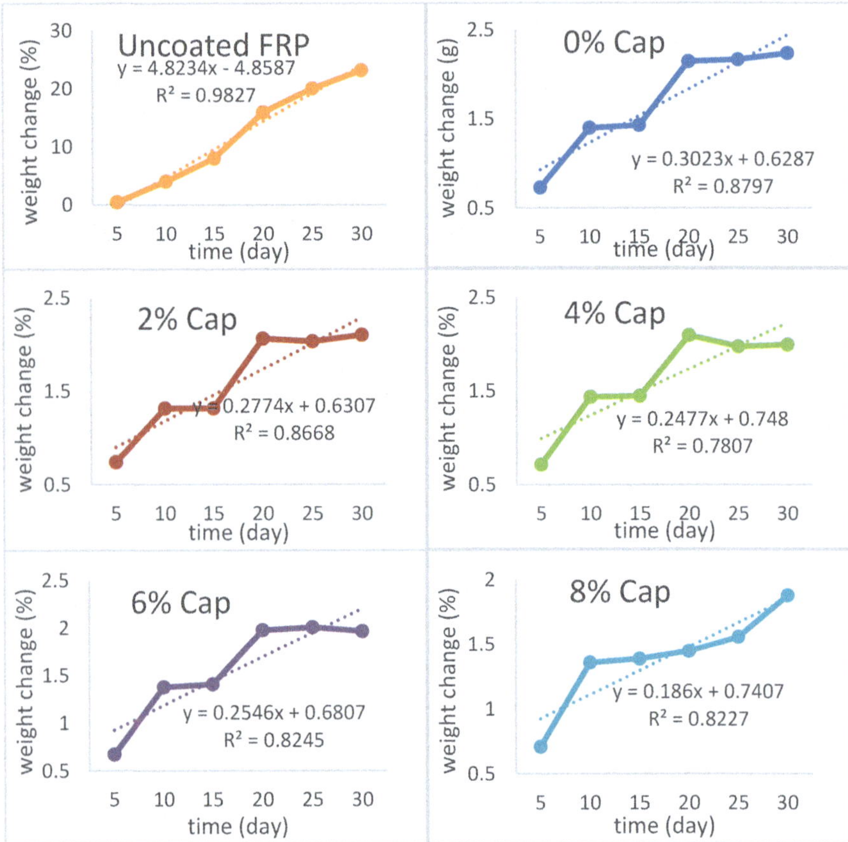

Fig. 24.3 Percentage weight change for composite fibreglass panels, **a** uncoated; **b** 0% cap; **c** 2% cap; **d** 4% cap; **e** 6% cap and **f** 8% cap

Table 24.2 Regression analysis summary for composite fibreglass panels, **a** uncoated; **b** 0% cap; **c** 2% cap; **d** 4% cap; **e** 6% cap and **f** 8% cap

Description	Percentage marine growth		Weigh change	
	Linear equation	R^2	Linear equation	R^2
Uncoated	$y = 1.1711x - 5.5571$	$R^2 = 0.8707$	$y = 4.8234x - 4.8587$	$R^2 = 0.9827$
0 wt% cap	$y = 0.0692x - 0.2539$	$R^2 = 0.8998$	$y = 0.3023x + 0.6287$	$R^2 = 0.8797$
2 wt% cap	$y = 0.0624x - 0.2654$	$R^2 = 0.8616$	$y = 0.2774x + 0.6307$	$R^2 = 0.8668$
4 wt% cap	$y = 0.042x - 0.1171$	$R^2 = 0.8403$	$y = 0.2477x + 0.748$	$R^2 = 0.7807$
6 wt% cap	$y = 0.0242x - 0.0618$	$R^2 = 0.831$	$y = 0.2546x + 0.6807$	$R^2 = 0.8245$
8 wt% cap	$y = 0.0242x - 0.0618$	$R^2 = 0.831$	$y = 0.186x + 0.7407$	$R^2 = 0.8227$

24.4 Conclusion

The natural Cap AF additive demonstrates interesting interaction on the fibreglass surface with low marine growth observed within 30 days of immersion. A higher ratio of Cap (6 and 8 wt%) shows better AF performance with a low percentage of weight change and marine growth observed. Thus, it is concluded that natural Cap additives AF have good interaction with the rosin-modified matrix and fibreglass surface. This finding also contributes to the immersion performance of Cap AF coating in the Lumut area. However, the performance of Cap and AF coating still needs further static and dynamic analysis. Further observation of the AF performance evaluation factor as discussed earlier also needed in various substrates.

Acknowledgements This work was financially supported by a Short-Term Grant (UniKL/Cori/STRG/21023). The author also appreciates the collaboration with Nanotechnology and Catalyst Research, UM Research Centre. The author is grateful and acknowledges all the support.

References

1. Guardiola, F.A., Cuesta, A., Meseguer, J., Esteban, M.A.: Risks of using antifouling biocides in aquaculture. Int. J. Mol. Sci. **13**(2), 1541–1560 (2012). https://doi.org/10.3390/ijms13021541
2. Demirel, Y.K., Turan, O., Incecik, A.: Predicting the effect of biofouling on ship resistance using CFD. Appl. Ocean Res. **62** (2017). https://doi.org/10.1016/j.apor.2016.12.003
3. Schultz, M.P., Bendick, J.A., Holm, E.R., Hertel, W.M.: Economic impact of biofouling on a naval surface ship. Biofouling **27**(1), 87–98 (2011). https://doi.org/10.1080/08927014.2010.542809
4. Kavanagh, J.: The future of antifouling. Mar. Profess. (2018)
5. Qian, P., Li, Z., Xu, Y., Li, Y.: Mini-review: marine natural products and their synthetic analogs as antifouling compounds: 2009–2014. Biofouling **31**(1), 101–122 (2015). https://doi.org/10.1080/08927014.2014.997226
6. Soliman, Y.A., Mohamed, A.S., NaserGomaa, M.: Antifouling activity of crude extracts isolated from two red sea puffer fishes. Egypt J. Aquat. Res. **40**(1), 1–7 (2014). https://doi.org/10.1016/j.ejar.2014.01.002
7. Chapman, J., et al.: Bioinspired synthetic macroalgae: examples from nature for antifouling applications. Int. Biodeterior. Biodegradation **86**, 6–13 (2014). https://doi.org/10.1016/j.ibiod.2013.03.036
8. Saat, A.M., et al.: Evaluation of antifouling coating performance with commercial and lemon natural antibacterial on composite fiberglass surface. Int. J. Innov. Technol. Explor. Eng. **8**(7S2), 253–256 (2019). https://doi.org/10.1088/1755-1315/291/1/012015
9. Hao, X., et al.: Antifouling and antibacterial behaviors of capsaicin-based pH responsive smart coatings in marine environments. Mater. Sci. Eng. C. **108** (2020). https://doi.org/10.1016/j.msec.2019.110361
10. Saat, A.M., et al.: Performance of commercial and natural antifouling embedded in rosin modified coated on wood surface. Int. J. Innov. Technol. Explor. Eng. **8**(12), 553–5556 (2019)
11. Liu, Y., Shao, X., Huang, J., Li, H.: Flame sprayed environmentally friendly high density polyethylene (HDPE)—capsaicin composite coatings for marine antifouling applications. Mater Lett. **238**, 46–50 (2019). https://doi.org/10.1016/j.matlet.2018.11.144

12. Hao, X., et al.: Antifouling and antibacterial behaviors of capsaicin-based pH responsive smart coatings in marine environments. Mater. Sci. Eng. C **108**, 110361 (2020). https://doi.org/10.1016/j.msec.2019.110361
13. Wang, W., et al.: PH-responsive Capsaicin@chitosan nanocapsules for antibiofouling in marine applications. Polymer **158**, 223–230 (2018). https://doi.org/10.1016/j.polymer.2018.10.067
14. Gu, Y., et al.: Research strategies to develop environmentally friendly marine antifouling coatings. Mar. Drugs **18**(7), 1–22 (2020). https://doi.org/10.3390/MD18070371
15. Liu, J., et al.: Capsaicin-modified fluorosilicone based acrylate coating for marine anti-biofouling. Coatings **12**(7), 988 (2022). https://doi.org/10.3390/coatings12070988
16. Tang, T., Effiong, K., Hu, J., Li, C., Xiao, X.: Chemical prevention and control of the green tide and fouling organism Ulva: key chemicals, mechanisms, and applications. Front. Mar. Sci. **8** (2021). https://doi.org/10.3389/fmars.2021.618950
17. Wang, X., Yu, L., Liu, Y., Jiang, X.: Synthesis and fouling resistance of capsaicin derivatives containing amide groups. Sci. Total Environ. **710**, 136361 (2020). https://doi.org/10.1016/j.scitotenv.2019.136361
18. Li, P., et al.: One-pot method for preparation of capsaicin-containing double-network hydrogels for marine antifouling. RSC Adv. **12**(24), 15613–15622 (2022). https://doi.org/10.1039/D2RA00502F
19. Saat, A.M., Alaauldin, S., Malik, A., Kamil, M.S., Zaini, A.L.M.: Hybrid composite fiberglass structure with embedded aluminum phosphate new fire retardants additive: effect of fiberglass types. Adv. Eng. Process. Technol. **II**, 31–40 (2021). https://doi.org/10.1007/978-3-030-67307-9_4
20. Saat, A.M., Yaakup, S., Razak, M.T., Johor, H., Azaim, F.Z.Z., Lokman, M.I.: Performance of rosin modified antifouling coated on mild steel surface at various immersion orientation. Int. J. Innov. Technol. Explor. Eng. **8**(12), 5562–5565 (2019)

Chapter 25
Evaluation of New Green Kapok and Cogon Fibre Composite Sorbent Material for Oil Spill Cleaning and Recovery

Muhammad Hanif Kamar, Ismila Ishak, Fairuzkhan Hassan, Nurul Safiah Yusoff, Muhammad Abdul Munaim bin Muhamad Idrus, Nur Aliya Hamizi, and Asmalina Mohamed Saat

Abstract Oil spill causes significant and long-term impacts on the ecology and socio-economic activities. Compared to other oil spill cleaning methods, sorption by sorbents is suggested as the most effective mechanism due to low operational cost. Kapok is already widely used as sorbent material due to its renewable, abundant, environment-friendly, low cost, high sorption, hydrophobicity, and reusability. However, kapok as a sorbent material has challenges in hydrophobicity water absorption, cleanliness of reuse kapok, and structural integrity of kapok. Cogon fibre also has the potential to be used as a low-cost sorbent for the sorption of various types of oil. Although researchers have found kapok fibre to be sorbent, no studies have been found that use a combination of both kapok and cogon fibres. On the laboratory

M. H. Kamar · I. Ishak · F. Hassan · N. S. Yusoff · M. A. M. M. Idrus · A. M. Saat (✉)
Malaysian Institute of Marine Engineering Technology, Universiti Kuala Lumpur, Lumut, Perak, Malaysia
e-mail: asmalina@unikl.edu.my

M. H. Kamar
e-mail: hanif.kamar@s.unikl.edu.my

I. Ishak
e-mail: ismila@unikl.edu.my

F. Hassan
e-mail: fairuzkhan@unikl.edu.my

N. S. Yusoff
e-mail: nurulsafiahy@unikl.edu.my

M. A. M. M. Idrus
e-mail: mamunaim@unikl.edu.my

N. A. Hamizi
Nanotechnology & Catalysis Research Centre, Universiti Malaya, 50603 Kuala Lumpur, Malaysia
e-mail: aliyahamizi@um.edu.my

A. Ismail et al. (eds.), *Exploring Diversity in Engineering and Technology for Knowledge and Innovation*, Advanced Structured Materials 215,
https://doi.org/10.1007/978-3-031-64330-9_25

scale, the fibres are studied through scanning electron microscope, contact angle, sorption capacity, retention profile, and pH slurry, and we found that a combination ratio of 10:90 provides the best performance of oil sorbent. The presence of lumen microstructures exhibited by kapok fibre and cogon fibre can promote buoyancy, oil sorption, and oil storage capabilities of the sorbent.

Keywords Oil spill · Sorbent · Kapok · Cogon fibre · Green absorbent

25.1 Introduction

The oil spill is a type of pollution, which releases liquid petroleum hydrocarbon into the aquatic environment and requires extensive cleanup activities that are very costly. Treatment procedures by readily available oil spill response equipment such as oil boom consume high operational orders. Alternatively, sorbents are used more than 80% of the time to overcome the problem. The important properties of sorbents are retention over time, oil sorption capacity, recovery of oil from sorbents, and reusability and biodegradability of sorbents. However, the use of a sorbent should be handled with caution to minimize uses that may contribute to secondary pollution and should be used as a supplementary method at the final stage of oil spill cleanup. Concerning sorption, adsorption, and absorption have a different meaning, the former describes the liquid molecular accumulation at the surface rather than in the sorbent, while the latter describes the liquid molecular assimilation throughout the sorbent, and these terms have been used interchangeably and loosely. Therefore, the sorption term will be used throughout this research because it encompasses adsorption and absorption. There are various types of oil spill sorbent materials reported earlier such as inorganic materials, natural base materials, and synthetic sorbents. However, to overcome these limitations, simple, effective, and environmentally friendly modifications are proposed to increase their potential in oil spill treatment. A growing amount of research focuses on natural sorbent materials due to their environmentally friendly advantage. The natural sorbent is also low in cost and the oil collected can be further utilized to be recycled. The natural base sorbent materials are also reported to have better sorption properties in comparison to synthetic sorbent materials. However, the main challenge is to improve the properties of oleophilic or hydrophobic properties that affect the sorption properties. The other challenges in natural sorbent materials are oil sorption performance and buoyancy properties [1]. Natural sorbent materials are commonly modified by using chemical, thermal, and mechanical methods. The mechanical method is the simplest and can ensure the environment-friendly and cost-effective value of the natural sorbent materials. Examples of common natural sorbent materials are palm fibres [2], rice husk [3], sawdust [4], wheat straw [5], kapok [6], and other agricultural waste [7]. Kapok fibre is a natural fibre that shows interesting characteristics as an oil-absorbent material [8]. It is abundantly available and exhibits high oil absorption capacity, leading to an attractive option for oil spill cleanup. Kapok fibre can be used as a raw sorbent material for oil spill cleanup, and it can be modified to improve its performance. For example, calcium stearate-coated kapok fibres have been shown to have improved hydrophobicity and oil sorption capacity [9]. Kapok fibre can also be combined with other materials to create oil collectors with superhydrophobic coatings for efficient oil layer sorption and oil-in-water emulsion separation [10]. Carbonized kapok-cotton sponges have also been developed for oil recovery due to their low fabrication costs, high surface area, high sorption capacity, and improved oil sorption kinetics [11]. There are some challenges in using kapok fibre as an oil-sorbent material, including hydrophobicity, reusability, structural integrity, and cost. The hydrophobic nature of kapok fibre makes it an excellent oil-sorbent material, and it can also make it difficult to wet the fibre with water, which is necessary for oil spill cleanup. Meanwhile, the reusability of kapok fibre requires proper cleaning and drying to maintain its oil sorption capacity [12]. The structural integrity of kapok fibre breaks down over time, reducing its oil sorption capacity [8]. Modification methods have been developed to enhance the structural integrity of kapok fibre, such as oxidation using sodium chlorite followed by sol–gel coating with tetraethyl orthosilicate and sponge formation via the freeze-drying method [12]. Kapok fibre is a renewable and abundant material, and the cost of modifying kapok to improve oil sorption may be higher than other synthetic materials. The modification process needs further research to evaluate the effectiveness of the modification process. This research aims to develop a sorbent material that combines kapok fibre and cogon fibre as an oleophilic sorbent to clean oil spills at sea. To demonstrate the achievement of the stated aim, the most efficient combination of oil sorbent by studying its morphological structure, the contact angle of water and oil, and its ability of sorbate retention is identified.

25.2 Methodology

Two specimens are prepared, namely kapok fibre (*Ceiba pentandra*) and cogon fibre (*Imperata cylindrica*). Kapok fibre is purchased from a local market in Lumut, Perak, while cogon fibre is collected from a lake in UniKL MIMET, Perak. The stems of cogon fibre are removed, keeping only their flowers (spikelets) for experimental purposes. All specimens are air-dried in a dryer at 40–50 °C overnight to remove moisture content. After drying, they are stored in Petri dishes and kept in the laboratory at room temperature. The cogon fibre and kapok fibre were mounted on a universal base using conductive carbon tape and placed in a vacuum chamber at high pressure to examine its morphology using a scanning electron microscope. The pH of the slurry is prepared by a weight of 0.25 g and ground into finer particles. The particles were then mixed with 450 ml distilled water and stirred for 1 h at 100 rpm. The pH of the slurry is recorded using a pH metre Eutech PC700. The sorption materials are made from a combination of cogon fibre and kapok prepared with various ratios of 75/25, 50/50, 25/75, and 10/90. Each ratio is packed into an enclosed net made of wire which will be used for the oil sorption test. The total mass for each combination is fixed at 0.25 g. The wetting behaviour of the specimen is characterized by using simple contact angle measurements and analyses with the ImageJ software. The smartphone is set up (1–2 cm) at a fixed position near the sample and a small droplet of liquid (distilled water, seawater, and lube oil) is dropped on the prepared absorbent surface. The torchlight is used to illuminate the surroundings to capture the liquid droplet. The sorption capacity of the specimen was assessed by performing an oil sorption test under a laboratory scale. Two systems are prepared consisting of lubricating oil (dry system), a mixture of lubricating oil with either distilled water or seawater (wet system). The absorbent material (0.25 g) is placed in a wire pad ($\approx$ 150 μm pore size, 6.2 cm in height, and 4.5 cm in diameter). The wet and dry systems are poured into a one-litre beaker and the wire pad is immersed in the beaker. After saturated condition, the wire pad is removed from the beaker and the oil is left to drain by gravitation process for 1 min. The weight of the samples is recorded. All processes are carried out at 30 °C ± 0.5 °C as a representation of ambient temperature in the tropics. The processes are repeated to measure the amount of oil deposited onto the wire pad as a constant, without a 0.25 g specimen.

25.3 Results and Discussion

Table 25.1 shows the contact angles of cogon–kapok ratios with various liquids for six combination ratios of kapok–cogon sorbent. The sorbate liquids are distilled water, seawater, and lubricating oil. Generally, distilled water and seawater are suspended on the surface of all sorbents, while lubricating oil penetrates through. The contact angle of seawater is the highest for all sorbents, followed by distilled water, and lubricating oil. Since seawater contains a mixture of minerals, it gives greater surface tension to the sorbent compared to distilled water. The lube oil is easily absorbed by the sorbent, thus giving minimum surface tension. The contact angle of seawater and distilled water increases as the kapok ratio increases from 100:0 to 0:100 while lubricating oil plateaus. As the amount of kapok increases at the expense of cogon, the contact angle for distilled water and seawater also increases.

Figure 25.1 describes the relationship between the sorption capacity of different sorbates and the sorbent ratio. The sorbates consist of either a mixture of lubricating oil and distilled water, or lubricating oil and seawater, or only lubricating oil. In terms of sorbent ratio, 10:90 has the highest sorption capacity for all sorbates while 100:0 has the lowest. From 100:0 to 75:25, the graph slightly inclines before declining when it reaches 50:50. From 50:50, the graph inclines steadily before decreasing at 0:100. When the composition of kapok increases, the sorption capacity also increases. This means that kapok has a more dominant effect than cogon. In terms of sorbate, lubricating oil has the highest uptake by sorbents, while the mixture of oil and distilled water is the lowest. Since both kapok and cogon fibres are hydrophobic and oleophilic, they can repel water to allow for oil sorption.

Figure 25.2 describes the effect of the sorbent ratio on the equilibrium retention profile. The least equilibrium time means that the sorbent stops leaking the fastest.

Table 25.1 Contact angle of cogon–kapok ratio with various liquids distilled water, seawater, and lubricant oil

Ratio (Cogon : Kapok)	Distilled Water (θ°)	Seawater (θ°)	Lube Oil (θ°)
100:0	(a) 115.980 	(b) 116.423 	(c) 0.000° 
75:25	(d) 116.110 	(e) 117.700 	(f) 0.000° 
50:50	(g) 116.220 	(h) 117.789 	(i) 0.000° 

(continued)

Table 25.1 (continued)

25:75	(j) 116.220 	(k) 117.789 	(l) 0.000° 
10:90	(m) 116.222 	(n) 117.890 	(o) 0.000° 
0:100	(p) 116.271 	(q) 117.898 	(r) 0.000° 

The regression analysis of the effect of sorbent ratio for contact angle, sorption capacity, and equilibrium retention profile is summarized in Table 25.2. In terms of sorbent ratio, the 10:90 ratio has the shortest equilibrium time for all sorbent materials, while 100:0 has the highest. The graphs decrease from a 100:0 to 75:25 ratio before increasing again to a 50:50 ratio. From 50:50, the graphs decrease steadily before increasing from 10:90 to 0:100. Kapok also has a more dominant effect to retain oil. In terms of sorbent, lubricating oil has the least value of equilibrium time for all sorbents, while the mixture of oil and distilled water is the most. Since both fibres of kapok and cogon are hydrophobic, it will try to eliminate water from their content thus taking a longer time to stop leaking and reach equilibrium. Compared to lubricating oil, only the oil sorption mechanism was involved, so the sorbents reach equilibrium faster.

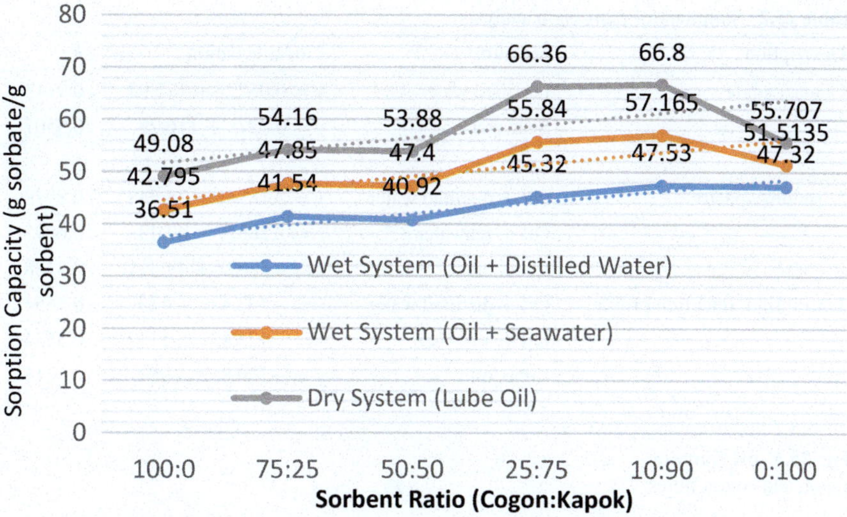

Fig. 25.1 Sorption capacity of different sorbates and sorbent ratios

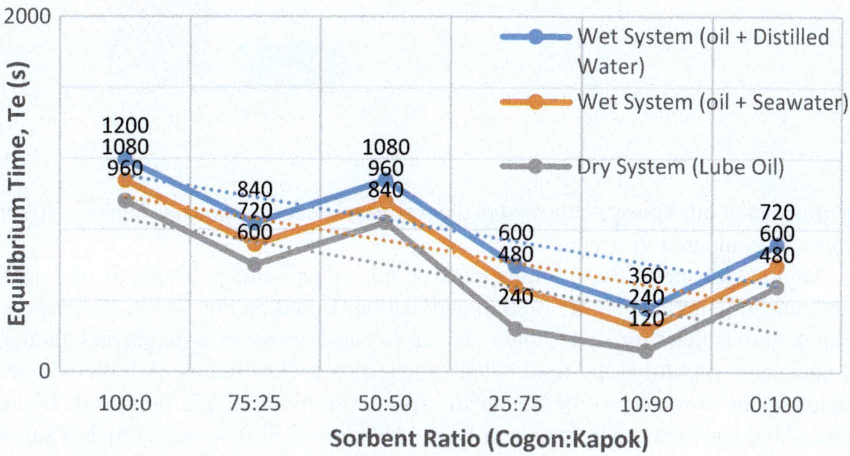

Fig. 25.2 Effect of sorbent ratio on equilibrium retention profile

Figure 25.3 describes the pH value of cogon and kapok fibre in aqueous solution. pH of kapok is higher which suggests that it is slightly alkaline, while cogon pH is lower and neutral. According to [13], influent pH changes alter the binding sites and surface and binding sites of sorbent thus making it an important factor to determine oil removal. During filtration, highly acidic condition causes oil to destabilize and forms larger into larger oil droplets, while highly basic condition causes saponification and

Table 25.2 Summary of regression analysis

Description	Medium	Linear equation	R^2
Contact angle	Distilled water	$y = 0.0529x + 115.97$	0.9113
	Seawater	$y = 0.227x + 116.79$	0.5507
	Lube. oil	–	–
Sorption capacity	Oil + distilled water	$y = 2.1834x + 35.548$	0.8978
	Oil + seawater	$y = 2.2851x + 42.43$	0.6095
	Lube. oil	$y = 2.3867x + 49.311$	0.3788
Equilibrium retention profile	Oil + distilled water	$y = -123.43x + 1232$	0.5554
	Oil + seawater	$y = -123.43x + 1112$	0.5554
	Lube. oil	$y = -126.86x + 984$	0.5215

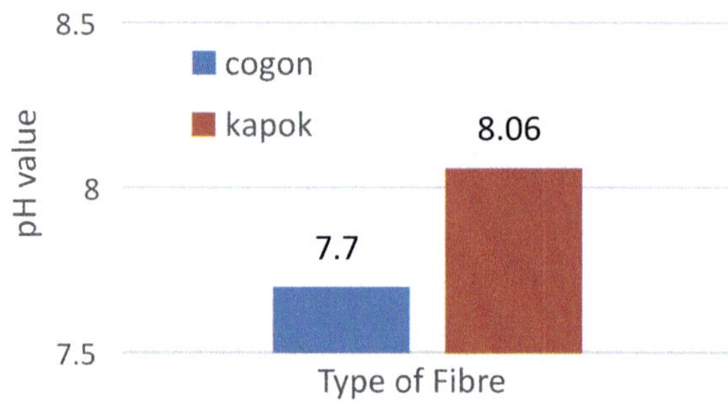

Fig. 25.3 pH slurry of kapok and cogon fibres

hydrolysis of oil. This is confirmed in the research because kapok has higher sorption capacity compared to cogon.

Figure 25.4 shows the scanning electron microscope image for kapok of ×1800 magnification and cogon of ×400 magnification. Based on Fig. 25.1a, the shape of kapok fibre is flat, thin, and follows an oval-to-round cross-section. Second, there is a noticeable rupture at the fibre, which suggests a hollow lumen. A hollow lumen functions to store oil, so the bigger the space, the more oil can be stored. It also gives fibre buoyancy support. According to [14], this hollow structure makes kapok distinctive from other plant fibres because of its high porosity of up to 80%. Third, its surface is smooth and has snowy deposits which suggests wax coating. Wax may cause the fibre to repel water while allowing oil sorption. These properties are known as hydrophobic and oleophilic, which are among the basic factors that affect oil sorption capacity [15]. Whereas Fig. 25.1b shows cogon fibre, its surface is uneven, jagged, and hairy which gives it a very rough surface. Unlike kapok, cogon fibre uses this rough surface as a mechanism to bind oil [16]. Second, it also has snowy deposits that suggest wax coating, which confirms the hydrophobic properties.

Figure 25.5 shows the image of cogon fibre and kapok before and after oil sorption. Before being subjected to oil sorption, both fibres are generally thinner. After oil sorption, both fibres appear to be swollen which indicates that oil is successfully

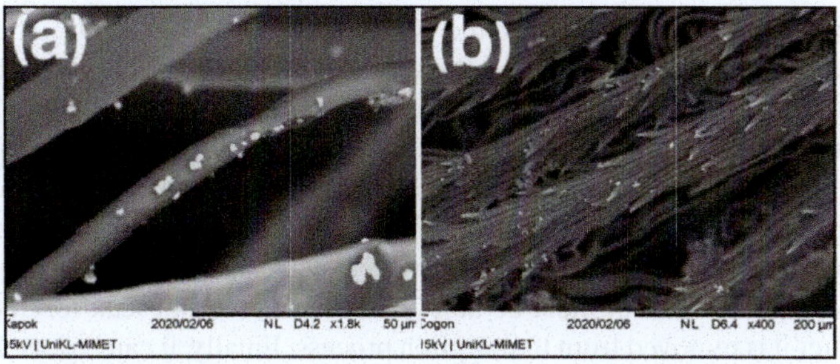

Fig. 25.4 Scanning electron microscope image for **a** kapok and **b** cogon fibres

Fig. 25.5 Digital microscope image for cogon grass (fibre) and kapok, before and after oil sorption

stored in their structure. Since kapok has a hollow structure while cogon fibre has a rough surface, kapok appears to swell more. Based on SEM images, it is found that kapok has more wax coating thus increasing its hydrophobicity.

25.4 Conclusion

The new green-based oil sorbent is prepared by using six combination ratios of kapok and cogon, namely 100:0, 75:25, 50:50, 25:75, 10:90, and 0:100. The sorbents were prepared by simple method that involves heating and compression and packed into a wire net. The performances of oil sorbents were evaluated by using contact angle, sorption capacity, and equilibrium time. We determined that a combination ratio of 10:90 produced the highest sorbent properties compared to other ratios. Its contact angle of distilled water, seawater, and lubricating oil is 116.22°, 117.890°, and 0.000° respectively, which suggests hydrophobicity and oleophilic properties. These properties are the basic factors that determine the efficiency of an oil sorbent, which repels water while allowing the sorption of oil into its body. The sorption capacity for the mixture of distilled water and lubricating oil, the mixture of seawater and lubricating oil and the pure lubricating oil is 66.8 g/g, 57.165 g/g and 47.53 g/g respectively, which are the highest values compared to the other ratios. Its equilibrium time for distilled water and lubricating oil mixture, seawater and lubricating oil mixture, and lubricating oil only is 360 s, 240 s, and 120 s, respectively. These results are the shortest compared to other ratios, which suggests that it reaches equilibrium fastest. A good oil sorbent must be able to retain oil and not leak too much of its content after it is removed from the sorption process. Finally, it can be concluded that sorbent ratio 10:90 is the best oil sorbent because it is the most hydrophobic, has the highest sorption capacity for oil uptake, and is able to achieve retention equilibrium fastest.

Acknowledgements This work was financially supported by UniKL MIMET Final Year Project. The author also appreciates the collaboration with Nanotechnology & Catalysis Research Centre, University of Malaya, Kuala Lumpur 50603, Malaysia. The author gratefully acknowledges both of their support.

References

1. Zamparas, M., Tzivras, D., Dracopoulos, V., Ioannides, T.: Application of sorbents for oil spill cleanup focusing on natural-based modified materials: a review. Molecules **25**(19), 4522 (2020). https://doi.org/10.3390/molecules25194522
2. Abdelwahab, O., Nasr, S.M., Thabet, W. M.: Palm fibers and modified palm fibers adsorbents for different oils. Alexandria Eng. J. **56**(4), 749–755 (2017). https://doi.org/10.1016/j.aej.2016.11.020
3. Ismail, U., Kamaruzaman, W.N., Wil, Z.C.: The effectiveness of oil spill remover in marine offshore from rice husk, mango peels and durian husk. Int. Res. J. Eng. Technol. (2020). https://www.irjet.net
4. Hussein, S.N.C.M., et al.: Study of acid treated mixed sawdust as natural oil sorbent for oil spill. Mater Today Proc. **19**, 1382–1389 (2019). https://doi.org/10.1016/j.matpr.2019.11.156
5. Omar, B.M., Abdelgalil, S.A., Fakhry, H., Tamer, T.M., El-Sonbati, M.A.: Wheat husk-based sorbent as an economical solution for removal of oil spills from sea water. Sci. Rep. **13**(1), 2575 (2023). https://doi.org/10.1038/s41598-023-29035-8
6. Quek, C.S., Ngadi, N., Ahmad Zaini, M.A.: The oil-absorbing properties of kapok fibre—a commentary. J. Taibah Univ. Sci. **14**(1), 507–512 (2020). https://doi.org/10.1080/16583655.2020.1747767
7. Wang, F., Lei, S., Ou, J., Li, C., Li, W.: Novel all-natural material for oil/water separation. Ind. Eng. Chem. Res. **58**(5), 1924–1931 (2019). https://doi.org/10.1021/acs.iecr.8b05535
8. Zhang, X., Wang, C., Chai, W., Liu, X., Xu, Y., Zhou, S.: Kapok fiber as a natural source for fabrication of oil absorbent. J. Chem. Technol. Biotechnol. **92**(7), 1613–1619 (2017). https://doi.org/10.1002/jctb.5155
9. Blaquera, A.L.M., Herrera, M.U., Manalo, R.D., Maguyon-Detras, M.C., Futalan, C.C.M., Balela, M.D.L.: Oil adsorption kinetics of calcium stearate-coated kapok fibers. Polymers **15**(2), 452 (2023). https://doi.org/10.3390/polym15020452
10. Wang, J., Zheng, Y., Wang, A.: Effect of kapok fiber treated with various solvents on oil absorbency. Ind. Crops Prod. **40**, 178–184 (2012). https://doi.org/10.1016/j.indcrop.2012.03.002

11. Datiles, W.C.P., Herrera, M.U., Manalo, R.D., Maguyon-Detras, M.C., Futalan, C.C.M., Balela, M.D.L.: Kapok-cotton carbon sponges for oil recovery. IOP Conf. Ser. Earth Environ. Sci. **812**(1), 012014 (2021). https://doi.org/10.1088/1755-1315/812/1/012014
12. Lugito, G., Kustiana, A., Martuani, R., Wenten, I.G.: Kapok fibre as potential oil-absorbing material: modification mechanism and performance evaluation. IOP Conf. Ser. Mater. Sci. Eng. **823**(1), 012033 (2020). https://doi.org/10.1088/1757-899X/823/1/012033
13. Wahi, R., Abdullah, L.C., Choong, T., Yaw, S., Ngaini, Z., Mobarekeh, M.N.: Removal from aqueous state by natural fibrous sorbent: an overview. Sep. Purif. Technol. **113**, 51–63 (2013). https://doi.org/10.1016/j.seppur.2013.04.015
14. Oleju, Z., Włókno, P.: Kinetics and thermodynamics of dispersed oil sorption by kapok fiber. Ecol. Chem. Eng. S **26**(4), 759–772 (2019). https://doi.org/10.1515/eces-2019-0053
15. Khalid, F.E., et al.: Application of cogon grass (*Imperata cylindrica*) as biosorbent in diesel-filter system for oil spill removal. Agronomy **11**(11), 2273 (2021). https://doi.org/10.3390/agronomy11112273
16. Ibrahim, S., Baharuddin, S.N.I.B., Ariffin, B., Hanafiah, M.A.K.M., Kantasamy, N.: Cogon grass for oil sorption: characterization and sorption studies. Key Eng. Mater. **775**, 359–364 (2018). https://doi.org/10.4028/www.scientific.net/KEM.775.359

Chapter 26
Finite Element Analysis of a Bulk Vessel: A Comprehensive Study

Zulzamri Salleh and Muhammad Akmal Afiq Zullastri

Abstract The marine construction sector is currently undergoing substantial expansion, driven by a focus on enhancing production efficiency and generating higher revenues through the integration of advanced technologies. To align with this advancement, it is crucial to establish cost-effective production objectives for critical aspects of ship and vessel manufacturing and management. However, there exists a knowledge gap, particularly concerning the mechanical damage behavior of marine structures, especially within structural analysis. To bridge this disparity, the study aims to furnish precise data simulations and output results pertaining to the structural analysis behavior and performance. The research will employ finite element analysis (FEA) formulations tailored for plate and shell structures, specifically utilizing the first-order shear deformation theory. The validation process will encompass various locations, considering both the extent and size of damage, relative to marine structures. To replicate the model experimentally in a real marine setting, hydrodynamic measurements will be conducted, alongside the introduction of a newly devised damage parameter and the utilization of von Mises strain distribution in FEA.

Keywords Finite element analysis (FEA) · Bulk vessel · Structural analysis · Simulation · Design optimization

Z. Salleh (✉) · M. A. A. Zullastri
Malaysian Institute of Marine Engineering Technology, Universiti Kuala Lumpur, Dataran Industri Teknologi Kejuruteraan Marin, Bandar Teknologi Maritim, Jalan Pantai Remis, 32200 Lumut, Perak, Malaysia
e-mail: zulzamri@unikl.edu.my

M. A. A. Zullastri
e-mail: makmal.zullastri@s.unikl.edu.my

A. Ismail et al. (eds.), *Exploring Diversity in Engineering and Technology for Knowledge and Innovation*, Advanced Structured Materials 215,
https://doi.org/10.1007/978-3-031-64330-9_26

26.1 Introduction

The meticulous design and analysis of bulk vessels demand thorough attention to structural integrity and performance across diverse operational scenarios. This paper offers a detailed exploration of employing finite element analysis (FEA) simulations for bulk vessels. The primary aim is to assess structural behavior, scrutinize critical areas vulnerable to stress and deformation, and suggest strategies for optimizing design. Through this study, there is a concerted effort to deepen understanding regarding bulk vessel performance and offer insights for enhancing both design and operational efficiency.

Bulk vessels contend with intricate loading conditions, encompassing factors such as cargo weight, hydrostatic pressure, and dynamic forces during transit and handling. This section discusses the key design considerations for bulk vessels, such as hull geometry, material selection, and structural elements [1]. Finite element analysis (FEA) simulation emerges as a potent tool for assessing the structural response of bulk vessels. The methodology of simulation entails several steps: initiating by generating a virtual model of the vessel, specifying material properties, applying loads and boundary conditions, and subsequently conducting a comprehensive structural analysis. This systematic approach enables engineers to accurately predict and understand how bulk vessels will behave under varying operational conditions, aiding in the refinement of their design and performance optimization. Various aspects, including stress distribution, deformation patterns, and fatigue life estimation, can be analyzed through FEA simulations [2]. This section outlines the modeling and analysis process for bulk vessel structures using FEA software. It covers the mesh generation, material assignment, and load application techniques. The FEA simulation investigates the stress concentration areas, deformation behavior, and overall structural response of the vessel under different operational scenarios [3].

In this research endeavor, a particular-sized bulk vessel model is both designed and scrutinized employing FEA software. The aim is to investigate how stresses are dispersed within the bulk vessel model when subjected to diverse forces. Through this analysis, insights into stress distribution under varying conditions are gained, contributing to a deeper understanding of the vessel's structural behavior and performance.

26.2 Methodology

Firstly, the creation and design of the bulk vessel model are imperative. While designs from prior research can be considered, preference is given to developing an original design. Subsequently, material selection for this simulation can be conducted directly within the FEA software. Lastly, the bulk vessel must undergo analysis using the FEA simulation software and the outcomes compared with those of previous studies.

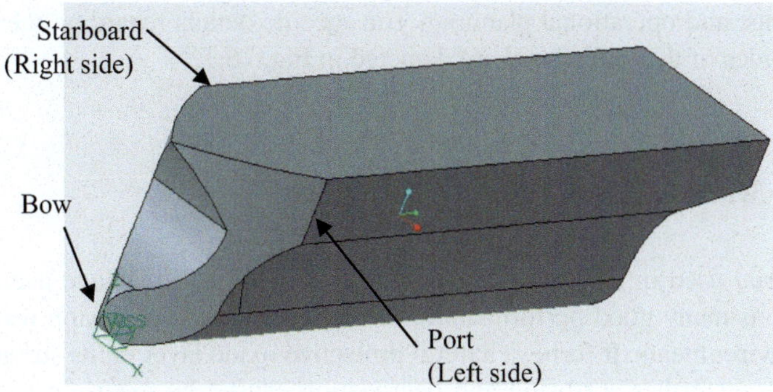

Fig. 26.1 Overview bulk vessel

Table 26.1 Bulk vessel model specification

Type	Model size (mm)
Length	2300
Height	600
Width	500

26.2.1 Design Stage

In the design stage, drawing software has been employed to streamline the design process, utilizing provided references as a foundation. A detailed visual representation of the sample model, referred to as the bulk model, is presented in Fig. 26.1. Additional specifications for these models are available in Table 26.1.

26.2.2 Analysis

The analysis was carried out and scrutinized utilizing application simulations accessed via FEA software. The designed model underwent an analytical process, with specific sections chosen as subjects for testing. These sections include:

(a) The bow of the bulk vessel model.
(b) The port and starboard sides of the bulk vessel model.

A bulk vessel model represents or simulates a bulk carrier ship, which is a specialized type of merchant vessel designed specifically for transporting loose bulk cargo such as grains, coal, ore, and similar materials. The bulk vessel model encompasses the structural features, dimensions, and characteristics typical of a standard bulk carrier, enabling analysis, simulations, and studies pertaining to its performance, stability, behavior, and other pertinent aspects. These models serve various purposes within the maritime industry, including engineering design, optimization, safety assessments, and operational planning. The specific details regarding the starboard and port sides of the bulk vessel are depicted in Fig. 26.1.

26.2.3 Materials

The material used in this study is the AlSiCu alloy. The advantage using AlSiCu is that it has many good performances, including excellent corrosion resistance in marine environments. It forms a natural protective oxide layer on its surface, which helps prevent further corrosion and degradation caused by exposure to saltwater, moisture, and other corrosive elements [4]. Furthermore, the AlSiCu alloy has a high strength-to-weight ratio, making it an attractive choice for marine structures. It provides sufficient strength and structural integrity while keeping the overall weight of the structure relatively low. This can contribute to improved fuel efficiency, higher payloads, and increased operational efficiency of marine vessels [5]. Aluminum silicon copper alloy offers good weldability, allowing for efficient fabrication and construction of marine structures. Welding is an essential process in shipbuilding and marine construction, and the weldability of the AlSiCu alloy simplifies the joining of various structural components. Another advantage that the alloy possesses is its good fatigue resistance, enabling it to withstand cyclic loading and dynamic stresses commonly experienced by marine structures [6]. This property is particularly important in applications where the structure is subjected to repeated waves, vibrations, or other dynamic forces.

Among its various advantages, the material has a good lightweight potential. It also has high strength-to-weight ratio which contributes to improved fuel efficiency and maneuverability of the boat. AlSiCu alloy also is highly resistant to corrosion, rot, and degradation caused by water exposure [7]. They can withstand harsh marine environments, making them a durable choice for marine construction. The materials also can be fabricated into various shapes and designs, allowing for greater flexibility in maritime design. This versatility enables the creation of complex and streamlined hull shapes, enhancing performance and hydrodynamics [8]. This characteristic ensures a higher level of safety for the vessel and its occupants. The cost maintenance for AlSiCu alloy is also less due to the required minimal maintenance compared to other materials. They are easy to clean, do not require frequent painting or varnishing, and generally have a longer lifespan, reducing overall maintenance costs [9]. For environmental effect due to moisture, generally materials have inherent moisture resistance properties. They do not absorb water, preventing issues such as swelling, warping, or rotting commonly associated with wooden boat models. The design of the bulk vessel can be versatile because it can be combined with other materials to create hybrid structures with enhanced properties [10]. For example, they can be mixed with carbon fibers to increase stiffness and strength in specific areas of the bulk vessel model. Overall, the AlSiCu alloy offers a compelling combination of strength, durability, lightweight construction, and design flexibility, making them an advantageous choice for building boat models [11].

26.3 Results and Discussion

During the initial phase of the simulation process, the model underwent assessment through the meshing stage. This step is crucial to verify that there are no discontinuities present in the design between lines or parts. The specifics of the meshing output are illustrated in Fig. 26.2, revealing a total of 483 elements comprising tetra—483, edge—859, and face—1142 entities.

There are two important categories of data or results available in the simulation software: stress and displacement. The data for each specimen were gathered and documented, referring to a previous experiment [10]. The specific outcomes for the bulk vessel can be observed in Fig. 26.3.

Figure 26.4 indicates that there is a contrast in stress and displacement values between the bulk model and boat model, allowing for a comparison between the two. As the force rate increases, the stress value decreases, while the maximum displacement increases [12]. This demonstrates that models subjected to higher force rates have a lower tolerance for pressure before breaking. Additionally, it can be inferred that increasing the force rate enhances the mechanical properties in terms of displacement, whereas applying a high force weakens the mechanical properties in terms of stress [13]. Hence, stress and displacement profoundly affect the mechanical properties. The data extracted from the simulation underscore that material choice significantly impacts both stress and displacement. In this investigation, when a force of 2000 kN is exerted on the bow, the bulk vessel exhibits the highest stress rate, measuring 793.8 kPa, which marginally decreases with the escalation of force up to 5000 kN. Likewise, the displacement at the bow of the bulk vessel exhibits a substantial increase as the force is raised to 5000 kN. These findings are corroborated by Fig. 26.3b, where it is evident that the bow of the bulk vessel displays a damaged area, highlighted in red. The characteristics of the AlCuSi alloy may exhibit rupture at a certain level of elastic behavior in the bow bulk vessel area.

In contrast, the results for the port and starboard regions indicate a slight decrease in stress concentration as the applied force exceeds 2000 kN. At this juncture, the

Fig. 26.2 AutoGEM meshing

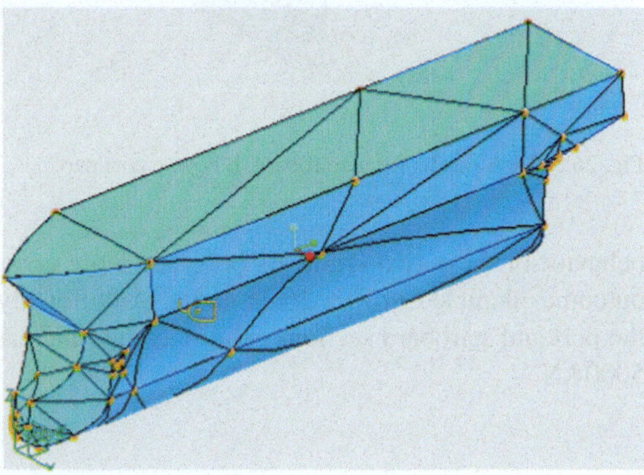

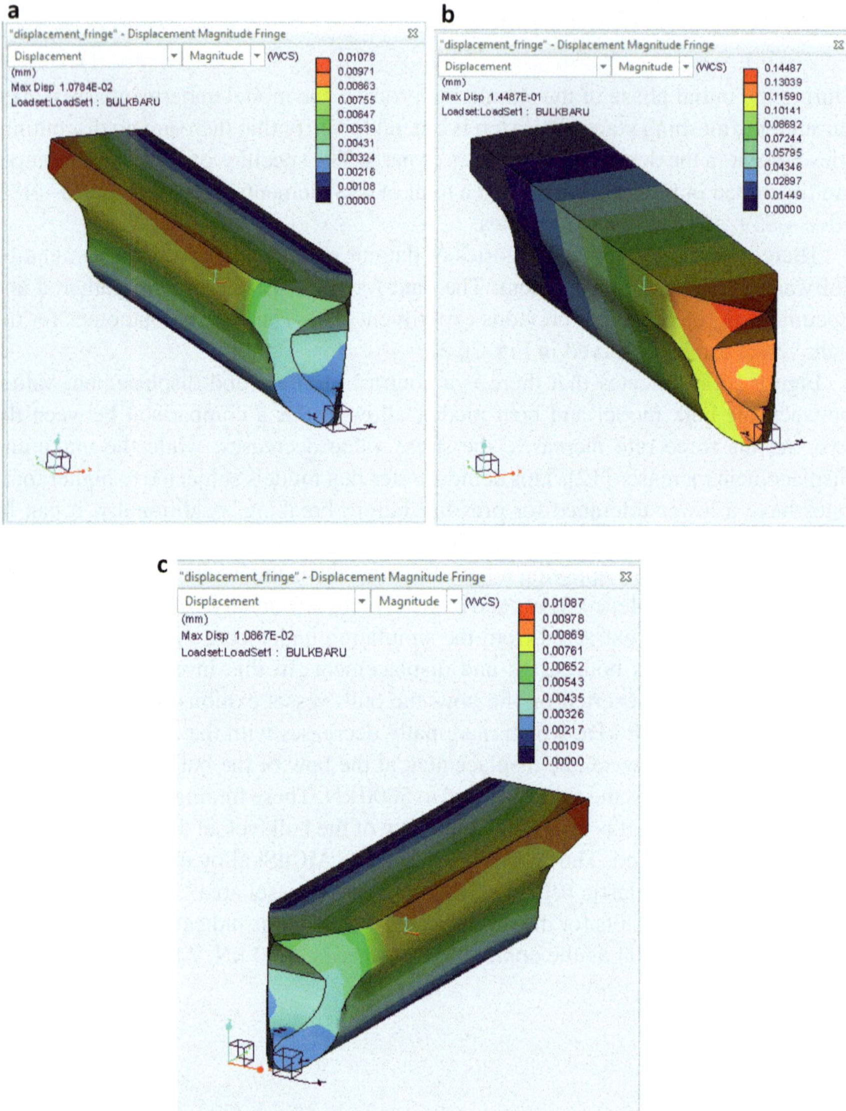

Fig. 26.3 FEA results for **a** starboard, **b** bow, **c** port part

behavior of these two segments becomes more apparent when compared with the outcomes depicted in Fig. 26.3a and c. Additionally, the displacement results for the port and starboard sections exhibit a slight increase as the applied force reaches 5000 kN.

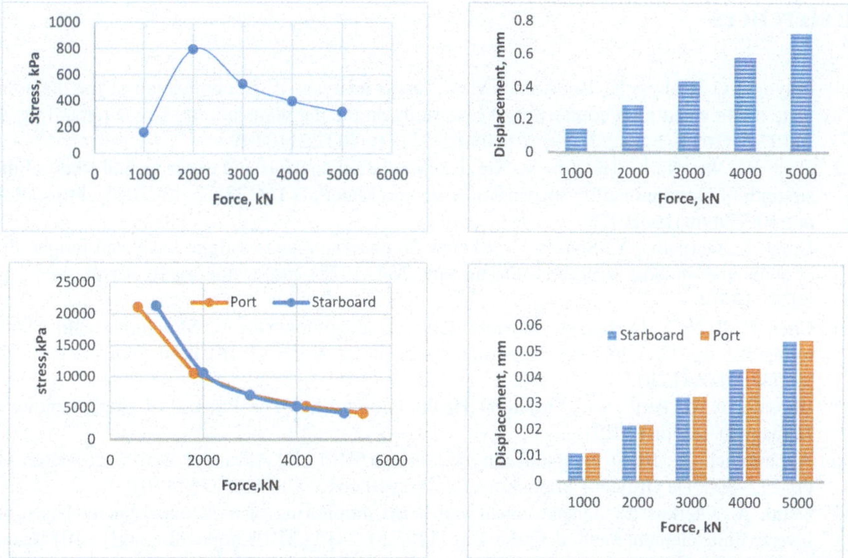

Fig. 26.4 Stress concentration result for **a** port and **d** starboard for bulk vessel and displacement **b** bow **c** port and starboard for bulk vessel

26.4 Conclusion

In conclusion, this research aimed to investigate the correlation between stress and displacement using finite element analysis. The data obtained will facilitate the examination of material properties by the production team prior to bulk vessel fabrication. Overall, the objectives of the project were successfully met. Initially, the design of the bulk vessel model was developed as discussed. Subsequently, the analysis of data derived from FEA aided in determining the optimal mechanical properties based on the selected materials, namely the AlSiCu alloy. In this study, under a force of 5000 kN, the highest stress value observed was 793.8 kPa for the bow bulk vessel, while for the port and starboard sections, the highest stress was 2138 kPa. Regarding displacement, under the same force, the bow bulk vessel recorded the highest displacement at 0.72437 mm, whereas the port and starboard sections recorded 0.054337 mm. The stress experienced by the specimen varies according to the applied force rate. The results also had a similar had stress distribution to all join members slightly reduced as mentioned also in a previous work [14].

Acknowledgements The author would like to thank the Ministry of Higher Education (MoHE), Malaysia, for providing a grant (FRGS/1/2021/TK0/UNIKL/02/20) to the first author in carrying out and completing this work.

References

1. Prayogo, G., Cahya, E., Budiyanto, M.A.: Stress analysis of the application of the diamond plate on the quad-joint connection: a case study on the flat plate hull of ships. Cogent Eng. **9**, 2110705 (2022). https://doi.org/10.1080/23311916.2022.2110705
2. Qian, W., Wu, N., Li, H., Yin, X., Xie, L.: Simulation analysis and experimental study of the strength of aluminum alloy suspension structure. Materials **16**(128), 2–13 (2023). https://doi.org/10.3390/ma16010128
3. Dong, Y., Garbatov, Y., Soares, G.: Review on uncertainties in fatigue loads and fatigue life of ships and offshore structures. Ocean Eng. **264** (2022). https://doi.org/10.1016/j.oceaneng.2022.112514
4. Chen, H.-P., Ni, Y.-Q.: Introduction to structural health monitoring. In: Structural Health Monitoring of Large Civil Engineering Structures, pp. 1–14. Wiley (2018). https://doi.org/10.1002/9781119166641.ch1
5. Bhosale, A.A., Patil, S.J.: Structural Health Monitoring: The Process of Implementing a Damage Detection (2019)
6. Abramovich, H.: 10. Structural health monitoring (SHM). In: Advanced Aerospace Materials, pp. 272–305. De Gruyter (2019). https://doi.org/10.1515/9783110537574-010
7. Kefal, A., Oterkus, E.: Displacement and stress monitoring of a chemical tanker based on inverse finite element method. Ocean Eng. **119**(15), 2–15 (2016). https://doi.org/10.1016/j.oceaneng.2016.04.025
8. Kefal, A., et al.: Three dimension shape and stress monitoring of bulk carrier based on iFEM methodology. Ocean Eng. **147**(13), 256–267 (2018). https://doi.org/10.1016/j.oceaneng.2017.10.040
9. Ji, S., et al.: Automated subject-specific, hexahedral mesh generation via image registration. Finite Elem. Anal. Des. **47**(10), 1178–1185 (2011). https://doi.org/10.1016/j.finel.2011.05.007
10. David Polly: G563 Quantitative Paleontology [Online] (2019). Available at: http://www.lusas.com/products/options/imdplus.html
11. Durand, R., Pantoja-Rosero, B.G., Oliveira, V.: A general mesh smoothing method for finite elements. Finite Elem. Anal. Des. **158**, 17–30 (2019). https://doi.org/10.1016/j.finel.2019.01.010. Finite Element Analysis Methods 1(1) (2009)
12. Birman, V., Genin, G.M.: Linear and nonlinear elastic behavior of multidirectional laminates. In: Comprehensive Composite Materials II, pp. 376–398. Elsevier (2017). https://doi.org/10.1016/B978-0-12-803581-8.09881-7
13. Carrera, E., Fazzolari, F., Cinefra, M.: Chapter 4—Fundamental of mechanics of beams, plates and shells (2017). https://doi.org/10.1016/B978-0-12-849892-7.00006-6
14. Salleh, Z., Wei Kee, G.: Simulation of Three-Point Bending Sandwich Composite Panel Through Finite Element Analysis (2022)

Chapter 27
Characterization of Hybrid Composite Polyester with Recycled HDPE Core by Hand Lay-Up Method

Muhammad Aiman Ahmad Danial, Fatin Zawani Zainal Azaim, Asmawi Ismail, Nor Ashikin Muhamad Khairussaleh, and Asmalina Mohamed Saat

Abstract High-density polyethylene (HDPE) waste is used in composite materials due to the increasing abundance of domestic polymer waste worldwide. Mostly pure HDPE has been used as a structural material; meanwhile, recycled HDPE (r-HDPE) is commonly transformed into particulate and used as a modified matrix in composite material. The use of r-HDPE was rarely converted into a panel and used as a core in the application of structural composites. Therefore, this research focused on converting the r-HDPE into a panel using a simple resizing and hot press method. The panel was used as core material and combined with fiberglass to fabricate a structural composite. The r-HDPE shows improved mechanical properties and fire performance with the combination of fiberglass layer. The new r-HDPE structural composite was shown to have good tensile, flexural, and water absorption properties. The fire performance also shows improvement with no dripping occurred after combination with fiberglass layers.

M. A. A. Danial · F. Z. Z. Azaim · A. Ismail · A. M. Saat (✉)
Universiti Kuala Lumpur, Malaysian Institute of Marine Engineering Technology, 32200 Lumut, Perak, Malaysia
e-mail: asmalina@unikl.edu.my

M. A. A. Danial
e-mail: mhdaimandanial@gmail.com

F. Z. Z. Azaim
e-mail: fatinzawani@unikl.edu.my

A. Ismail
e-mail: asmawiis@unikl.edu.my

N. A. M. Khairussaleh
Faculty of Civil Engineering Technology, Universiti Malaysia Pahang, 26300 Gambang, Kuantan, Pahang, Malaysia
e-mail: aasyiq@ump.edu.my

A. Ismail et al. (eds.), *Exploring Diversity in Engineering and Technology for Knowledge and Innovation*, Advanced Structured Materials 215,
https://doi.org/10.1007/978-3-031-64330-9_27

Keywords Recycled HDPE core · Structural composite · Mechanical test · Composite fiberglass boat · Fire retardant

27.1 Introduction

High-density polyethylene (HDPE) is a thermoplastic polymer made from petroleum that is commonly used for packaging, such as milk bottles, detergent bottles, bleach bottles, shampoo bottles, and Tupperware. HDPE is one of the most widely used plastics globally due to its versatility, durability, and resistance to various chemicals. However, like all plastics, HDPE can contribute to environmental pollution if not managed properly after its useful life. HDPE is commonly recycled and has the number "2" as its resin identification code. HDPE is one of the plastic waste packagings that contributes to 40% of plastic waste packaging [1]. HDPE waste refers to discarded or post-consumer HDPE products and packaging that are no longer in use and are destined for disposal or recycling. Recycled HDPE can be used for a variety of applications, including drainage pipe systems, floor tiles, fencing, wire covering, and wood-plastic composites. HDPE waste must be recycled and recovered to avert environmental pollution [1]. Recycled HDPE is gaining interest in composite material applications due to the increasing of HDPE waste in the world. The increase in polymer waste will lead to environmental micro and nano-plastic pollution and affect both the ocean and land. A better approach to the life cycle of HDPE and plastic waste is essential to control the waste and maintain its value [2]. Virgin HDPE was commonly reported to be used as a matrix for many natural recycle resources from leather dust waste [3], clay [4], sugarcane waste [5], hemp waste [6], rubber waste [7], jute fiber [8], date palm [9], and others. HDPE waste is commonly recycled using a mechanical recycling approach, and the utilization of recycled HDPE is in the form of particles [3]. The size of waste HDPE is reduced and formed into a particle before combining it with other additives by using the injection or compression molding method [10]. Structural composite with a pure HDPE panel as a layer is reported to use the fabrication method of hand lay-up [8]. However, none has yet been found for recycling HDPE fabrication using the hand lay-up method. Common HDPE waste is recycled in particulate form, modified with other additives and fabricated into a product. The use of r-HDPE as core material was also not reported elsewhere. Thus, this research prepares a new hybrid composite polyester with r-HDPE core by the most economical and easiest method of hand lay-up. The HDPE waste is recycled by resizing and hot compression methods. The resulting panel was then fabricated into a structural composite as a core material by the hand lay-up method. The design layers of composite with fiberglass layers are varied to evaluate the compatibility and improvement of mechanical, water absorption, and fire-retardant performance of the new sandwich panels. The mechanical properties and fire performance of HDPE and r-HDPE are summarized in Table 27.1.

Table 27.1 Summary of selected mechanical properties and fire performance data on HDPE composite

Author	Description	Finding
Sayem et al. [8]	• HDPE sheet/jute fiber (multilayer) • Hand lay-up method • Tensile test ASTM D3039 • Flexural test ASTM D790	• The tensile and flexural strengths as fiber content increases (due to good bonding) • Tensile strength, max. value: 36.37 MPa (six-layer HDPE/jute fiber), min. value: 22.39 MPa (pure HDPE) • Flexural strength, highest: 38.73 MPa (six-layer HDPE + jute fiber), lowest: 24.64 MPa (pure HDPE)
Mousa et al. [7]	• Pure HDPE, waste tire rubber particles (WTRP)-HDPE-marble and granite dust (GM) hybrid composite • Hot pressed in a mold (200 mm × 200 mm) WTRP-HDPE-GM hybrid composite plate • Charpy impact test (ASTM D6110)	• The toughness decreases with the addition of GM and WTRP in a hybrid composite • Pure HDPE impact strength max.: 181 kJ/m^2, min.: 28 kJ/m^2
Abd El-Fattah and Abd ElKader [4]	• Wood plastic composite (WPC) (r-HDPE sheet/ wood flour/organic modified clays) • Identify the most effective clay-type modified bentonite (mBNT) and modified layered double hydroxide (mLDH) for WPC • Melt mixing and compression molding method • Mechanical, thermal, and water absorption tests	• The mBNT clay (2 wt%) most effective formulation with enhanced WPC properties • The tensile strength increased as 2 wt% mBNT clay was added. Further, an increase in clay led to a decrease in tensile strength • The Izod impact test decreased (10.5%) with 2 wt% of clay • The water resistance increased by 27.5% after immersion for five days • The mLDH addition decreases the tensile strength WPC

(continued)

Table 27.1 (continued)

Author	Description	Finding
Kavianiboroujeni et al. [6]	• Composite panels' three layers [pure HDPE, Azodicarbonamide (ACA), and Hemp] • Compression molding 170 °C and a 3-ton load to generate the composite layers • Dimensions (250 × 250 × 2) mm • Flexural test ASTM D790	• Hemp content improves flexural properties • Specific modulus (30% increase) and strength (36% higher) when using a foam core • Pure HDPE flexural strength, max.: 23 MPa, min.: 18.8 MPa
Lin et al. [11]	• HDPE (pure)/clay/glass fiber woven • Thermally compressed at 180 °C with a pressure of 10 MPa and laminated • Composites thickness of 1 mm • Tensile strength in ASTM D638	• Tensile strength max. 80 MPa and min 65 MPa in HDPE/clay/glass fiber woven • HDPE/clay composite. Tensile strength max. 30 MPa and min. 25 MPa
El-Fattah et al. [5]	• R-HDPE and sugarcane bagasse fibers • Sheet prepared by compression molding (20mmx20mmx3mm) • Impact test ASTM D 256-06 R-HDPE and sugarcane bagasse fibers • Water absorption test (ASTM D 570) immersed in distilled water	• The impact strength decreases as fiber increases • Pure HDPE impact strength. Max.: 10 kJ/m^2, min.: 6 kJ/m^2 • The water absorption increased with the increase of fiber content • Highest—3% water absorption. Lowest—1% water absorption
Aldousiri et al. [9]	• HDPE layers combine with natural date palm fiber (DPF) and fiberglass chopped strand mat (CSM) layer • Design layers' arrangement HDPE/ fiberglass/natural fiber/ fiberglass/natural fiber (five layers) • Fabrication by hot press	• Tensile strength; min. value 28 MPa (HDPE); max. value 50 MPa (HDPE/ 6% DPF/6% CSM) • Compressive strength: vertical; max. value 119 MPa (HDPE + 6% CSM), min. value 15 MPa (HDPE + 6% DPF + 2% CSM); horizontal: max. value 200 MPa (HDPE + 6% DPF + 4% CSM), min. value 10 MPa (HDPE)

(continued)

Table 27.1 (continued)

Author	Description	Finding
Wang et al. [12]	• HDPE layer combines with kenaf	• Fire performance of HDPE composite recorded that pure HDPE observed a dripping with a high burning rate • Addition of kenaf HDPE, the dripping is missing, and the burning rate is reduced
Zhang et al. [13]	• HDPE modified with untreated and treated celluloses (doped P, N)	• HDPE composites' reinforced cellulose also observed that the fire performance is reduced • Mixture with P, N-doped cellulose HDPE composite shows improvement in fire performance

27.2 Methodology

The fabrication of a structural composite with recycled HDPE was conducted by first preparing the recycled HDPE panel. The process started by cutting the HDPE waste into small pieces, then distributing it evenly on an aluminum plate and heated at 180 °C for 30 min until it melts. Then, the samples were compressed at fixed pressure and held for 1 h until they cooled down. The recycled HDPE panel was used as a core panel and then fabricated as a sandwich structure. Each specimen was fabricated with various design layers with combination layers of chopped strand mat (CSM), and woven roving (WR) by using the hand lay-up method is summarized in Table 27.2. Polyester resin and a catalyst of methyl ethyl ketone peroxide (MEKP) were used as a binder in the sandwich composite panels' fabrication. The prepared panels were then characterized by mechanical testing, water absorbance, and fire-retardant test. The American Standard of Testing Material (ASTM) and International Standard Organization (ISO) were used in the respective testing, tensile test (ASTM D3039), flexural test (ASTM D790), impact test (ISO 179), absorbance test (ASTM D570), and fire-retardant test (ASTM D635). The tensile and flexural testings were conducted using a universal testing machine Instron 1195. Meanwhile, impact testing (Charpy method) was conducted using the Charpy Izod Impact Tester 300 J (LS-22006-300J). The fire-retardant performance was conducted by referring to UL94 as explained in another paper. The mechanical testing conducted was also similar as presented earlier in the same paper [14].

27.3 Results and Discussion

Figure 27.1a shows the tensile stress and strain analysis for samples of Panels R, A, B, and C. Panel A shows a tensile strength of 7.13 MPa. The combination of various types of CSM shows an increase in the tensile strength. Samples of Panel C show the highest tensile strength with 75 MPa. The lowest tensile strength is observed in

Table 27.2 Summary of various design panel layers for recycled HDPE sandwich composite panels

Design layers	Sketching
Panel R	r-HDPE
Panel A	CSM / r-HDPE / CSM
Panel B	WR / r-HDPE / WR
Panel C	CSM / r-HDPE / WR

Panel B. This finding shows that the interaction between CSM and r-HDPE is better than the interaction between WR and r-HDPE.

Figure 27.1b shows the flexural strength for Panels R, A, B, and C. Similar to tensile, the addition of fiberglass increases the flexural strength, however not significant as tensile data. The r-HDPE observed 20.78 MPa. The highest value was observed in Panel C with a value of 23.56 MPa. The fiberglass types did not significantly improve the flexural strength. However, the flexural strain observed the same pattern as in tensile strain. The use of CSM shows better interaction for the panels compared to WR. The use of a combination of CSM and WR shows the highest strain value in both tensile and flexural strains. This finding is similar to the reported journal earlier which discusses the higher value of flexural strength due to good bonding between R-HDPE and fiberglass [8].

Figure 27.2 shows the impact strength analysis for Panels R, A, B, and C. The addition of a fiberglass layer improves the impact strength of r-HDPE (Panel R). The addition of the CSM layer in Panel A shows the highest value of impact strength at 0.2086 kJ/m^2 and the lowest values in Panel R with 0.103 kJ/m^2. The use of WR shows a decrease in impact strength compared to CSM; meanwhile, the combination of CSM/WR shows intermediate value. The presence of chopped strand mat enhances nesting and resists cracks due to greater bonding between layer interfaces [15]. Table 27.3 shows the fracture samples of three types of mechanical testing conducted on a new structural r-HDPE composite. The tensile samples observed most of the

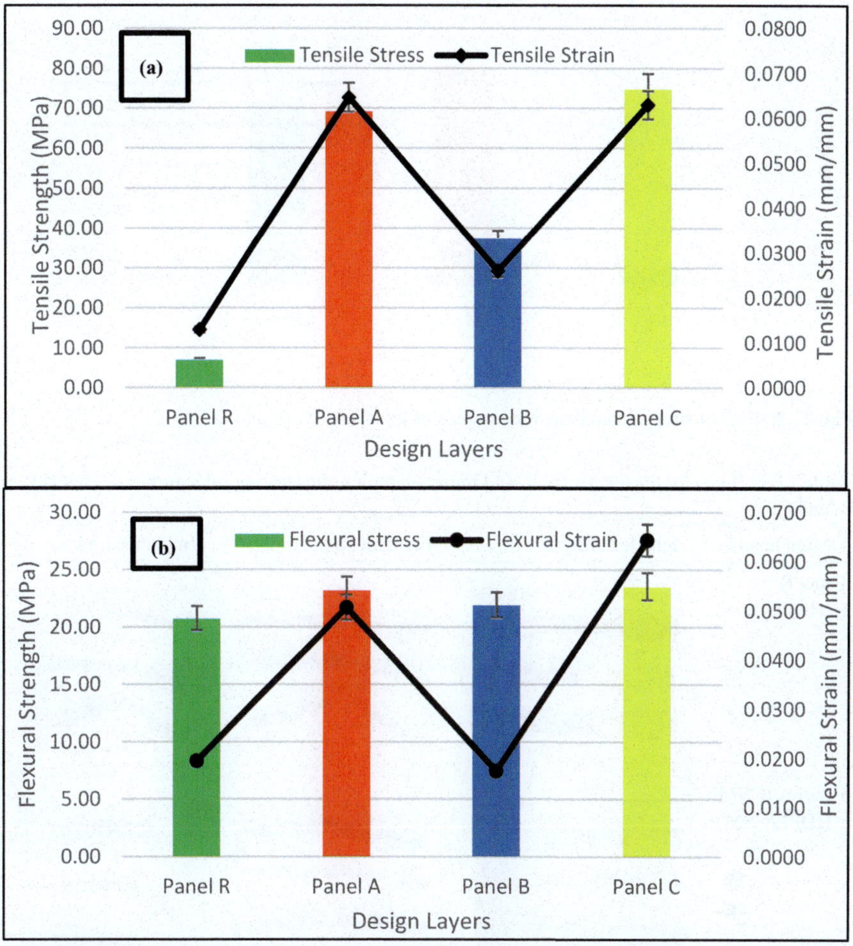

Fig. 27.1 **a** Tensile strength and strain; **b** flexural strength and strain data analysis for samples of Panels R, A, B, and C

sample fracture at the gauge length. All fractured samples observed experienced delamination once mechanical force was applied to the samples due to the protection reaction mechanism. Delamination of WR occurred more than the delamination of CSM fiberglass. This is due to the good interaction binding of r-HDPE with CSM compared to WR.

Figure 27.3 shows the water absorption analysis for samples of Panels R, A, and C. Panel B with WR shows the highest water absorption in distilled water. Meanwhile, the lowest water absorption value is observed in Panel C with the combination of CSM and WR layers in the panels. The use of CSM shows better interaction with r-HDPE compared to WR. This finding is similar to the observation of delamination occurring

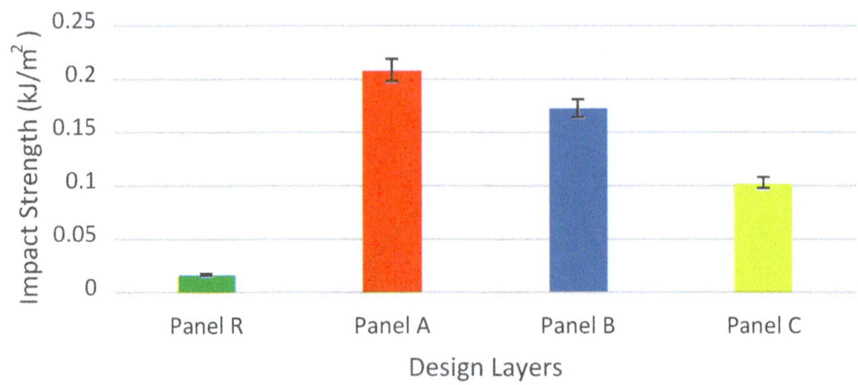

Fig. 27.2 Impact strength analysis for samples of Panels R, A, B, and C

Table 27.3 Image of tensile, flexural, and impact samples after testing for samples of Panels R, A, B, and C

Design layers	Tensile samples	Flexural samples	Impact samples
Panel R (r-HDPE)	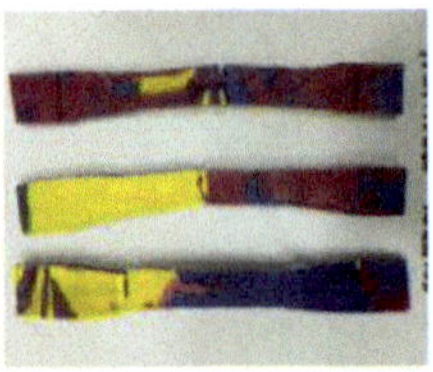		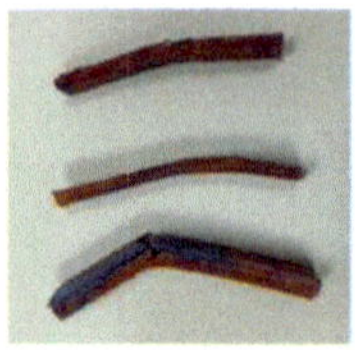
Panel A (CSM/ r-HDPE/CSM)	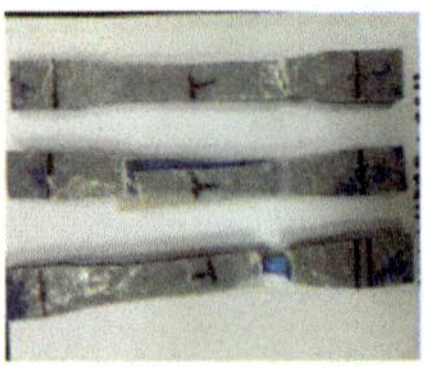	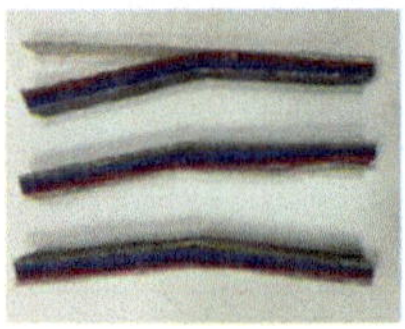	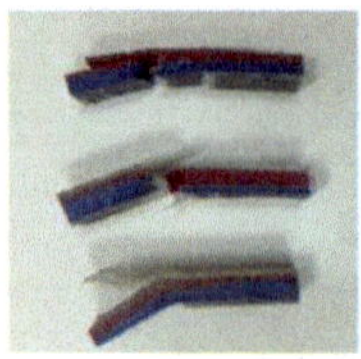
Panel B (WR/ r-HDPE/WR)	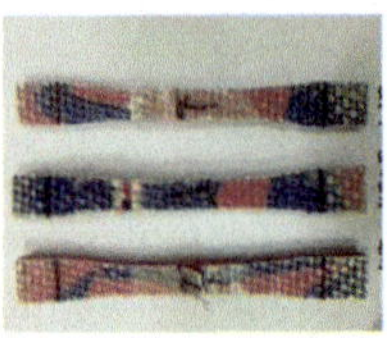	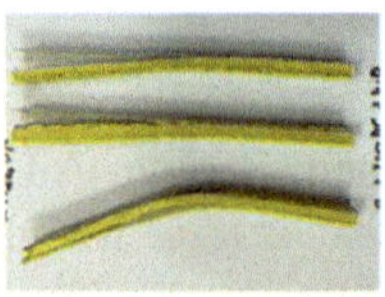	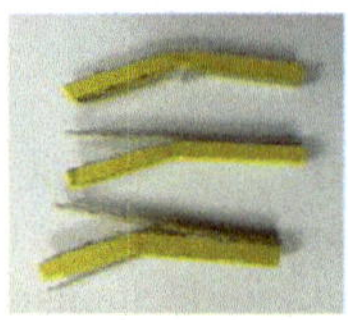
Panel C (CSM/ r-HDPE/WR)	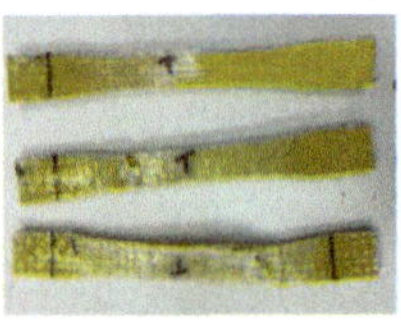	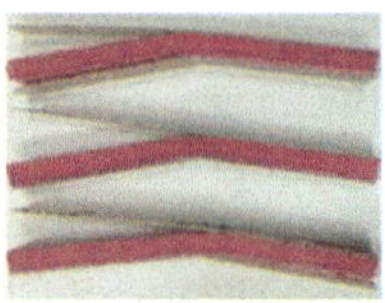	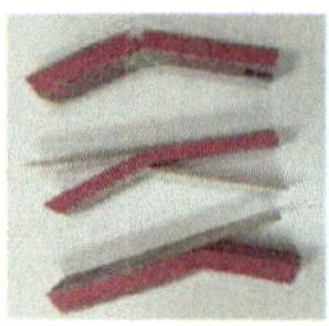

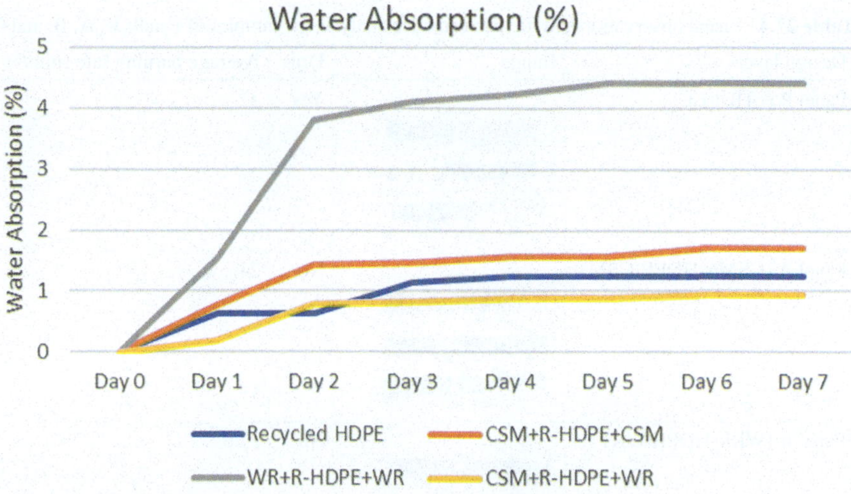

Fig. 27.3 Water absorption analysis for samples of Panels R, A, B, and C

in all of the fractured samples. The water absorption tendency was increased due to the combination of hybrid composite with fiberglass [5].

Table 27.4 shows the visual observation and fire performance analysis of Panels R, A, B, and C. The combination of fiberglass with r-HDPE shows a decrease in the burning rate of the polymer. The dripping problem was also solved with the introduction of fiberglass in the r-HDPE. The burning rate for recycled r-HDPE shows the highest value at 42 mm/s, while Panel B shows the lowest value at 22.6 mm/s. The WR fiberglass layer shows good interaction with r-HDPE to improve the fire-retardant performance. The fire performance of additional fiberglass improved the fire-retardant properties showing similar findings as recorded earlier [16, 17].

Table 27.4 Visual observation and fire performance analysis for samples of Panels R, A, B, and C

Design layers	Image	Drip	Average burning rate (mm/s)
Panel R (r-HDPE)	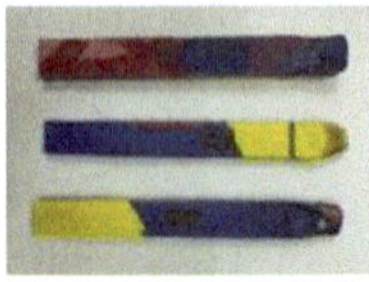	Yes	42
Panel A (CSM/r-HDPE/CSM)		No	24
Panel B (WR/r-HDPE/WR)		No	22.6
Panel C (CSM/r-HDPE/WR)	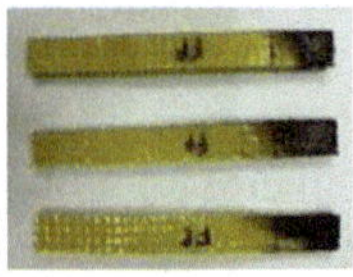	No	31.13

27.4 Conclusion

The new r-HDPE structural composite panels are successfully prepared by using HDPE waste. The fabrication process of the r-HDPE core and structural composite is simplified by resizing, hot press, and hand lay-up methods. The mechanical properties and fire-retardant performance show the highest value in Panel C with the combination of CSM and WR fiberglass. The CSM shows better interaction with r-HDPE compared to WR. However, the combination of CSM and WR in the structural composite r-HDPE produced a better outcome as proven in the tensile, flexural, and water absorption analyses. The data can further be analyzed with another analysis method such as the design of experiment (DOE) and design analysis software to discuss further the interaction between r-HDPE cores with CSM and WR. This structural composite panel has the potential to be used as a composite fiberglass boat structure thus helping in protecting the environment.

Acknowledgements This work was financially supported by the UniKL MIMET Final Year Project. The author also appreciates the University Kuala Lumpur Malaysian Institute of Marine Engineering Technology facilities, equipment, and commitment of technical staff. The author also appreciates the collaboration with the Faculty of Civil Engineering Technology, Universiti Malaysia Pahang. The author gratefully acknowledges all of the support.

References

1. Evode, N., Qamar, S.A., Bilal, M., Barceló, D., Iqbal, H.M.N.: Plastic waste and its management strategies for environmental sustainability. Case Stud. Chem. Environ. Eng. (2021). https://doi.org/10.1016/j.cscee.2021.100142
2. Schyns, Z.O.G., Shaver, M.P.: Mechanical recycling of packaging plastics: a review. Macromol. Rapid Commun. (2021). https://doi.org/10.1002/marc.202000415
3. Kiliç, E., Tarrés, Q., Delgado-Aguilar, M., Espinach, X., Fullanaipalmer, P., Puig, R.: Leather waste to enhance mechanical performance of high-density polyethylene. Polymers **12**(9), 1–15 (2020). https://doi.org/10.3390/polym12092016
4. Abd El-Fattah, A., Abd ElKader, E.: Influence of different clays on the mechanical, thermal, and water absorption properties of recycled high-density polyethylene/wood flour hybrid composites. J. Compos. Mater. **52**(9), 1215–1226 (2018). https://doi.org/10.1177/0021998317723180
5. El-Fattah, A.A., El Demerdash, A.G.M., Alim Sadik, W.A., Bedir, A.: The effect of sugarcane bagasse fiber on the properties of recycled high density polyethylene. J. Compos. Mater. **49**(26), 3251–3262 (2015). https://doi.org/10.1177/0021998314561484
6. Kavianiboroujeni, A., Cloutier, A., Rodrigue, D.: Mechanical characterization of asymmetric high density polyethylene/hemp composite sandwich panels with and without a foam core. J. Sandw. Struct. Mater. **17**(6), 748–765 (2015). https://doi.org/10.1177/1099636215597667
7. Mousa, B.H., El Gamsy, R., Abdel Latif, M.H.: Mechanical behaviour of rubber hybrid composites. IOP Conf. Ser. Mater. Sci. Eng. **610**(1), 012064 (2019). https://doi.org/10.1088/1757-899X/610/1/012064
8. Sayem, A.S.M., Haider, J., Sayeed, M.A.: Development and characterisation of multi-layered jute fabric-reinforced HDPE composites. J. Compos. Mater. **54**(14), 1831–1845 (2020). https://doi.org/10.1177/0021998319885440
9. Aldousiri, B., Alajmi, M., Shalwan, A.: Mechanical properties of palm fibre reinforced recycled HDPE. Adv. Mater. Sci. Eng. (2013). https://doi.org/10.1155/2013/508179
10. Fazli, A., Stevanovic, T., Rodrigue, D.: Recycled HDPE/natural fiber composites modified with waste tire rubber: a comparison between injection and compression molding. Polymers **14**(15), 3197 (2022). https://doi.org/10.3390/polym14153197
11. Lin, J.H., Chen, C.K., Chen, W.C., Tung, Y.C., Lou, C.W.: Lamination of glass fiber woven fabrics and high density polyethylene/clay composition: preparation, tensile properties, and crystallization properties. Appl. Mech. Mater. **749**, 257–260 (2015). https://doi.org/10.4028/www.scientific.net/amm.749.257
12. Wang, Q., Jones, J., Lu, N., Johnson, R., Ning, H., Pillay, S.: Development and characterization of high-performance kenaf fiber–HDPE composites. J. Reinforced Plast. Compos. **37**(3), 191–200 (2018). https://doi.org/10.1177/0731684417739127
13. Zhang, S., Chen, H., Zhang, Y., Zhang, Y., Kan, W., Pan, M.: Flame retardancy of high-density polyethylene composites with P,N-doped cellulose fibrils. Polymers **12**(2), 336 (2020). https://doi.org/10.3390/polym12020336
14. Saat, A.M., et al.: Effect of aluminum phosphate on mechanical and flame retardant properties of composites fiberglass. ARPN J. Eng. Appl. Sci. **12**(4), 1315–1318 (2017). https://doi.org/10.1016/j.rinp.2017.04.025
15. Heckadka, S.S., Nayak, S.Y., Narang, K., Pant, K.V.: Chopped strand/plain weave E-glass as reinforcement in vacuum bagged epoxy composites. J. Mater. **2015**, 1–7 (2015)
16. Saat, A. M. et al.: Effect of thermal aging on the mechanical properties hybrid multilayer composite with various core structure. In: Advancement in Emerging Technologies and Engineering Applications, Lecture Notes in Mechanical Engineering, pp. 125–132 (2020)
17. Saat, A.M., Malik, A.A., Azmi, A., Latif, M.F.A., Ramlee, N.E., Johan, M.R.: Effect of aluminum phosphate on structural and flame retardant properties of composites fiberglass. ARPN J. Eng. Appl. Sci. **12**(4), 1315–1318 (2017)

Chapter 28
Performance of Enhanced Natural Kapok/Coconut Coir Composite Absorbent for Oil Spill Recovery

Wan Farhan Zakaria, Mohamad Fahmi Shaharuddin, Mohd Izuan Ishak, Wan Ahmad Safwan Zakaria, Norazizah Che Mat, Muhammad Abdul Mun'aim Mohd Idrus, Nurul Safiah Yusoff, Irwan Nurdin, and Asmalina Mohamed Saat

Abstract Catastrophic oil spills emphasized the importance of fast response to ensure marine environmental sustainability, as well as cost-effective technology for oil spill cleaning and recovery. However, the existing method for oil spill response causes damage to the ecosystem and does not recover the oil. The use of organic oil

W. F. Zakaria · M. F. Shaharuddin · M. I. Ishak · W. A. S. Zakaria · N. C. Mat · M. A. M. M. Idrus · N. S. Yusoff · A. M. Saat (✉)
Universiti Kuala Lumpur, Malaysian Institute of Marine Engineering Technology, Lumut Perak, Malaysia
e-mail: asmalina@unikl.edu.my

W. F. Zakaria
e-mail: wanfarhan9@yahoo.com

M. F. Shaharuddin
e-mail: mfahmi.shaharuddin@s.unikl.edu.my

M. I. Ishak
e-mail: izuan.ishak@unikl.edu.my

W. A. S. Zakaria
e-mail: safwan.zakaria@unikl.edu.my

N. C. Mat
e-mail: norazizah@unikl.edu.my

M. A. M. M. Idrus
e-mail: mamunaim@unikl.edu.my

N. S. Yusoff
e-mail: nurulsafiahy@unikl.edu.my

I. Nurdin
Chemical Engineering Department, Politeknik Negeri Lhokseumawe, 24301 Lhokseumawe, Indonesia
e-mail: irwan_nurdina@yahoo.com

A. Ismail et al. (eds.), *Exploring Diversity in Engineering and Technology for Knowledge and Innovation*, Advanced Structured Materials 215,
https://doi.org/10.1007/978-3-031-64330-9_28

sorbent tends to absorb both water and oil and leads to the sinking of sorbent materials. The organic sorbent material is in loose particles and makes it difficult to collect after spreading on the water. The use of natural organic sorbent raised attention due to its low cost, high effectiveness, and attractive properties such as high recovery, reusability, and biodegradability. Thus, this research focuses on evaluating the best composition of a coconut coir and kapok composite absorbent material that can clean, collect, and recycle oil spills. The observation showed the composition of coconut coir and kapok with a ratio of 0:4 in the water absorbance test, oil absorbance test, and reusability test. However, the addition of coconut coir improved the water, oil absorbance, and reusability properties of lubricant oil. This is due to the hydrophilic properties of coconut coir. The combination of coconut coir and kapok shows better absorbance properties for lubricant oil and has improved the buoyancy of composite sorbent materials in real oil spill cleaning conditions.

Keywords Oil spill · Oil recovery · Kapok · Coconut coir · Natural composite absorbance

28.1 Introduction

The rising incidents of oil spills have resulted in the continual growth of oil spill response solutions to treat them effectively and reduce their negative effects [1]. Common oil spill cleaning and recovery methods recorded are using oil booms [2], skimmers [3], sorbent [4], burning in situ [5], dispersant [6], hot water/high-pressure washing [7], manual recovery and natural attenuate [8]. The choice of method depends on the location and type of spill, as well as other factors such as weather conditions and water depth. Responding to oil spills can be logistically challenging, particularly in remote areas or in adverse weather conditions. Some oil spill response technologies may have limited effectiveness, particularly in certain environmental conditions. Another challenge is that the oil spill response can be expensive, and the cost of cleanup can be significant. Sorbents are used more than 80% of the time [9] for oil spill cleanup. Sorbents (insoluble materials) are used to collect the oil by the mechanism of absorption and adsorption, or both. Absorbents swell up more than 50 percent to collect and preserve liquid throughout its molecular structure. The absorbents are coated by a liquid on their surface, including pores and capillaries that support the solid swelling. The sorbents need to have both hydrophobic (water-repellent) and oleophilic (oil-attracting) properties. The sorbents are used as a final stage to clean the oil spills. It is also the best option to clean the oil spill in the area not accessible by skimmer.

Materials used as oil sorbents are divided into three categories, organic natural products, organic synthetic products, and inorganic mineral products. The organic natural product from vegetables degrades naturally compared to organic synthetic products (polyurethane and polypropylene) that degrade very slowly [6]. Meanwhile, examples of inorganic minerals are such as sorbent clay, diatomite graphite, and perlite. Synthetic organic oil sorbents are commonly used due to their oil-absorbing performance (absorb 70 times its weight) and can be reused. However, due to the limitation of being very hard to degrade [10] and the high cost of synthetic oil sorbent [9], natural oil absorbance becomes the best solution. The use of natural oil sorbent is an interesting method due to attractive properties such as recovery, reusability, and biodegradability. It is also a promising, highly effective oil binding agent that reduces costs and provides an environmentally friendly method of cleaning up oil spills [11]. Although the properties are not as good as synthetic oil absorbent, the properties of natural sorbent materials can be improved by using chemical enhancement such as Para ID="Par4">The natural oil sorbent materials faced a challenge due to properties of buoyancy, oil sorption performance, and hydrophobicity/oleophilic [12]. Hydrophobicity is important to describe the performance of oil sorption. The hydrophobic–oleophilic properties of an oil sorbent depend on the physical appearance of the fiber, the porosity, the surface wax, the surface roughness, the twists and the crimps, and the chemical constituent of the sorbent. The oil sorbent performance also depends on the oil properties such as the amount of oil, the oil concentration, specific gravity, and the temperature [11]. These characteristics also contribute to the buoyancy of natural absorbance.

Kapok fiber is a natural oil sorbent that is known as cellulosic fiber. Kapok fibers have both properties, superhydrophobic (due to the high proportion of acetyl groups) and oleophilic, they do not get wet in water [8] and have a low density. Kapok also has an air cavity (lumen) with a waxy fiber wall that leads to good buoyancy and high oil sorbent performance [7]. The water absorbent and oil absorbent capacity is evaluated by referring to ASTM F 726 and conducted at room temperature. The water absorbent is designed to measure the water molecules attached to the sorbent materials. Meanwhile, the oil absorbent properties are a test conducted to determine the optimum absorbent capacity without comparing the molecules of water. The reusability test repeats the measurements of the oil absorbency test using the same fiber several times by recording the weight change. After the test, the sample is compressed to remove all excess oils. The summary of kapok sorbent selected properties such as water absorbent, oil absorbent, and reusability properties is presented in Table 28.1. Meanwhile, another interesting natural organic sorbent from agriculture waste is coconut coir. The fiber is extracted from the fibrous mesocarp layer of the coconut fruit. The major composition of coconut is polysaccharides (50.9%) and lignin (32.8%). The minor constituents are nitrogenous substances, inorganic salts, coloring substances, pectin, wax and others [13]. Coconut coir is reported as an absorbent material with the combination of human hair. A combination of coconut coir with human hair observed improvement of the absorbance in diesel oil from 6530 mg/g to 7916 mg/g in coconut husk-to-human hair ratio 1:1 [14]. Coconut coir fiber surface has carbonaceous with an -OH group. These groups tend to react at specific pH values and enable the uptake of metal ions from aqueous solutions. The characteristics lead to good oil sorbent properties and high water sorbent properties [9]. Kapok and coir fibers are naturally renewable materials with a high lumen and hydrophobicity, allowing them to absorb a lot of oil [15]. Both kapok and coconut coir have interesting sorbent properties. Their ability to absorb oil spills depending on the combination composition is still not found to be reported earlier. Thus, this research focuses on producing environment-friendly and cost-effective materials for oil cleaning and recovery. The aim is to produce an absorbent pad to collect and clean the oil spillage with natural oil absorbent as their filtration system so that it can recover the oil easily. The performance of the new combination of coconut coir kapok natural sorbent was evaluated for their oil absorbency, water absorbency, and reusability performance.

28.2 Methodology

There are three stages of processes throughout the project as shown in Fig. 28.1. The project began with the preparation of kapok and coir as a pad of absorbent (Fig. 28.2). A total of 4 g of weight is used to compress the absorbent to make it a pad. Next, the combination of coir and kapok absorbents consisting of (4:0, 3:1, 2:2, 1:3, and 0:4) is prepared. The process took three days to make the structure of the absorbent become a pad. The absorbent pad was then evaluated by conducting the oil absorption test, seawater absorption test, and reusability test.

Seawater (400 ml) is poured into a one-liter beaker and added with 4 g of absorbent material (W_1). The absorbent materials are left for 2 min to settle. The same process is repeated for oil absorbency properties. However, the seawater is mixed with 200 ml of oil first before absorbent materials are immersed. The condition of the water and oil absorbency is visually observed, and the change in weight is recorded. (Note: If 10% or more of the absorbent material has sunk, the absorbent is considered to have failed the test.) Strain the contents of the beaker using a mesh to catch the absorbent. After 30 s of the drain, the sample is weighed (W_2). The water and oil absorbencies are calculated by using Eq. (28.1).

$$\text{Absorbency} = \frac{W_2 - W_1}{W_1} \times 100 \quad (28.1)$$

28.3 Results and Discussion

Table 28.2 shows the image of the seawater absorbency test for various coconut coir and kapok ratios. No significant change in physical appearance is observed. In the wet condition, the fiber tends to stick to each other. The weight change is presented in Fig. 28.3. Figure 28.3 shows that 100% of coconut coir observed the highest seawater absorbency performance with 190 wt.%; meanwhile, the lowest values observed in samples 100 wt.% of kapok (0:4) which is 0.83%. This finding was due to the composition and hydrophobic or oleophilic properties that kapok has. Meanwhile, coconut coir has hydrophilic properties due to the large amount of

Table 28.1 Summary of selected kapok and natural absorbance properties

Authors	Properties	Material	Finding
[9]	Water absorbency test	Kapok, coir, human hair	• Water absorbency rate (kapok < human < coir) • Kapok (not dissolved in the seawater, floats on water) • Kapok water absorbency (2.575%)
[16]		Untreated kapok, HCl-treated, NaOH-treated, NaClO-treated, NaClO-NaOH-NaClO-treated, $NaClO_2$-NaOH-$NaClO_2$-treated based	• Six types of kapok, untreated kapok (smallest moisture sorption; 45.29%) • Treated kapok with NaClO-NaOH-NaClO-treated (highest moisture sorption; 285.29%)
[17]		Coconut (shell and husk), pineapple (leaf), mangosteen (rind), banana (skin), corn (cob), rambutan (rind), kapok durian (skin)	• % water absorbed increases as coconut husk fiber content increases. • Coconut husk absorption in distilled water (3–17%)
[9]	Oil absorbency test	Kapok, coir, human hair	• Kapok absorbing diesel oil (46.401 g/g) • Coir absorbing diesel oil (7.0904 g/g) • Absorption properties in crude and diesel oil; Kapok > coir > human Hair
[18]		Raw kapok fiber, chemically treated kapok fiber	• Kapok absorbency of aviation oil increases due to treatment with HCl, chloroform, and NaOH, but decreases with treatment NaOH at increasing concentrations of NaOH (1–5%) • Raw kapok absorbs engine oil (38 g/g), jet A1-fuel (26 g/g), and Skydrol (49 g/g)
[17]		Coconut (shell and husk), pineapple (leaf), mangosteen (rind), banana (skin), corn (cob), rambutan (rind), kapok, durian (skin)	• Composites with pineapple and rambutan fibers (highest absorption of the engine and cooking oil) compared to fiber-filled composites • Raw coconut fiber oil absorption (5.4 g/g) • Kapok's oil absorption (36–45 g/g)

(continued)

Table 28.1 (continued)

Authors	Properties	Material	Finding
[18]	Reusability test	Raw kapok fiber Chemically treated kapok fiber	• Raw kapok oil absorbs (21 g/g: 1st cycle), as the number cycle increases the absorbance properties decrease (10 g/g: 5th cycle)
[19]		Kapok, Kapok with NaOH	• Raw kapok (32 g/g: 1st run) (15 g/g:5th run) • The reusability of fiber decreases immediately after the 1st run (the sorption capacity of the fiber reduces with several cycles) • The absorption capacity of treated does not reduce much compared to raw kapok due to more stable fibrous material

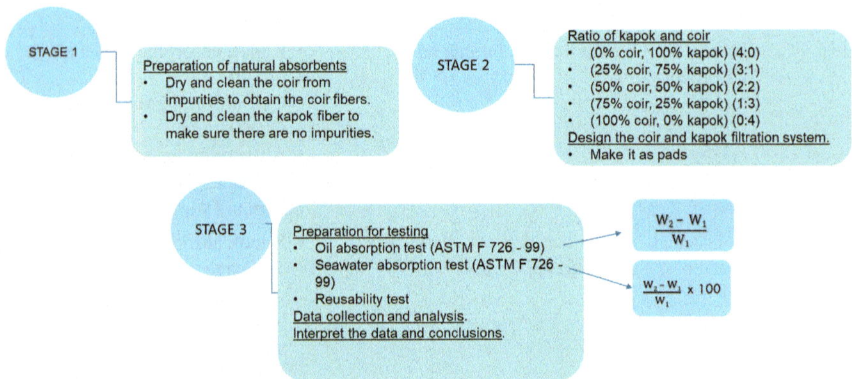

Fig. 28.1 Flowchart of sample preparation, data collection, and analysis

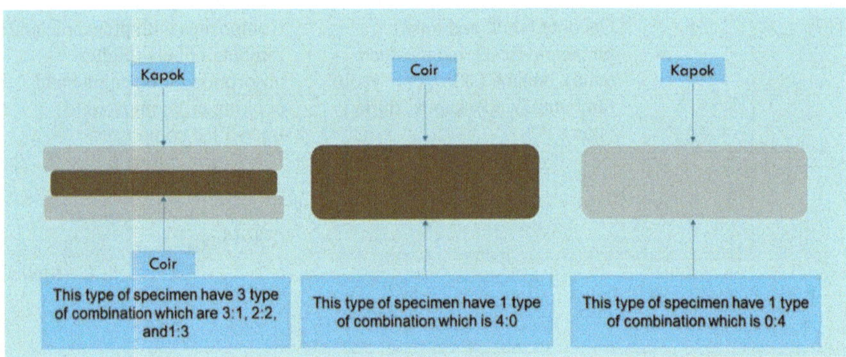

Fig. 28.2 Sample preparation for kapok and coconut coir absorbance materials

water absorbed during the test. Kapok fibers contain cellulose, lignin, and pentosan [20]. But coir contains cellulose, hemicellulose, and lignin [14]. This will affect the absorption percentage of water. Based on previous research, the amount of seawater that has been absorbed is higher compared to this result, and this happens because coir which had been granulated and dissolved in the seawater made it hard to retrieve and lost its buoyancy [21]. For this research, long coir fibers with a range of length between 15 and 28 cm are used resulting in the reduction of the contact surface between the coconut coir and kapok.

Table 28.3 shows the visual observation of oil absorbency for two different oils, lubricant oil and diesel oil. The image shows that lubricant oil is observed a darker color compared to diesel oil. Absorption of lubricant oil in composite filter containing coconut coir shows better absorption. Figure 28.4 shows that increasing the ratio of kapok observed an increased absorption rate of oil. Oil absorbency shows the highest values in the sample 100% of kapok (0:4) which is 16.85 g/g for main engine lubricant oil and 24.1 g/g for diesel oil. Meanwhile, the lowest values were observed in samples with 100 wt.% coir (4:0) which is 7.1 g/g for main engine lubricating oil and 2.35 g/g for diesel oil. From the graph, the increasing kapok ratio leads to an increase in the absorption rate for both diesel and lubricating oil. However, at a higher coconut coir ratio, the lubricant oil absorption is higher than diesel oil. This observation shows that the interaction of coconut coir with lubricant oil is better than diesel oil. The

Table 28.2 Visual observation for seawater absorbency test for various ratios of coconut coir:kapok (4:0, 3:1, 2:2, 1:3, 0:4)

Ratio	Before	After
4:0		
3:1		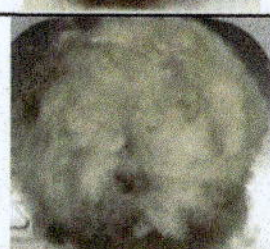
2:2	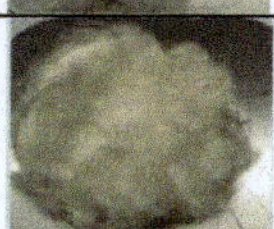	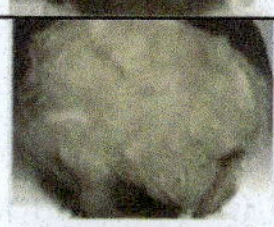
1:3	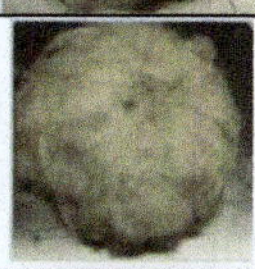	
0:4		

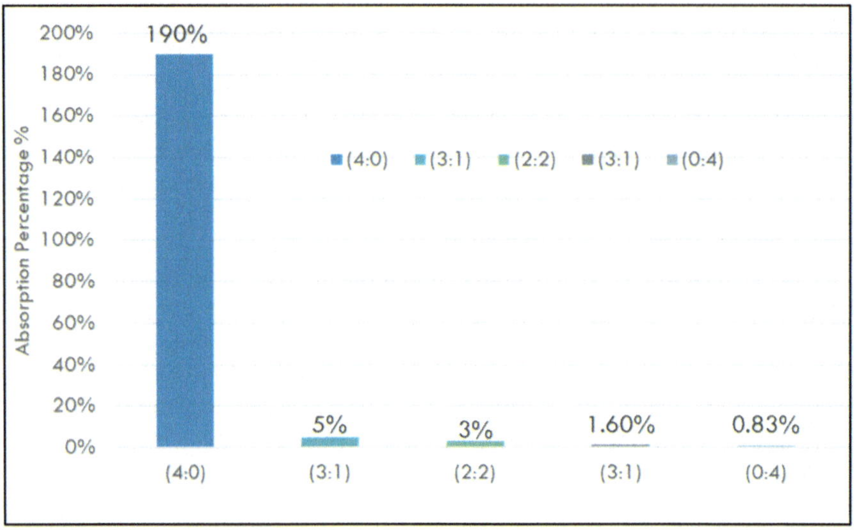

Fig. 28.3 Seawater absorbency performance for various ratios of coconut coir:kapok (4:0, 3:1, 2:2, 1:3, 0:4)

observation was due to the relationship between the water molecule and lubricant acid value [22]. Typically, lubricant oil allows a higher concentration of dissolved water compared to diesel water. According to the previous journal, the mechanism of oil absorption is generally attributed to the hollow lumen and surface wax of kapok fiber [15], and for the coconut coir, it has a layer of the surface stacked on each other. These layers could have a capillary that enables them to collect and trap oil [9].

Table 28.4 shows the reusability visual observation for various ratios of coconut coir and kapok. The sample (4 g) in a beaker was immersed in oil and water at room temperature for 2 min. Then, the mesh was taken up from the oil and drained for 30 s, wiped with filter paper to remove excess oil from the bottom of the mesh. Finally, the oil-loaded sample was transferred into another beaker and compressed by putting 5 kg of weight pressure on the piston to remove any oil remaining and left for 2 min The test was repeated and the oil absorbency of the sample after five desorption cycles was evaluated. The reusability of samples with high ratios kapok shows a significant change in physical appearance and color. The kapok color changes from pearl white to dark brown. The before condition is after the first cycle of absorption. The higher the ratio of kapok, the darker the color changes. This is due to the ability of kapok to absorb more oil than coconut coir. Furthermore, kapok's properties are oleophilic as compared to coconut coir which is hydrophilic. This finding supports the observation of the absorbance properties earlier. The color for absorption in lubricant oil appeared much darker compared to diesel oil due to the natural color of the oil. Other than that, the absorbent composite absorbs more lubricant oil compared to diesel oil, especially for composition with higher coconut coir.

Table 28.3 Visual observation of oil absorbency (diesel and lubricant oil) for various ratios of coconut coir:kapok ratio (4:0, 3:1, 2:2, 1:3, 0:4)

Ratio	Before (diesel)	After (diesel)	Before (lubricating oil)	After (lubricating oil)
4:0	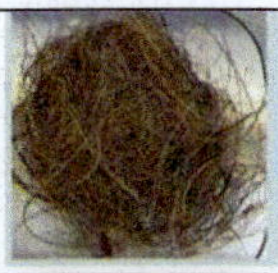			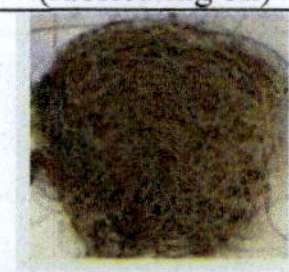
3:1				
2:2	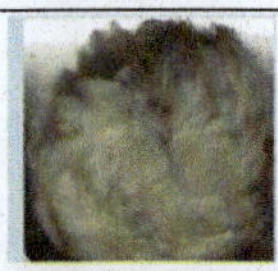			
1:3	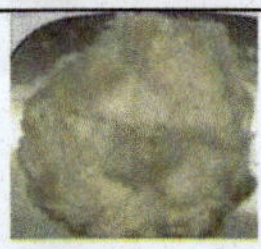			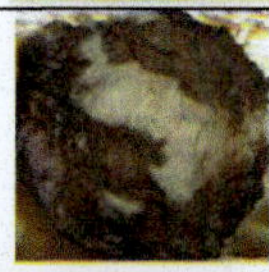
0:4		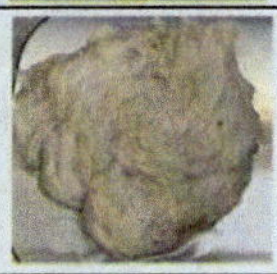		

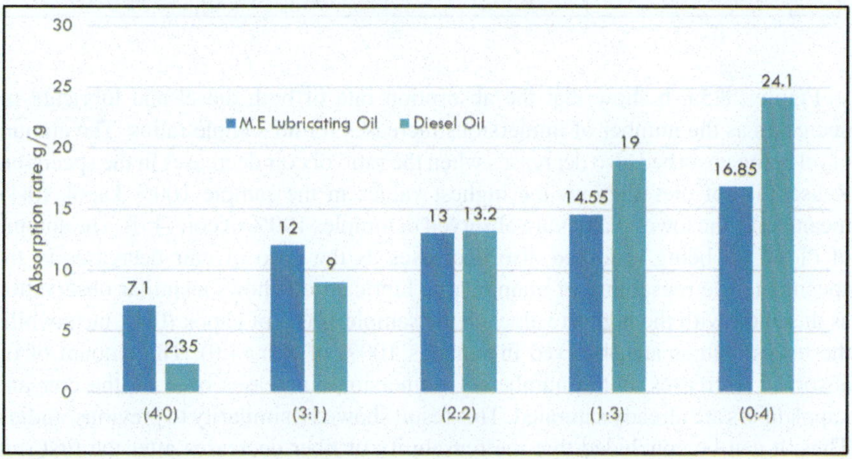

Fig. 28.4 Oil absorbency performance for various ratios of coconut coir:kapok ratio (4:0, 3:1, 2:2, 1:3, 0:4)

Table 28.4 Visual observation of reusability test (diesel oil and marine engine lubricant oil) for various ratios of coconut coir:kapok (4:0, 3:1, 2:2, 1:3, 0:4)

Ratio	Before (diesel)	After (diesel)	Before (lubricating oil)	After (lubricating oil)
4:0				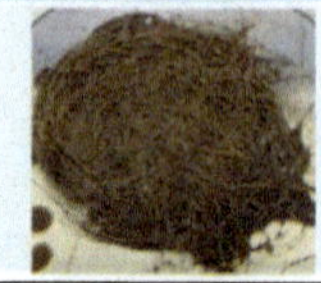
3:1				
2:2				
1:3				
0:4				

Figure 28.5a, b show that the absorption rate of both diesel and lubricant oil decreases as the number of immersions increases for all sample ratios. The amount of oil being absorbed also decreases when the ratio of coir decreases in the specimen. Reusability of diesel shows the highest values in the sample 100% kapok (0:4); meanwhile, the lowest values are observed in samples 100% of coir (4:0). The amount of diesel oil being absorbed also decreases as the ratio of coir decreases in the specimen. The reusability of main engine lubricant oil shows a similar observation as diesel oil with the highest values in the sample 100% of kapok (0:4); meanwhile, the lowest values are observed in samples 100% of coir (4:0). The amount of oil absorbed decreases as the number of immersions increases because the coir and kapok fibers are already saturated. This result showed a similarity to previous studies. Thus, it can be concluded that the reusability of fiber decreases after the first run. This proves that the absorption capacity of the fiber reduces after several cycles [1]. However, lubricant oil shows better absorbance compared to diesel oil with a higher absorbance weight. For samples with a higher ratio of coconut coir (4:0), the lubricant oil absorbance weight is higher than diesel oil. The combination of coconut

coir with kapok improved the absorbance properties of lubricant oil. Since coconut coir absorbs better in a system combined with water, the combination of coconut coir and kapok is expected to have better absorbance properties and buoyancy in real seawater oil spill cleaning. The composite coconut coir and kapok also show better performance of oil cleaning compared to treated kapok which reduces the kapok oil absorbance properties. In this natural composite absorbance material, the coconut coir reacts as a stabilizer and optimized kapok's ability to absorb the oil and clean the water. The modification of kapok by producing a composite with coconut coir is much cheaper and easier compared to chemical treatments reported earlier.

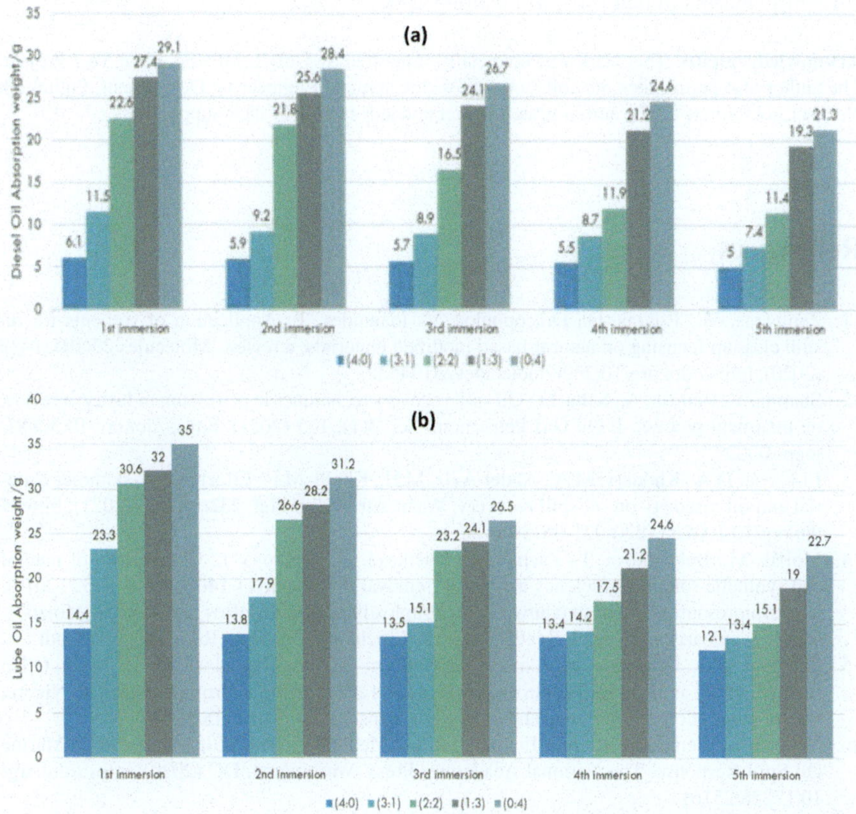

Fig. 28.5 Reusability performance of **a** diesel oil; **b** main engine lubrication oil for various ratios of coconut coir:kapok ratio (4:0, 3:1, 2:2, 1:3, 0:4)

28.4 Conclusions

The performance of coconut coir and kapok composite absorbance oil collectors was successfully prepared and evaluated. The composition of 100% kapok and 0% coconut coir with a ratio of 0:4 is the best in terms of water sorption efficiency (0.83%), oil sorption efficiency of lubricant oil (16.85 g/g), diesel oil (24.1 g/g), and reusability efficiency. However, the addition of coconut coir improved the water, oil absorbance, and reusability properties of lubricant oil. This is due to the hydrophilic properties of coconut coir. The combination of coconut coir and kapok shows better absorbance properties for lubricant oil and has improved the buoyancy of composite sorbent materials in real oil spill cleaning cases.

Acknowledgements This work was financially supported by UniKL MIMET Final Year Project. The author also appreciates the collaboration with Chemical Engineering Department, Politeknik Negeri Lhokseumawe. The author gratefully acknowledges both of their support.

References

1. Zamparas, M., Tzivras, D., Dracopoulos, V., Ioannides, T.: Application of sorbents for oil spill cleanup focusing on natural-based modified materials: a review. Molecules **25**(19), 1–22 (2020). https://doi.org/10.3390/molecules25194522
2. Sharma, V., Ramish, A., Sahu, O.: Oil spill recovery techniques in petroleum industry: a review on treatment process. J. Oil Gas Petrochem. Sci. **3**(1), 1–5 (2021). https://doi.org/10.30881/jogps.00029
3. El-Gayar, D.A., Khodary, M.A., Abdel-Aziz, M.H., Khalil, M.F.: Effect of disk skimmer material and oil viscosity on oil spill recovery. Water Air Soil. Pollut. **232**(5), 193 (2021). https://doi.org/10.1007/s11270-021-05150-y
4. Mojžiš, M., Bubeníková, T., Zachar, M., Kačíková, D., Štefková, J.: Comparison of natural and synthetic sorbents efficiency at oil spill removal. Bioresources **14**(4), 8738–8752 (2019). https://bioresources.cnr.ncsu.edu/wp-content/uploads/2019/09/BioRes_14_4_8738_Mojzis_BZKS_Comparison_Natural_Synth_Sorbents_Efficiency_Oil_Spill_Removal_15989.pdf
5. Ekperusi, A.O., Onyena, A.P., Akpudo, M.Y., Peter, C.C., Akpoduado, C.O., Ekperusi, O.H.: In-situ burning as an oil spill control measure and its effect on the environment. In: SPE Nigeria annual international conference and exhibition. https://doi.org/10.2118/198777-MS
6. National Academies of Science, Engineering and Medicine: The Use of Dispersants in Marine Oil Spill Response. The National Academies Press, Washington, DC (2020). https://doi.org/10.17226/25161
7. Broman, D., Ganning, B., Lindblad, C.: Effects of high pressure, hot water shore cleaning after oil spills on shore ecosystems in the Northern Baltic proper. Mar. Environ. Res. **10**(3), 173–187 (1983). https://doi.org/10.1016/0141-1136(83)90009-0
8. Péquin, B., Cai, Q., Lee, K., Greer, C.W.: Natural attenuation of oil in marine environments: a review. Mar. Pollut. Bull. **176**, 113464 (2022). https://doi.org/10.1016/j.marpolbul.2022.113464
9. Xiang, J.F.K.: Investigation of oil spill absorbents using different materials. Final Year Project, Universiti Teknologi Petronas, pp 1–58 (2011). http://utpedia.utp.edu.my/id/eprint/10524
10. Radvanska, A.: The environmental impacts of inland water transport and possibilities of oil spills cleaning. Acta Technica Corviniensis— Bull. Eng. **3**(3), 73–78 (2010)

11. Anuzyte, E., Vaisis, V.: Natural oil sorbents modification methods for hydrophobicity improvement. Energy Procedia **147**, 295–300 (2018). https://doi.org/10.1016/j.egypro.2018.07.095
12. Zamparas, M., Tzivras, D., Dracopoulos, V., Ioannides, T.: Application of sorbents for oil spill cleanup focusing on natural-based modified materials: a review. Molecules **25**(19), 4522. https://doi.org/10.3390/molecules25194522
13. Mishra, L., Basu, G.: Coconut fibre: its structure, properties and applications. In: Handbook of Natural Fibres, pp 231–255. Elsevier (2020). https://doi.org/10.1016/B978-0-12-818398-4.00010-4
14. Osamor, A., Momoh, Z.: An evaluation of the adsorptive properties of coconut husk for oil spill cleanup. In: International conference on advances in applied science and environmental technology—ASET 2015, pp 33–37. Institute of Research Engineers and Doctors (2015). https://doi.org/10.15224/978-1-63248-040-8-38
15. Quek, C.S., Ngadi, N., Ahmad Zaini, M.A.: The oil-absorbing properties of kapok fibre—a commentary. J. Taibah Univ. Sci. **14**(1), 507–512 (2020). https://doi.org/10.1080/16583655.2020.1747767
16. Wolok, E., Lahay, I.H., Machmoed, B.R., Pakaya, F.: Modification and characterization of Ceiba Pentandra (L.) Gaertn. (Kapok) fiber: physical properties. Int. J. Res. GRANTHAALAYAH **7**(7), 381–390 (2019). https://doi.org/10.29121/granthaalayah.v7.i7.2019.791
17. Ahad, N.A.: Oils and water absorption behavior of natural fibers filled TPU composites for biomedical applications. In press: Accepted manuscript—Note to users oils and water absorption behavior of natural fibers filled TPU composites for biomedical applications. J. Eng. Res. Educ. **10**, 16–21 (2018)
18. Khairuddin, M.A.: Effect of Treated Kapok Fiber on Aviation Oils Absorbency. Universiti Kuala Lumpur Malaysian Institute of Aviation Technology, Kajang (2015)
19. Muniandy, K.: Investigation of Modified Kapok Fibre on the Sorption of Oil in Water. Universiti Teknologi PETRONAS, Tronoh, Perak (2015)
20. Chaiarrekij, S., Apirakchaiskul, A., Suvarnakich, K., Kiatkamjornwong, S.: Characteristics of kapok fiber. Bioresources **7**(1), 475–488 (2011)
21. Lim, T-T., Huang, X.: Evaluation of kapok (Ceiba pentandra (L.) Gaertn.) as a natural hollow hydrophobic–oleophilic fibrous sorbent for oil spill cleanup. Chemosphere **66**(5), 955–963 (2007). https://doi.org/10.1016/j.chemosphere.2006.05.062
22. Zhou, F., Yang, K.: The influence of water content on the acid number of diesel engine lubricant in electrochemical measurement. IOP Conf. Ser. Mater. Sci. Eng. **1043**(5), 052054 (2021). https://doi.org/10.1088/1757-899X/1043/5/052054

Chapter 29
Semi-submersible Trimaran Hull Form Improvement to Reduce Ship Fuel Consumption

Wasis Dwi Aryawan, Hesty Anita Kurniawati, Rizky Chandra Ariesta, Haikal Anjasmara, Nur Aini Amalia Dinda Agustina, Steven Caramoy, and Md Redzuan Zoolfakar

Abstract Indonesia has many tourist areas; therefore, it requires facilities to support the development of these regions. One of them is a semi-submersible trimaran ship that can explore the underwater scenery. Previous trimarans faced difficulties during operation due to the occurrence of forward flow in the area around the bridge, caused by water hitting the bridge area. This issue led to increased fuel consumption due to increased resistance. Therefore, optimization of the trimaran's hull is necessary to reduce the ship's resistance and, in turn, reduce fuel consumption. The optimization is carried out by modifying the bridge distance (the distance between the main hull and the demi hull) and by changing the length of the demi hull. The slender body method is chosen to calculate the trimaran's resistance because of the slender shape of

W. D. Aryawan (✉) · H. A. Kurniawati · R. C. Ariesta · H. Anjasmara · N. A. A. D. Agustina · S. Caramoy
Department of Naval Architecture, Faculty of Marine Technology, Sepuluh Nopember Institute of Technology, Surabaya, Indonesia
e-mail: wasis@na.its.ac.id

H. A. Kurniawati
e-mail: tita@na.its.ac.id

R. C. Ariesta
e-mail: chandra@its.ac.id

H. Anjasmara
e-mail: haikal.anjasmara98@gmail.com

N. A. A. D. Agustina
e-mail: dindaamalia5698@gmail.com

S. Caramoy
e-mail: stevencaramoy98@gmail.com

M. R. Zoolfakar
Universiti Kuala Lumpur, Malaysian Institute of Marine Engineering Technology, Jalan Pantai Remis, 32200 Lumut, Perak, Malaysia
e-mail: redzuan@unikl.edu.my

A. Ismail et al. (eds.), *Exploring Diversity in Engineering and Technology for Knowledge and Innovation*, Advanced Structured Materials 215,
https://doi.org/10.1007/978-3-031-64330-9_29

the trimaran's hull. This slender body method is then validated through towing tests to obtain the trimaran's resistance value. The results show that for various bridge extensions, the lowest resistance value is obtained at a speed of 11 knots with a resistance of 31.6 kN. The resistance value for various reductions in the length of the demi hull at a speed of 10 knots is 36.5 kN. This indicates a reduction in the ship's resistance, which can reduce fuel consumption.

Keywords Bridge · Demi hull · Resistance · Slender body · Trimaran

29.1 Introduction

Indonesia is the largest archipelagic country, with a land area of 2.55 million km^2 and a sea area of 5.8 million km^2 spanning from Sabang to Merauke. With the sea area comprising nearly two-thirds of Indonesia's total territory, the government has focused on developing the tourism sector, particularly maritime tourism, which has significant potential to contribute to economic growth and benefit the government, general public, and private sector. One of the popular tourist attractions is the breathtaking underwater beauty of the Bunaken National Park. As one of the National Strategic Tourism Areas, according to the Government Regulation of the Republic of Indonesia Number 50 of 2011 and the guidance of the President of the Republic of Indonesia, the proposal for the development of maritime tourism facilities in the form of tourist vessels to explore the underwater wonders by using semi-submersible tourism vessels. Therefore, a study on the semi-submersible trimaran has been conducted to support tourism in the Bunaken National Park area [1].

The construction of a glass-bottomed ship to support tourism in Bunaken National Park, Manado, North Sulawesi, has been constructed and could be seen in Fig. 29.1. The design of this tourist vessel features a trimaran hull shape, which consists of three hulls: a main hull and two smaller secondary hulls. It is noted that during the operation of the ship, there is a partial water flow returning to the front, caused by the phenomenon of water hitting the bridge. This results in an increase in resistance, which affects the fuel consumption required by the engine and has an impact on the operational costs of the ship. Therefore, hull optimization is needed to reduce ship resistance with the aim of reducing fuel consumption.

Hull form optimization aims to decrease resistance for enhancing ship efficiency. Studies suggest that the profile type and angle of attack of anti-rolling hydrofoils significantly influence resistance [2]. Moreover, considerations must be made regarding the spacing and center distance between the main hull and the secondary hull [3]. This article employs three-dimensional potential theory and Green's function to provide valuable insights into wave loads, encompassing vertical shear forces, vertical bending moments at various main hull locations, and transverse bending moments at the cross-structure-main hull junction. Additionally, parametric optimization of wave loads is conducted for diverse side hull layouts and lengths, factoring in various wave speeds and headings [4].

Fig. 29.1 Semi-submersible trimaran

The present study focuses on enhancing the semi-submersible trimaran hull form through the slender body method to reduce resistance and consequently decrease fuel consumption. By employing the slender body method for resistance calculation and utilizing data from towing tests, the required power for the ships can be determined. The objective of this paper is to serve as a guide for optimizing trimaran hull forms to mitigate wave-making resistance, thereby reducing fuel consumption and carbon emissions.

29.2 Methodology

29.2.1 Slender Body Method

This approach can be employed for various hull configurations, including multihulls. However, it is essential for individual hulls to possess slenderness and symmetry around their local centerline. Ideally, a high slenderness ratio (*L*/*B*) is preferred. Therefore, this method is particularly suitable for trimaran hulls due to their slender bodies. The formulation of thin ship theory is outlined below. Total drag consists of two elements: viscous drag caused by the ship's movement through a viscous fluid, and wave-making drag caused by the ship's movement on the free surface. Wave-making drag arises from the dissipation of energy in surface waves. The total drag coefficient is expressed as follows:

$$C_T = C_v + C_W. \tag{29.1}$$

In this equation, C_v represents the coefficient of viscous drag, while C_w stands for the coefficient of wave-making drag. Viscous drag consists of two components: frictional drag and form drag, expressed as $C_v = (1 + k)C_f$, where C_f represents the coefficient of frictional drag and k denotes the form factor. The coefficient of frictional drag is determined according to the ITTC57 formula, which is as follows:

$$C_f = \frac{0.075}{(\log \mathrm{Rn} - 2)^2}, \tag{29.2}$$

where Rn is the Reynolds number given by

$$\mathrm{Rn} = \frac{UL}{\upsilon}, \tag{29.3}$$

where U represents the ship's velocity, L signifies the ship's length, and υ. denotes the kinematic viscosity of water. Typically, the length of the central hull of a trimaran vessel differs from that of the side hulls. Consequently, this dissimilarity yields distinct Reynolds numbers for the central hull and the side hulls, leading to varying frictional resistance coefficients. Thus, the overall viscous drag of a trimaran ship can be formulated as follows:

$$R_v = 0.5(C_{vc}S_c + 2C_{vs}S_s)\rho U^2. \tag{29.4}$$

The optimization process centers on refining the shape of the trimaran hull, specifically targeting the issue of excessive wake formation in the bridge region beten the main hull and the side hulls. Adjustments are made to mitigate wave resistance by altering the configuration of the side hulls. Two configurations are considered: modifying the spacing between the centerline of the main hull and the se hulls and adjusting the distance between the midship of the main hull and the side hulls while maintaining the same displacement. The trimaran optimization parameters are illustrated in Fig. 29.2, where parameter '*a*' denotes the distance between the midship of the main hull and the demi hull, while parameter '*b*' represents the distance between the centerline of the main hull and the demi hull. The model slated for modification adheres to the dimensions outlined in Table 29.1.

29.2.2 Towing Test Experiment

The trimaran semi-submersible ship has undergone resistance calculations using the towing test experimental method. The towing test is a reliable means of determining the ship's resistance. Subsequently, modifications will be made to the trimaran hull form to achieve an improved ship configuration that minimizes wave-making resistance. The resistance calculations are conducted using the slender body approach to validate the resistance values obtained from the towing test. Result from the towing

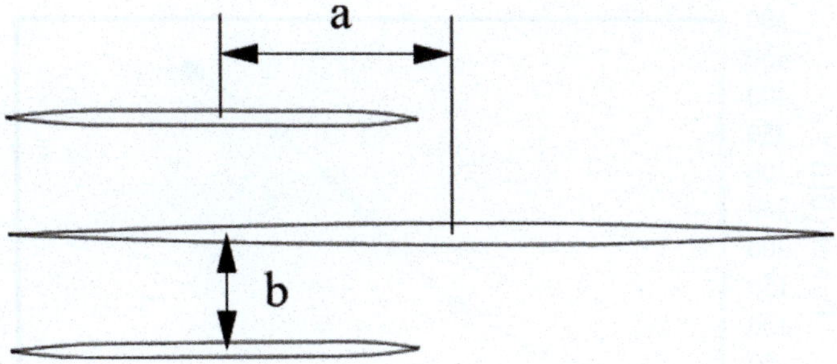

Fig. 29.2 Parameters for trimaran optimization

Table 29.1 Main Dimensions of semi-submersible trimaran hull form

Parameter	Variable	Value	Unit
Length, Overall	LoA	23.10	meter
Length, Perpendicular	Lpp	20.80	meter
Length, Waterline	Lwl	21.68	meter
Beam, Overall	B	8.00	meter
Beam, Main hull	B_0	2.80	meter
Beam, Demi hull	B_1	1.60	meter
Beam, Main hull to side hull	B_2	3.20	meter
Depth to main deck, at side	H	3.00	meter
Draft, Full load	T	2.22	meter
Service speed	V_s	10	knot
Maximum speed	V_{max}	12	knot

test can be seen in Table 29.2 and Fig. 29.3, respectively display the occurring resistance values at speeds ranging from 7 to 12 knots. This trimaran ship has a service speed of 10 knots and a maximum speed of 12 knots. The resistance experienced at the service speed is 34.3 kN, while the resistance at the maximum speed amounts to 51.75 kN.

Table 29.2 Towing test result

Ship						
Run	*V* (knots)	Re ($\times 10^8$)	C_F ($\times 10^{-3}$)	C_T ($\times 10^{-3}$)	R_T (kN)	P_E (EHP)
1	7	0.8454	2.135	5.156	7.1177	34.8559
2	8	0.9662	2.094	6.449	11.6287	65.0822
3	9	1.0870	2.058	8.805	20.0939	126.5167
4	10	1.2077	2.028	12.187	34.3362	240.2104
5	11	1.3285	2.000	13.487	45.9807	353.8412
6	12	1.4493	1.976	12.756	51.7514	434.4531

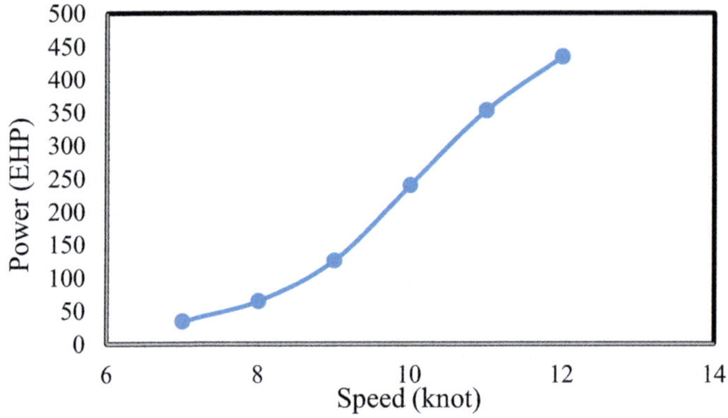

Fig. 29.3 Towing test resistance graph

29.3 Results and Discussion

The research study was conducted to investigate the effects of variation designs of the semi-submersible trimaran vessel's on fuel oil consumption rate (FOC). The study focused on two variations: variation 1, bridge lengthening, which involved extending the bridge between the demi hulls, and variation 2, demi hull shortening, which entailed shortening the demi hulls (Fig. 29.4). Detail of the variation could be seen in the FOC of a ship which is closely linked to its resistance and power (Fig. 29.5). Generally, a reduction in resistance and power results in a lower FOC. In Table 29.3, the variations' impact on the vessel's resistance and power at different operational speeds is documented.

The findings indicate that variation 1, involving the extension of the bridge between the demi hulls, resulted in lower total resistance compared to the existing design across all speeds examined. Additionally, variation 1 demonstrated a decrease in power at 85% maximum continuous rating (MCR) compared to the existing design. This suggests that variation 1 may contribute to a potential reduction in FOC under service and maximum speed.

Conversely, variation 2, which involved shortening the demi hulls, exhibited higher total resistance values compared to the existing design. Variation 2 also showed increased power at 85% MCR across the speed range. These findings suggest that variation 2 may lead to lower FOC due to decreased resistance under service speed and conversely the demi hull variation may lead to higher FOC due to increased resistance and power requirements with ship maximum speed.

If the ship operates primarily at a service speed of 10 knots, variation 2 may be considered feasible for reducing the fuel oil consumption (FOC). At the service speed of 10 knots, the existing design exhibits a total resistance of 37.5 kN and a power of 526.35 HP at 85% maximum continuous rating (MCR). In comparison, variation 2 shows a slightly lower total resistance of 35.3 kN and a power of 495.86 HP at the same speed. The lower total resistance and power values associated with variation 2

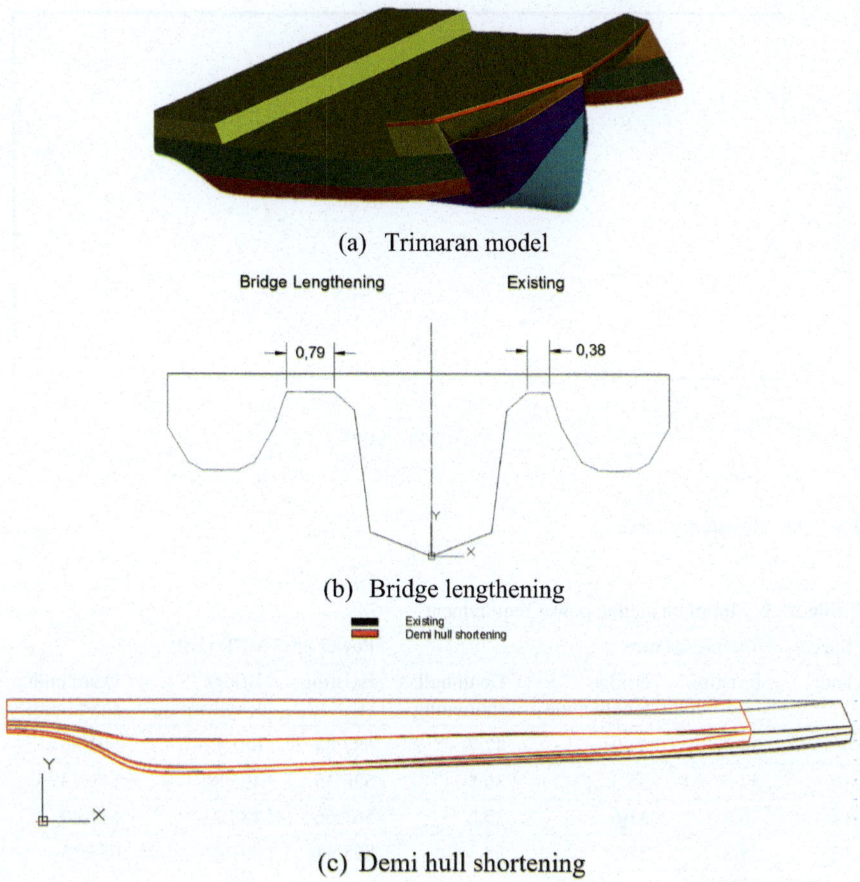

Fig. 29.4 Trimaran optimization **a** whole model; **b** bridge lengthening; **c** demi hull shortening

indicate potential improvements in fuel efficiency compared to the existing design. With reduced resistance, the vessel can move through the water more easily, requiring less power to maintain the desired speed.

It is important to consider the technical aspects of these variations when analyzing their impact on FOC. Extending the bridge between the demi hulls in variation 1 can potentially enhance the vessel's hydrodynamics, resulting in reduced drag and lower resistance. On the other hand, shortening the demi hulls in variation 2 may have adverse effects on the vessel's hydrodynamic performance, leading to increased resistance and higher power demands.

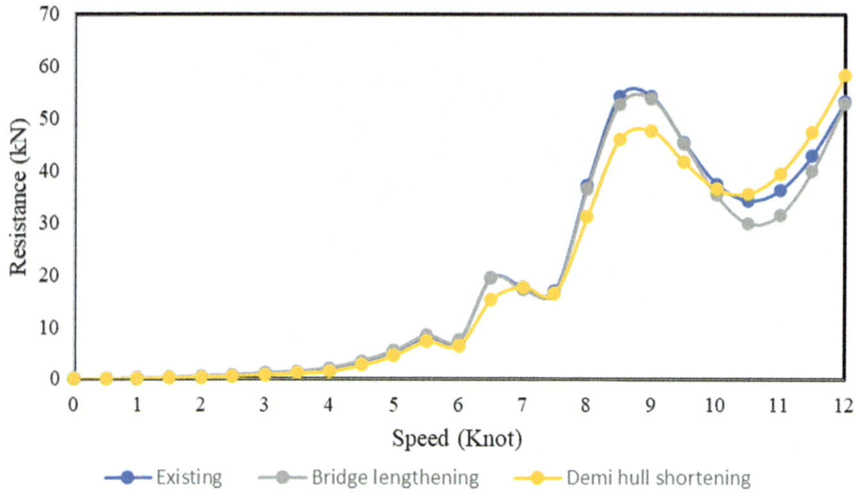

Fig. 29.5 Resistance curve

Table 29.3 Minimum engine power requirement

Speed	Total resistance			Power 85% MCR (HP)		
knot	Existing	Bridge lengthening	Demi hull shortening	Existing	Bridge lengthening	Demi hull shortening
9	54.3	53.9	47.7	687.27	682.17	603.86
10	37.5	35.3	36.5	526.35	495.86	513.23
11	36.3	31.6	39.6	561.66	488.23	611.89
12	53.3	52.9	58.3	898.31	891.27	983.55

29.4 Conclusion

Semi-submersible trimaran hull form improvement has been carried out. The conclusion formed on the improvement of the new hull form that has been accomplished, including.

1. The distance between main hull and side hull must be larger than before to avoid larger wave-making resistance, which affect the ship's total resistance.
2. The ratio of the main hull and side distances to the ship's length must be adjusted to get as minimum resistance as possible.

Acknowledgements This research was funded by Department of Naval Architecture, Faculty of Marine Technology, Institut Teknologi Sepuluh Nopember research scheme.

References

1. Aryawan, W.D., Ariesta, R.C.A., Sukma, N.P.L., Achmadi, T., Utama, D.: Fatigue life prediction on the side hull structure of semi-submersible tourism. IOP Conf. Series: Earth and Environmental Science, vol 1198, pp 1–8 (2023)
2. Zhang, Y., Hu, J., Ma, S., Wang, P.: Anti-rolling analysis and resistance optimization of a new anti-rolling hydrofoil for the trimaran vessel. Ocean Eng. **272**, 113837 (2023). https://doi.org/10.1016/j.oceaneng.2023.113837
3. Yang, C., Noblesse, F., Lohner, R.: Practical hydrodynamic optimization of a trimaran. SNAME Trans. **109**, 185–196 (2001)
4. Xu, M., Zhang, S.L.: A numerical study on side hull optimization for trimaran. J. Hydrodyn. **23**, 265–272 (2011). https://doi.org/10.1016/S1001-6058(10)60112-6

SPRINGER NATURE

GPSR Compliance

The European Union's (EU) General Product Safety Regulation (GPSR) is a set of rules that requires consumer products to be safe and our obligations to ensure this.

If you have any concerns about our products, you can contact us on ProductSafety@springernature.com

In case Publisher is established outside the EU, the EU authorized representative is:

Springer Nature Customer Service Center GmbH
Europaplatz 3
69115 Heidelberg, Germany

The manufacturer's authorised representative in the EU is Springer Nature Customer Service Centre GmbH, Europaplatz 3, 69115 Heidelberg, Germany. If you have any concerns regarding our products, please contact ProductSafety@springernature.com

Printed and bound by CPI Group (UK) Ltd, Croydon, CR0 4YY

30/09/2026

02225586-0011